MODELLING, ESTIMATION AND AI APPLICATIONS FOR LITHIUM-ION BATTERY MANAGEMENT SYSTEMS

Modelling, Estimation and AI Applications for Lithium-Ion Battery Management Systems

SHUNLI WANG

College of Electrical Engineering, Sichuan University, Chengdu, China; College of Electric Power, Inner Mongolia University of Technology, Hohhot, China

QI HUANG

School of Information and Control Engineering, Southwest University of Science and Technology, Mianyang, China

LIYA ZHANG

College of Electric Power, Inner Mongolia University of Technology, Hohhot, China; College of Electrical Engineering, Sichuan University, Chengdu, China

GUANGCHEN LIU

College of Electric Power, Inner Mongolia University of Technology, Hohhot, China

CARLOS FERNANDEZ

School of Pharmacy, Applied Sciences and Public Health, Robert Gordon University, Aberdeen, United Kingdom

FREDE BLAABJERG

School of Energy, Aalborg University, Aalborg, Denmark

ELSEVIER

Elsevier
Radarweg 29, PO Box 211, 1000 AE Amsterdam, Netherlands
125 London Wall, London EC2Y 5AS, United Kingdom
50 Hampshire Street, 5th Floor, Cambridge, MA 02139, United States

ISBN: 978-0-443-45347-2

Publisher: Priscilla Braglia
Acquisitions Editor: Ambika Kapoor
Editorial Project Manager: Avni Gupta
Production Project Manager: Karthikeyan Durairaj
Cover Designer: Arumugam Kothandan

Typeset by TNQ Tech

Contents

Preface

Under the strategic guidance of "peak carbon emissions and carbon neutrality," energy storage systems have emerged as critical enablers for stabilizing renewable energy fluctuations and optimizing energy allocation within next-generation power systems. Consequently, innovative developments and engineering applications of performance testing theories and technologies for energy storage systems have attracted significant attention. A scientifically sound and comprehensive system of performance testing theory and technology serves as a fundamental basis for the real-time monitoring of system operating states and the optimization of energy management strategies. It provides robust technical support for enhancing system operational efficiency, ensuring equipment safety and longevity, and advancing intelligent operation and maintenance, thereby driving the construction of next-generation power systems and fostering the development of smart energy economies.

Lithium-ion batteries are the dominant energy storage medium in grid-scale energy storage systems, new energy vehicles, and portable electronic devices due to their high energy density, long life cycle, low self-discharge rate, and other outstanding advantages. The performance quality of lithium-ion batteries directly determines the overall reliability and operational integrity of energy storage systems. Therefore, research on performance testing, modeling analysis, and intelligent management technologies for lithium-ion batteries holds substantial theoretical significance as well as practical engineering value.

Building on years of theoretical insights, practical experience, and research methodologies in lithium-ion battery management systems (BMSs), this book focuses on lithium-ion batteries as its core subject. It integrates the technical requirements for battery systems in power grid storage applications and new energy vehicles to present a comprehensive overview. The book systematically examines key technologies, including the working principles of lithium-ion batteries, performance testing methodologies, modeling approaches, and intelligent state-estimation techniques. It establishes a comprehensive technical framework spanning fundamental theory to engineering applications, providing a reference for the design, development, experimental validation, and practical implementation of intelligent battery management systems.

The book consists of nine chapters with the following content: Chapter 1 provides an overview of the working principles of lithium-ion batteries, performance testing methods, key performance characteristics, and typical application scenarios—establishing a holistic understanding among readers. Chapter 2 introduces foundational concepts in artificial intelligence (AI) and algorithms—including core theories in machine learning (ML) and deep learning (DL), along with data preprocessing techniques—laying the groundwork for intelligent modeling. Chapter 3 presents a systematic review of equivalent circuit models (ECMs), electrochemical models (EMs), and AI-based modeling approaches for lithium-ion batteries—offering diverse solutions tailored to different application needs. Chapters 4 through 6 focus on the core state-estimation techniques: Chapter 4 details state-of-charge (SOC) estimation; Chapter 5 addresses the state-of-health (SOH); Chapter 6 discusses the state-of-power (SOP). Each chapter elaborates on the definitions, significance, algorithmic principles—including traditional filtering methods, such as Kalman filters—and advanced deep learning–based approaches—with an emphasis on implementation strategies and improvement mechanisms. Chapter 7 investigates joint estimation across multiple battery states by analyzing the coupling characteristics among SOC/SOH/SOP, while highlighting the limitations inherent in independent estimation approaches. Chapter 8 centers on the experimental design methodology, including data cleaning procedures, preprocessing techniques, and multidimensional visualization tools—all aimed at supporting model validation processes as well as algorithm optimization efforts. Chapter 9 explores the future directions in battery management technology along with the broader societal impacts.

During manuscript preparation, Wang Shunli, Carlos Fernandez, and Frede Blaabjerg jointly developed the overall structure and core content framework; Zhang Liya contributed significantly to the content supplementation revision work leading up to the finalization; Professor Huang Qi from the National Research Center for Science and Technology Development reviewed the entire manuscript; and Liu Guangchen provided extensive engineering practice references essential to this work. Members of our research group—Beining Hu, Bohan Hu, Chang Liu, Chao Ma, Chao Wang, Daijiang Mo, Donglei Liu, Eben Kououyi, Faryal Bourkache, Guo Chen, Guofang Li, Jiahao Chai, Jiakang Zhou, Jiale Qin, Jiaqi Chen, Jie Wei, Josep M. Guerrero, Junkai Wang, Kaibo Ma, Kuo Yang, Lei Zhou, Liangdong Sun, Liangwei Cheng, Linzhi Li, Mai Geng, Mamadou Fall, Miao Yu, Mingjie Zhang, Minlong Bi, Qin Zhang, Qingshi Geng, Quan

Dang, Shaoyang Zeng, Shuyang Ji, Wenxia Zhang, Xiao Deng, Xinyu Tian, Xinyu Yan, Yang Wang, Yanxin Xie, Yingyang Wu, Yixuan Wu, Yongke Li, Yu Fu, Yuchen Huo, Yuhao Qin, Zhe Chen, Zhengqing Gao, Zipeng Yu, and Ziwen Guo—actively participated in experimental data acquisition tasks, graphical representation generation activities, and drafting portions of several chapters, contributing substantially toward ensuring both rigorous scientific accuracy and practical applicability. This research was supported by the National Key Research and Development Program (No. 2024YFB2408400), the Support Program for Introduced Talents of Inner Mongolia Autonomous (DC2500004284), and the "Inner Mongolia Autonomous Region Graduate Research Innovation Program" (Grant no. KC2025040S). We extend our sincere gratitude to all the funding agencies involved, along with all the individuals who contributed directly or indirectly throughout this endeavor.

Given that this book covers multiple interdisciplinary domains, including electrochemistry, electrical engineering, and artificial intelligence, certain limitations inevitably exist due to the authors' varying academic backgrounds and practical experience. We welcome constructive criticism from the readers regarding any inaccuracies or areas requiring improvement. Correspondence may be addressed via email: wangshunli@imut.edu.cn. We hope this book will serve not only as a scholarly resource, but will also foster deeper academic exchange among peers worldwide toward advancing innovation in the key technologies related to battery storage systems while promoting their effective deployment across real-world applications.

CHAPTER ONE

Overview and applications of lithium-ion batteries

1.1 Operating principles of lithium-ion batteries

In modern society, portable electronic devices, electric vehicles, and renewable energy storage systems have become indispensable components of our daily lives and industrial development. Among various energy storage technologies, Lithium-ion batteries (LIBs) stand out due to their outstanding performance characteristics. The widespread application of lithium-ion batteries stems from their multiple key advantages. They exhibit high energy density, meaning they can store substantial amounts of energy relative to their weight and volume, making them particularly suitable for portable devices with limited space and weight constraints. Additionally, compared to other rechargeable battery technologies such as Nickel-cadmium (Ni–Cd) and Nickel-metal hydride (Ni-MH), lithium-ion batteries have a lower self-discharge rate. This ensures that they retain their charge for longer periods when not in use.

Another significant advantage is its long cycle life, allowing hundreds to thousands of charge/discharge cycles before substantial performance degradation occurs. Moreover, lithium-ion batteries are more environmentally friendly than some conventional battery types, as they do not contain heavy metals such as cadmium that are harmful to the environment [1]. These combined advantages make lithium-ion batteries the preferred energy storage solution for applications ranging from smartphones and laptops to electric vehicles and grid-scale energy storage systems.

A lithium-ion battery is a complex electrochemical device composed of multiple critical components that work together to enable the storage and release of electrical energy. Each component plays a vital role in determining the battery's overall performance and safety. The primary structural elements of a lithium-ion battery include the cathode (positive electrode), anode (negative electrode), electrolyte, and separator. These components

Modelling, Estimation and AI Applications for Lithium-Ion Battery Management Systems
ISBN: 978-0-443-45347-2
https://doi.org/10.1016/B978-0-443-45347-2.00001-7

are typically arranged in a layered or wound configuration according to the battery's physical form, such as cylindrical, prismatic, or pouch cells.

Fig. 1.1 demonstrates the charging and discharging process of lithium-ion batteries. The cathode is the positive electrode of a lithium-ion battery, responsible for storing lithium ions during charging and releasing them during discharging. It is typically composed of lithium-containing transition metal oxides. Common cathode materials include Lithium cobalt oxide ($LiCoO_2$), Lithium manganese oxide ($LiMn_2O_4$), Nickel manganese cobalt oxide ($LiNiMnCoO_2$, commonly referred to as NMC), and Lithium iron phosphate ($LiFePO_4$, LFP). These materials exhibit distinct characteristics in terms of energy density, power density, cycle life, and safety.

The cathode material is typically mixed with a conductive additive (such as carbon black) and a binder (such as Polyvinylidene fluoride, PVDF) to form a composite electrode. This mixture is coated onto a current collector, which is usually a thin aluminum foil. The current collector facilitates electron flow between the cathode and the external circuit.

The negative electrode is the cathode of the battery, serving as the host for lithium ions during charging. When the battery charges, lithium ions

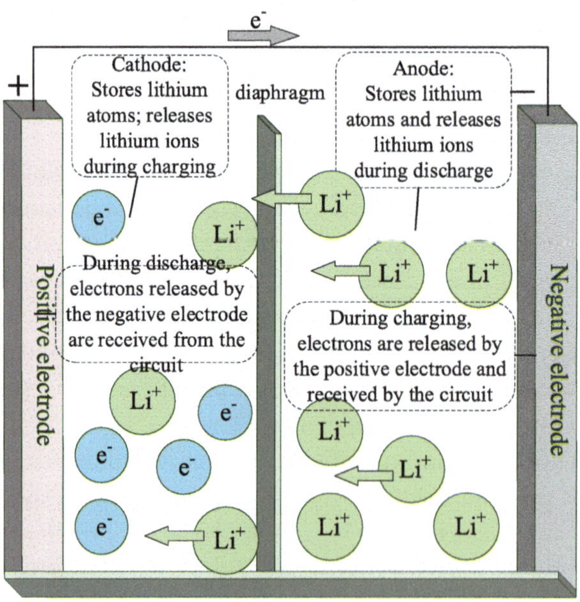

Figure 1.1 Schematic diagram of lithium-ion battery structure.

migrate from the positive electrode to the negative electrode and intercalate into the negative electrode material. During discharge, lithium ions deintercalate from the negative electrode and return to the positive electrode. Graphite is the most commonly used negative electrode material in commercial lithium-ion batteries due to its high lithium intercalation capacity, good electrical conductivity, and relatively low cost.

Similar to the cathode, the anode material is mixed with a conductive additive and a binder (typically PVDF or Styrene-butadiene rubber, SBR) and then coated onto a current collector. The anode current collector is usually a thin copper foil, just as the aluminum foil is used in the cathode, enabling electron flow between the anode and the external circuit.

Electrolytes are critical components in lithium-ion batteries, serving as the medium for lithium-ion transport between the cathode and anode while preventing direct electron passage. Typically, lithium-ion battery electrolytes are liquid in nature, composed of lithium salts dissolved in organic solvent mixtures. Common lithium salts include Lithium hexafluorophosphate ($LiPF_6$), Lithium perchlorate ($LiClO_4$), and Lithium tetrafluoroborate ($LiBF_4$), with $LiPF_6$ being widely adopted due to its favorable combination of ionic conductivity and chemical stability.

The selection of organic solvents must strike a balance between high dielectric constant (to promote dissociation of lithium salts) and low viscosity (to facilitate lithium-ion transport). Commonly used organic solvents include Ethylene carbonate (EC), Propylene carbonate (PC), Dimethyl carbonate (DMC), and Diethyl carbonate (DEC), typically employed in mixed formulations to optimize electrolyte performance. Additionally, minor additives may be incorporated into the electrolyte, such as film-forming additives (which generate a stable Solid electrolyte interphase, SEI, on electrode surfaces) and flame-retardant additives (to enhance battery safety), thereby improving cycle life and safety performance.

A separator is a porous film that physically isolates the positive and negative electrodes in a lithium-ion battery. Its primary function is to prevent direct contact between the electrodes, thereby avoiding short circuits, while allowing lithium ions to migrate through the pores within the electrolyte [2]. The performance of the separator directly affects the safety and rate capability of the battery; therefore, it must possess excellent mechanical strength, chemical stability, thermal stability, and high porosity.

Commonly used separator materials in commercial lithium-ion batteries are polyolefin polymers, such as Polyethylene (PE), Polypropylene (PP), or PE-PP composite membranes. Polyethylene separators melt at elevated temperatures, thereby closing pores and preventing lithium-ion migration, effectively interrupting current flow. This serves as one of the critical safety mechanisms in battery operation. In recent years, to meet the demands of high-energy-density batteries, novel separator materials such as ceramic-coated separators have also been widely adopted, offering enhanced thermal stability and mechanical strength.

The operation of lithium-ion batteries is based on the intercalation and deintercalation of lithium ions between the cathode and anode, along with the directional flow of electrons through an external circuit. This entire process constitutes a "rocking-chair" mechanism, in which lithium ions shuttle back and forth within the crystal lattice structures of the cathode and anode materials during charge and discharge cycles. Since no metallic lithium deposition or dissolution occurs within the battery itself, this design confers high cycle stability.

When a lithium-ion battery is connected to an external power source for charging, the electrical energy supplied by the source drives redox reactions within the battery. Under the influence of the electric field, the cathode material undergoes decomposition, and lithium ions are extracted from the crystal lattice of the cathode material and migrate into the electrolyte. Simultaneously, the cathode material loses electrons (oxidation), and these electrons flow through the external circuit toward the anode. The extracted lithium ions then migrate through pores in the separator toward the anode via the electrolyte. Upon reaching the anode through the external circuit, electrons are transferred to the anode material, which gains electrons. The lithium ions that have migrated to the vicinity of the anode subsequently intercalate into the crystal lattice of the anode material, forming a lithium intercalation compound.

As the charging process proceeds, lithium ions are continuously released from the positive electrode and continuously intercalated into the negative electrode, causing the battery voltage to gradually increase. Electrical energy is thus stored within the battery in the form of chemical energy. Charging should be terminated when the negative electrode material reaches its maximum lithium intercalation capacity or when the positive electrode material reaches its maximum lithium deintercalation capacity; otherwise, overcharging may occur, posing safety risks.

When a lithium-ion battery is connected to an external load (such as a mobile phone or motor), the chemical energy stored within the battery is converted into electrical energy to power the load, placing the battery in a discharge state. The discharge process is the reverse reaction of charging. Lithium ions, previously intercalated into the crystal lattice of the negative electrode material, deintercalate and enter the electrolyte, while the negative electrode material loses electrons. These electrons travel through the external circuit toward the positive electrode. The deintercalated lithium ions migrate through pores in the separator toward the positive electrode via the electrolyte. Upon reaching the positive electrode through the external circuit, electrons are accepted by the positive electrode material. Simultaneously, lithium ions near the positive electrode reintercalate into its crystal lattice structure, forming a stable compound.

During discharge, the anode continuously releases lithium ions, while the cathode continuously intercalates lithium ions, causing the battery voltage to gradually decrease. The discharge process ends when lithium ions are nearly completely deintercalated from the anode material or when the cathode material reaches its maximum lithium intercalation capacity. At this point, the battery must be recharged before it can be used again.

During the entire charge/discharge cycle, the energy conversion process in lithium-ion batteries is achieved through the reversible intercalation and deintercalation of lithium ions between the positive and negative electrodes. Macroscopically, this process manifests as battery charging and discharging; microscopically, it involves a complex, coordinated sequence of electrochemical reactions, ion transport, and electron conduction. It is precisely this reversible ion migration mechanism that enables lithium-ion batteries to maintain high energy output and efficiency over numerous charge/discharge cycles.

From the perspective of electrochemical reactions, the charging and discharging processes of batteries essentially involve the mutual coupling of two half-reactions. Taking the typical $LiCoO_2$/graphite system as an example, its reaction process can be simplified into the following form:

$$\begin{aligned} &\text{cathode(charge)}: LiCoO_2 \rightarrow Li_{1-x}CoO_2 + xLi^+ + xe^- \\ &\text{anode(discharge)}: xLi^+ + xe^- + C \rightarrow Li_xC \end{aligned} \tag{1.1}$$

The "rocking and chair" ion migration mechanism is a distinctive feature of lithium-ion batteries, setting them apart from conventional batteries, and also the fundamental source of their high cycle stability and high energy density.

On the thermodynamic level, the open-circuit voltage of a battery can be described by the Nernst equation as:

$$E = E^0 - \frac{RT}{nF} \ln \frac{\alpha_{\mathrm{Red}}}{\alpha_{Ox}} \tag{1.2}$$

where E_0 is the standard electrode potential, R is the gas constant, T is the absolute temperature, n is the number of electrons transferred in the reaction, F is the Faraday constant, and α_{Red} and α_{Ox} represent the activities of the reduced and oxidized species, respectively. This equation elucidates the relationship between battery voltage and conditions such as reactant concentration and temperature, forming the foundation for thermodynamic analysis of lithium-ion batteries.

The migration pathway of lithium ions traverses the electrolyte and separator, and the ionic conductivity of the electrolyte directly affects the speed of their transport between electrodes. Ionic conductivity can be expressed by the following equation:

$$\sigma = nq\mu \tag{1.3}$$

where σ is the ionic conductivity, n is the carrier concentration, q is the elementary charge, and μ is the ionic mobility. This equation indicates that increasing the concentration of lithium salt and enhancing solvation effects can effectively improve the ionic conductivity of the system. Common electrolyte systems enable efficient migration of lithium ions between electrodes through solvation interactions between polar solvents and lithium ions.

During the initial charging process, electrolyte molecules undergo reductive decomposition at the surface of the negative electrode, forming a stable Solid electrolyte interphase (SEI) layer. This membrane exhibits ionic conductivity and electronic insulation, effectively preventing further electrolyte decomposition and thereby ensuring system stability. Similarly, under high voltage conditions, a Cathode electrolyte interphase (CEI) layer forms on the surface of the positive electrode to suppress electrolyte oxidation reactions.

During the diffusion of lithium ions from the surface to the interior of the electrode, the concentration gradient is a critical factor influencing mass transfer efficiency. The diffusion flux can be described by Fick's first law:

$$J = -D\frac{dC}{dx} \tag{1.4}$$

In Eq. (1.4), J represents the diffusion flux, D the diffusion coefficient, and $\frac{dC}{dx}$ the concentration gradient. If the ion diffusion rate is lower than the charge transfer rate, concentration polarization occurs, leading to a reduction in battery output voltage. Ion diffusion rates can be effectively enhanced by optimizing electrode microstructure, shortening diffusion pathways, or introducing nanostructured materials.

Moreover, significant polarization occurs in the system when the external current is too large or when the reaction rate is limited. The kinetics of electrode interface reactions can be described by the Butler–Volmer equation:

$$i = i_0\left[e^{\frac{\alpha_a F\eta}{RT}} - e^{\frac{\alpha_c F\eta}{RT}}\right] \tag{1.5}$$

In Eq. (1.5), i represents the current density, i_0 the exchange current density, η the overpotential, and α_a and α_c denote the charge transfer coefficients at the anode and cathode, respectively. This equation reveals the exponential relationship between electrode reaction rate and overpotential, forming a fundamental basis for studies in electrochemical kinetics. The charging and discharging processes of lithium-ion batteries involve not only electrochemical energy conversion but also heat generation. The primary sources of heat include Joule heating due to ohmic internal resistance, side reactions, and entropy changes. Accumulated heat leads to increased battery temperature, which in turn affects electrolyte stability and interfacial film structure. When temperature exceeds a critical threshold, the separator may undergo thermal shrinkage or even melting, resulting in internal short circuits and thermal runaway. To address these risks, modern battery designs typically incorporate multi-layer separator structures, high-temperature-resistant electrolyte systems, and active thermal management systems to ensure safe operation under high energy density conditions.

The performance of lithium-ion batteries, such as energy density, power density, cycle life, safety, and cost, is largely determined by the selection of cathode and anode materials. Different electrode materials possess distinct characteristics and are suited for various application scenarios. Table 1.1 is a comparative analysis of several mainstream cathode and anode materials.

As revealed by the comparison, lithium iron phosphate material is widely applied in power batteries and energy storage fields due to its

Table 1.1 Comparison of mainstream cathode materials for lithium-ion batteries.

Cathode material	Energy density (Wh/kg)	Cycle life (times)	Security	Cost	Primary application scenarios
Cobalt acid	150–200	500–1000	Lower	Higher	Portable electronic devices
Lithium manganese oxide	100–150	300–500	Medium	Medium	Electric tools, low-end electric vehicles
Lithium nickel manganese cobalt oxide	180–280	1000–2000	Medium	Higher	Electric vehicles, energy storage systems
Lithium iron phosphate	120–180	2000–5000	Higher	Lower	Energy storage systems, power batteries

excellent safety and long cycle life; nickel manganese cobalt oxide is preferred for high-end electric vehicles seeking extended driving range, owing to its higher energy density; while lithium cobalt oxide, despite its relatively high energy density, is mainly confined to small portable electronic devices due to cost and safety concerns.

Currently, the primary anode material in commercial lithium-ion batteries is graphite; however, its theoretical capacity is relatively limited and insufficient to meet the demands of high-energy-density batteries. Consequently, the development of novel anode materials has become a research hotspot in the field of lithium-ion batteries. Table 1.2 is a comparison of several major anode materials.

Silicon-based materials are considered one of the most promising candidates for next-generation high-energy-density lithium-ion battery anodes due to their exceptionally high theoretical capacity. Currently, through techniques such as nanostructuring, compositing (e.g., silicon-carbon composites), and surface coating, the volume expansion issue of silicon-based materials can be effectively mitigated, thereby enhancing their cycle stability. Some silicon-based anode batteries have already begun to be applied in high-end electronic devices and electric vehicles.

In summary, the operating principle of lithium-ion batteries can be reduced to a multiphysics coupled system involving ion migration, electron conduction, interfacial reactions, and energy conversion. Battery performance is not only determined by the chemical properties of electrode materials but also influenced by multiple factors such as electrolyte composition, interfacial film stability, diffusion kinetics, and thermal management strategies. A profound understanding of the electrochemical mechanisms and their mathematical descriptions provides essential theoretical support for the development of next-generation lithium-ion batteries with high energy density, long cycle life, and enhanced safety.

1.2 Performance testing methods

1.2.1 Energy test experiments

Energy is one of the core performance indicators of lithium-ion batteries, directly determining the endurance capability of terminal devices. Energy testing experiments precisely measure the actual available energy by simulating the battery's energy output/input processes under typical charging and discharging conditions, providing a critical basis for battery

Table 1.2 Comparison of mainstream anode materials for lithium-ion batteries.

Anode material	Theoretical capacity (mAh/g)	Cyclic stability	Conductivity	Volume expansion rate (%)	Development stage
Graphite	372	Excellent	Good	About 10	Commercial maturity
Silicon-based materials	4200	Poor	Poor	Approximately 300	Semi-commercial
Lithium titanate	175	Excellent	Poor	Approximately 0.3	Niche applications
Tin-based material	994	Poor	Good	Approximately 260	R&D stage

selection, calibration of battery management system parameters, and application scenario adaptation.

1.2.1.1 Test background and core objectives

1.2.1.1.1 Background

Conventional energy testing is primarily based on constant-current charging and discharging patterns. Although operationally simple, such methods fail to adequately reflect the dynamic energy variations of batteries under real-world applications. As application scenarios become increasingly complex, industry demands for energy testing have shifted from "static quantification" to "dynamic precision." Testing protocols must now be designed in alignment with actual operating conditions to ensure that test results accurately represent battery performance in real-world applications.

1.2.1.1.2 Core objective

The core objectives primarily involve three aspects: first, quantifying the rated energy and actual available energy of batteries under standard and real-world operating conditions. Second, analyzing the influence patterns of key parameters, such as temperature, charge/discharge rate, and cut-off voltage on battery energy output. Third, verifying whether the battery's energy performance meets the design requirements for end-user applications.

1.2.1.2 Test principles and definition of key parameters

1.2.1.2.1 Principle of energy calculation

The fundamental energy of a lithium-ion battery stems from the conversion of electrochemical energy during charging and discharging processes. Its core calculation formula is based on the integral relationship among "current, voltage, and time," as specifically expressed below:

Discharge energy (E_discharge): the total energy output by the battery during discharge from a fully charged state to the cutoff voltage, calculated using the following formula:

$$E_{\mathbf{discharge}} = \int_0^t U_d(t) * I_d(t) d_t \qquad (1.6)$$

Where in $U_d(t)$ is the instantaneous voltage during the discharge process, $I_d(t)$ is the discharge current, and t is the total duration of discharge.

Charging energy (E_charge): the total energy input during the process of charging the battery from a fully discharged state to a fully charged state, calculated by the formula:

$$E_{\mathbf{discharge}} = \int_0^t U_d(t)^* I_d(t) d_t \quad (1.7)$$

Where in, $U_c(t)$ is the real-time voltage during charging, $I_C(t)$ is the charging current, and t is the total charging duration.

1.2.1.2.2 Definition of key parameters

Cut-off voltage: the voltage limits that restrict battery charging and discharging, including charge cut-off voltage and discharge cut-off voltage. These must strictly comply with the specifications provided by the battery manufacturer to prevent performance degradation or safety hazards caused by overcharging or over-discharging.

Charge and discharge rate: a metric for evaluating the speed of charging and discharging. A 1C rate indicates that the battery can be fully charged or discharged within 1 hour.

Test temperature: Ambient temperature significantly affects the electrochemical reaction rate of batteries; therefore, testing must be conducted at standard temperature or the target application temperature to ensure result comparability.

1.2.1.3 Test equipment and sample preparation

1.2.1.3.1 Core testing equipment

Energy testing requires high precision and stability from the equipment, The requirements must meet the technical specifications shown in Table 1.3.

1.2.1.3.2 Sample preparation

Sample selection: Select 3 to 5 batteries of the same batch and model as test samples, such as 18,650 cylindrical batteries, ternary lithium batteries, or lithium iron phosphate batteries.

Sample pretreatment:

① Perform three standard charge-discharge cycles on the sample to eliminate the battery's "memory effect" and stabilize its performance.

② Before testing, allow the sample to remain in a temperature-controlled chamber set at the target temperature for more than 4 h to ensure thermal equilibrium between the battery's internal temperature and the ambient temperature.

1.2.1.4 Testing procedure and implementation steps

Energy testing must strictly follow the "standardized procedure" to avoid operational errors affecting result accuracy. The specific steps are as follows.

Table 1.3 Technical parameters of mainstream lithium-ion batteries.

Device name	Core features:	Technical requirements:
High-precision charge and discharge tester	Provides controllable charging and discharging current and voltage, and records real-time data.	Current accuracy $\leq \pm 0.1\%$ FS, voltage accuracy $\leq \pm 0.05\%$ FS, sampling frequency ≥ 1 Hz
Temperature-controlled environmental chamber	Maintain a constant test temperature to simulate various application environments.	Temperature control range: 40°C–85°C, temperature fluctuation $\leq \pm 0.5$°C
Data acquisition and analysis software	Real-time recording of voltage, current, and temperature data, calculation of energy values	Supports data export and includes a points calculation function.

1.2.1.4.1 Device initialization

Power on the charge/discharge tester and environmental chamber, set the target test temperature, and wait for the temperature to stabilize. Connect the battery sample to the tester. Launch the data acquisition software, configure the sampling frequency, and ensure that voltage, current, and temperature data are recorded in real time.

1.2.1.4.2 Charging phase

The charging mode adopted is constant current–constant voltage. The battery is charged at a constant current set at a specified rate until the battery voltage reaches the cutoff voltage. The charging mode then switches to constant voltage, maintaining the voltage at the cutoff level while the current gradually decreases until it drops to 0.05C, at which point charging ceases. After charging is completed, the battery is left to rest for 30 min, and the total energy input during charging is recorded.

1.2.1.4.3 Discharge phase

The battery is discharged at a constant current set by a predefined discharge rate until the battery voltage drops to the cutoff voltage. During discharge, the software records voltage and current data in real time. After discharge, the battery is left to rest for 30 min, and the total discharged energy is recorded.

1.2.1.4.4 Operating condition simulation testing

To simulate actual operating conditions such as those of electric vehicles or energy storage systems, custom operating curves must be imported. The tester charges and discharges according to the current variation patterns specified in the curve, calculating energy output under dynamic conditions. For example, the UDDS driving cycle for electric vehicles includes idle, acceleration, and deceleration phases, during which the current fluctuates within a range of 0–5C. The total energy over the entire cycle is determined through integration.

1.2.1.4.5 Data repetition verification

Each batch of samples was tested three times repeatedly, and the average energy value and standard deviation were calculated to ensure the reliability of the test results.

1.2.1.5 Data processing and results analysis

1.2.1.5.1 Data processing steps

① Raw data filtering: Remove abnormal data during the testing process to ensure data continuity.

② Energy calculation: The charge and discharge energy can be computed using the software's integration function, or by performing segmented integration on the "time, voltage, current" data using tools such as Excel.

③ Result standardization: If the test temperature or rate is not under standard conditions, the results must be converted to energy values under standard conditions to enable comparison among different samples.

1.2.1.5.2 Analytical dimensions for typical results

① Effect of discharge rate on energy: As the discharge rate increases, the actual energy output of the battery decreases. This is because, at higher discharge rates, internal polarization within the battery intensifies, leading to a faster voltage drop.

② Effect of temperature on energy: Under low-temperature conditions, the diffusion rate of lithium ions decreases, resulting in a significant reduction in discharge energy. Under high-temperature conditions, energy slightly increases; however, prolonged exposure to high temperatures accelerates battery aging.

③ Effect of cycle count on energy: The energy of a new battery is close to the rated value, and as the number of cycles increases, the energy degrades to below 80% of its initial value.

1.2.1.6 Advantages, disadvantages, and application scenarios of the method

1.2.1.6.1 Advantages

① High accuracy: The calculation method based on the integration of "current, voltage, time" can accurately reflect the actual energy output of the battery.

② High versatility: applicable to all types of lithium-ion batteries (e.g., NMC, LFP, and LMO), and adaptable to varying test requirements across different application scenarios.

③ Standardization of operations: The testing process can be automated through equipment, minimizing human error and ensuring high reproducibility of results.

1.2.1.6.2 Disadvantages

① Long testing cycle: low-rate testing requires several hours to complete a single charge/discharge cycle, making it unsuitable for rapid screening.

② Complex simulation under dynamic conditions: custom condition curves are required, demanding advanced software functionality and data processing capabilities from the equipment.

③ Ignoring the impact of self-discharge: The test did not account for the self-discharge energy loss that occurs during prolonged battery storage; in practical applications, this should be corrected based on results from self-discharge testing.

1.2.1.6.3 Application scenarios

① Battery R&D phase: Compare the energy performance of batteries with different material systems and structural designs, and optimize formulations and processes.

② Production quality inspection phase: randomly inspect the energy values of batteries leaving the factory to ensure compliance with rated standards.

③ Terminal application selection: Select batteries with appropriate energy matching for electric vehicles and energy storage systems.

1.2.2 Hybrid pulse power characterization

The Hybrid pulse power characteristic (HPPC) test is a fundamental method for evaluating the dynamic power output capability of lithium-ion batteries. Proposed by the United states advanced battery consortium (USABC), it was originally developed for testing power batteries in electric vehicles and has since become a standard approach for assessing power performance in fields such as energy storage and aerospace. The HPPC test applies "high-power pulses" to simulate the battery's response under transient high-load conditions, quantifying its maximum power output capacity across various states of charge. This provides critical data for designing power control strategies in Battery management systems (BMS).

1.2.2.1 Test background and core objectives

1.2.2.1.1 Background

Conventional constant-current power testing can only reflect the battery's power output under steady-state conditions and cannot simulate instantaneous high-power demands encountered in real-world applications. The core advantage of HPPC lies in its "dynamic pulse loading," which enables the capture of voltage responses before and after pulses, thereby allowing an assessment of the battery's power performance stability and limit values.

1.2.2.1.2 Core objective

The core objectives primarily involve three aspects. First, quantifying the maximum discharge and charge power of the battery at different State-of-charge (SOC) levels. Second, analyzing the voltage recovery characteristics of the battery following pulse loading to evaluate its internal polarization level. Third, establishing an "SOC maximum power" relationship curve to provide data support for the BMS power limitation strategies.

1.2.2.2 Test principles and definition of key parameters

1.2.2.2.1 Test principle: based on the "pulse response-polarization model"

The fundamental principle of HPPC is to apply a cyclic sequence of "discharge pulse, rest, charge pulse, rest," observe the battery voltage variations during the pulse application and rest phases, and then calculate the maximum power via an equivalent circuit model. Its key logic is as follows:

Pulse loading phase: A high-power pulse is applied, causing the battery voltage to rapidly drop (discharge) or rise (charge); the magnitude of the drop/rise reflects the battery's ohmic polarization and electrochemical polarization.

Rest stage: After the pulse ends, the voltage gradually recovers to its pre-polarization level. The recovery rate reflects the battery's ability to eliminate polarization.

Maximum power calculation: The maximum available power is calculated using the power formula based on the voltage at the end of the pulse U_{end}, the cutoff voltage U_{cutoff}, and the pulse current I_{Pulse}. The calculation formula is as follows:

$$P = \left(U_{\text{end}} - U_{\text{cutoff}}\right) * I_{\text{Pulse}} \tag{1.8}$$

1.2.2.2.2 Definition of key parameters

Pulse power/current: Set according to application scenarios. For example, electric vehicle battery packs typically have a discharge pulse power of 3C and a charge pulse power of 1C, with pulse durations of 10 s (discharge) and 5 s (charge), respectively.

SOC range: Covers the typical battery SOC range, with intervals generally set at 10%–20%. Low SOC ranges require special attention during testing, as voltage becomes more sensitive to power output in this region.

Polarization voltage: the voltage difference before and after pulse loading, categorized into ohmic polarization voltage and electrochemical polarization voltage. The smaller the value, the better the battery's power performance.

1.2.2.3 Test equipment and sample preparation

1.2.2.3.1 Core testing equipment

HPPC imposes extremely high requirements on the "dynamic response speed" of equipment, necessitating rapid loading and precise control of pulsed currents. The key equipment includes Table 1.4.

1.2.2.3.2 Sample preparation

① Sample selection: As in 1.2.1, select three to five batteries from the same batch, preferably new batteries with no visible swelling or leakage when fully charged.

② Sample pretreatment: Perform three standard charge-discharge cycles on the sample to eliminate initial polarization. Charge the battery to full capacity, then place it in a temperature-controlled chamber set at the target temperature and allow it to rest for more than 6 h to ensure thermal equilibrium between the interior and exterior of the battery.

Table 1.4 Parameters of HPPC core test equipment.

Device name	Core features:	Technical requirements:
Dynamic charge and discharge tester	Output rapid pulse current with short response time	Current rise time $\leq$1 ms, pulse current accuracy $\leq \pm$0.5% FS, supports user-defined pulse sequences
Temperature-controlled environmental chamber	Maintain a constant test temperature	Temperature range: -40–85°C, temperature stability $\leq \pm$0.5°C
High-precision data acquisition card	Voltage and current data during high-speed pulse recording processes	Sampling frequency $\geq$1 kHz (to ensure capture of instantaneous voltage variations), resolution $\geq$16 bits

1.2.2.4 Testing procedure and implementation steps

HPPC testing must be conducted according to the logic of "SOC gradient + pulse cycling." After completing a set of pulse tests at each SOC point, the system should be adjusted to the next SOC point. The specific steps are as follows.

1.2.2.4.1 Equipment and parameter configuration

Power on the dynamic charge/discharge tester, environmental chamber, and data acquisition card; set the test temperature and pulse parameters. Import the SOC test sequence and configure the charge/discharge cutoff voltages for each SOC point.

1.2.2.4.2 Pulse testing under full charge condition

① Quiescent stage: After the battery is fully charged, allow it to stand for 30 min and record the initial voltage.

② Discharge pulse: Apply a 3C discharge pulse for 10 s, record the real-time voltage during the pulse process, with particular emphasis on recording the voltage at the end of the pulse.

③ Standstill after discharge: After the pulse ends, allow a 40 seconds rest period, record the voltage recovery process, and note the final recovered voltage.

④ Charge pulse: Apply a 1C charge pulse for 5 s, and record the voltage at the end of the pulse.
⑤ Rest after charging: After the pulse ends, rest for 40 s and record the recovered voltage.
⑥ Data recording: Save the pulse voltage, current, and time data at this SOC point, and calculate the discharge/charge polarization voltage.

1.2.2.4.3 SOC adjustment and cycle testing

After completing the SOC = 100% test, discharge at a constant current of 1C until reaching the next SOC point, followed by a 30 minutes rest period to eliminate discharge polarization. Repeat the "pulse and rest" procedure described in Step 2 to sequentially complete testing at SOC points of 80%, 60%, 40%, 20%, 10%. Note: During the discharge pulse at low SOC points, continuously monitor the voltage. If the voltage rapidly declines to the cut-off voltage, terminate the pulse prematurely to prevent over-discharge.

1.2.2.4.4 Test completion and sample handling

After completion of all SOC point tests, charge the battery at a constant current of 0.5C until fully charged, then rest for 24 h and inspect the battery for abnormalities such as swelling or leakage. Export all test data for subsequent power calculations and characteristic analysis.

1.2.2.5 Data processing and results analysis

1.2.2.5.1 Core data computation

① Polarization voltage calculation:

The calculation formula for the discharge polarization voltage is as follows:

$$\Delta U_{d,\mathrm{SOC}} = U_{\mathrm{init,SOC}} - U_{d,\mathrm{end,SOC}} \tag{1.9}$$

The calculation formula for charging polarization voltage is as follows:

$$\Delta U_{c,\mathrm{SOC}} = U_{c,\mathrm{end,SOC}} - U_{\mathrm{unit,SOC}} \tag{1.10}$$

② Maximum power calculation:

Based on the difference between the pulse-end voltage and the cutoff voltage, combined with the pulse current, the maximum available power at each SOC point is calculated.

The calculation formula for maximum discharge power is as follows:

$$P_{d,\mathrm{max,SOC}} = \left(U_{d,\mathrm{end,SOC}} - U_{d,\mathrm{cutoff}}\right)^{*} I_{d,\mathrm{pulse}} \tag{1.11}$$

The formula for calculating the maximum charging power is as follows:

$$P_{c,\max,SOC} = \left(U_{c,\text{cutoff}} - U_{c,\text{end,SOC}}\right) * I_{c,\text{pulse}} \tag{1.12}$$

1.2.2.5.2 Analytical dimensions for typical results

① SOC-power curve: Plots the maximum discharge/charge power curves at different SOC levels, typically exhibiting a "high in the middle, low at both ends" trend. This behavior arises because low SOC results in reduced voltage, while high SOC leads to intensified polarization.

② Polarization voltage variation pattern: As SOC decreases, the discharge polarization voltage gradually increases, reflecting the rise in internal resistance and the weakened power output capability of the battery at low SOC.

③ Effect of temperature on power: At low temperatures, the maximum discharge power may be only 30%–50% of that at 25°C. At high temperatures, power output slightly increases; however, prolonged exposure to high temperatures accelerates battery aging.

1.2.2.6 Advantages, disadvantages, and application scenarios of the method

1.2.2.6.1 Advantages

① High dynamicity: simulates instantaneous high-power operating conditions encountered in real-world applications, resulting in test outcomes that closely match actual performance.

② Information-rich: Multiple-dimensional data such as power, polarization, and voltage recovery characteristics can be simultaneously obtained, enabling a comprehensive evaluation of the battery's dynamic performance.

③ High standardization: Clear industry standards exist, enabling test results to be compared across different laboratories.

1.2.2.6.2 Disadvantages

① High equipment cost: Dynamic charge/discharge testers, particularly those supporting high-power pulses, are expensive and entail high maintenance costs.

② Long testing duration: Each SOC point requires completion of pulse-rest cycles, and testing across the full SOC range may take several hours to 1 day.

③ High sample requirements: high-power pulses may cause battery temperature to rise, necessitating real-time temperature monitoring to prevent sample damage.

1.2.2.6.3 Application scenarios

① Selection of power batteries for electric vehicles: evaluate battery performance under high-power operating conditions, such as acceleration and hill climbing to ensure the fulfillment of the vehicle's overall power requirements.

② Frequency regulation/peak load regulation by energy storage systems: Testing the battery's response capability to instantaneous power regulation.

③ BMS Power Control Strategy Optimization: Based on the SOC-power curve, set power limit thresholds at different SOC levels to protect the battery and extend its lifespan.

1.2.3 Charge-discharge tests at various C-rates

The charge and discharge rate of a lithium battery is a multiplier indicating the magnitude of the charge and discharge current, denoted by C. The charge and discharge rate refers to the ratio between the working current during charging or discharging and the battery's nominal capacity. By employing different charge and discharge currents, the nominal capacity of energy storage lithium batteries can be estimated. The charge and discharge rate is related to charge and discharge efficiency, which varies with different C values. Typically, the magnitude of charging and discharging current is expressed in terms of the charge and discharge rate, defined as: charge and discharge rate = charge/discharge current/nominal capacity. For example, a battery with a nominal capacity of 100 mAh discharging at 20 mA has a discharge rate of 0.2C. The C-rate during battery discharge, such as 1C, 2C, or 0.2C, is a measure indicating the speed of discharge. The charge and discharge rate reflects the speed at which a battery charges or discharges, with its specific calculation method given by Eq. (1.13):

$$M = \frac{I}{C_n} \tag{1.13}$$

In the formula, M represents the charge/discharge rate, I denotes the charging or discharging current, and C_n is the rated capacity of the battery.

The capacity of lithium-ion batteries is influenced by numerous factors, and the charge/discharge rate significantly affects the output capacity under different charge and discharge conditions. Generally, as the charge/discharge rate increases, the available capacity decreases. At high charge/discharge rates, the chemical reaction kinetics within the battery accelerate, leading to noticeable structural changes in the electrode materials, sometimes even causing structural degradation, which adversely impacts battery lifespan. Furthermore, internal battery resistance causes a significant voltage drop during high-rate charge and discharge processes. This results in increased power dissipation due to internal resistance, reducing the output voltage and thereby diminishing the effective usable capacity of the battery. Therefore, when utilizing Lithium-ion batteries (LIBs), it is generally necessary to select an appropriate charge/discharge rate to protect the battery effectively and balance power output requirements with longevity. Fig. 1.2 illustrates voltage and time curves under different discharge rates. As shown in the figure, at a 2C discharge rate, discharge occurs more rapidly and results in lower capacity. In contrast, at a 0.5C discharge rate, longer discharge duration is observed with higher achievable capacity.

1.2.4 Capacity calibration experiments

Battery capacity refers to the total amount of electrical charge that a lithium-ion energy storage battery can deliver when fully charged, typically denoted by the symbol Q and expressed in units of mAh (milliampere-hours) or Ah (ampere-hours). Battery capacity is a crucial indicator of battery quality and is closely related to the battery's operational duration. In laboratory settings, constant current or constant voltage discharge methods are generally employed to discharge energy storage lithium-ion batteries. The stored capacity is determined either by measuring the total electrical energy released during the entire discharge process or by calculating the product of discharge current and time. The resulting capacity value is expressed in

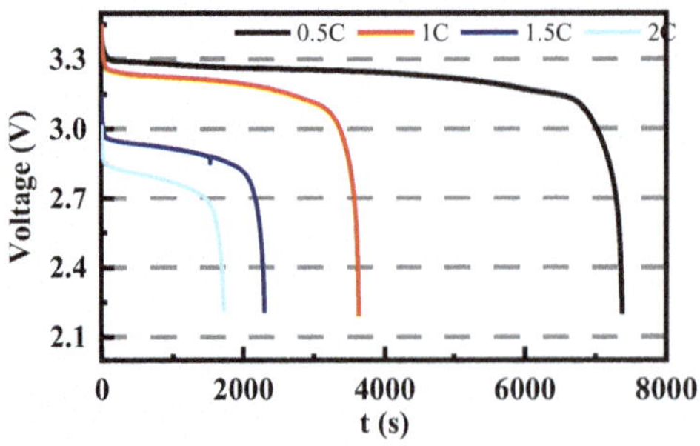

Figure 1.2 Voltage-time curves of the battery at different discharge rates.

Ah or mAh. Battery capacity can be categorized into rated capacity, actual capacity, and theoretical capacity.

The theoretical capacity is the calculated value of the total charge stored in a lithium-ion battery based on its internal electrochemical reactions, assuming complete participation of all lithium ions in the reaction, as determined by Faraday's law. Since the theoretical capacity represents an ideal value, it is greater than both the rated capacity and the actual capacity. The actual capacity is typically lower than the rated capacity due to degradation caused by battery usage. The rated capacity is measured by the battery manufacturer and directly marked on the battery's exterior surface. It serves as a key indicator of the amount of charge a battery can store and also reflects the battery's operational duration. Also known as nominal capacity, it refers to the minimum amount of charge that a battery must be able to deliver under specified discharge conditions, as mandated by relevant national authorities. The rated capacity is specified by the manufacturer in ampere-hours and constitutes one of the essential parameters for energy storage lithium-ion batteries.

The actual capacity of a lithium-ion energy storage battery refers to the maximum amount of charge that can be discharged under specific discharge conditions after the battery is purchased and brought home. The actual capacity is typically calculated as the product of the discharge current and discharge time. This actual capacity does not necessarily correspond to the capacity of a brand-new battery; it may also represent the capacity of a used energy storage lithium-ion battery, in which case the actual capacity will inevitably be lower than that of a new battery. The actual capacity reflects the amount of electrical energy a battery can deliver under real-world operating conditions. Numerous factors influence the actual capacity, including ambient temperature, the materials used in manufacturing the lithium-ion energy storage battery, and its duration of use. The charge delivered by a fully charged battery under certain conditions corresponds to the integral of discharge current over time, and this calculation process differs depending on whether it is conducted under constant current or alternating current conditions.

The discharge capacity at constant current is given by Eq. (1.14).

$$C = IT \tag{1.14}$$

The discharge capacity during communication is given by Eq. (1.15).

$$C = \int_{0}^{T} I(t)dt \quad (1.15)$$

Due to aging caused by the recycling of battery materials, the battery capacity may deviate significantly from its rated capacity at the time of manufacture. The actual discharge capacity of the battery is of great significance for SOC estimation in energy storage lithium-ion batteries. Therefore, the first step is to calibrate the capacity of the energy storage lithium-ion batteries. Three energy storage lithium-ion batteries were selected as experimental subjects, cobalt oxide batteries with a rated capacity of 1.4 Ah and a voltage of 3.7 V. A cycle charging and discharging experiment was conducted at room temperature. The specific procedures are as follows:

1) Charge the battery at a constant current of 1C, setting the charging cutoff voltage to 4.2 V. Then switch to constant voltage charging, with the charging cutoff current set at 0.02C, ensuring the battery reaches a fully charged state;
2) Since the selected energy storage lithium battery capacity is relatively small, a resting time of 30 min was chosen;
3) Constant-current discharge experiment was conducted on the energy-storage lithium-ion battery at a 1C rate, with a cutoff voltage of 2.75 V;
4) After discharge, allow to stand for 30 min;
5) Repeat steps 1) to 4) three times, and conclude the experiment.

The experimental results are presented in Table 1.5, where the unit of capacity is Ah.

Analysis of the data in the table reveals that the discharge capacity of individual battery cells is consistently lower than their corresponding charge capacity during the same cycle. This discrepancy arises due to the influence of internal battery resistance, which causes a portion of energy to be dissipated within the battery during discharge. Consequently, the amount of energy available for release during operation is always less than the energy input during charging.

1.2.5 Aging tests

The performance of lithium-ion batteries is not static. During storage and operation, their performance gradually degrades irreversibly over time, a process known as "aging." Aging ultimately manifests as capacity fade, increased internal resistance, declining power characteristics, and eventually, end-of-life. Aging tests constitute the fundamental methodology for researching, evaluating, and predicting the lifespan and reliability of

Table 1.5 Experimental data for capacity calibration.

Monomer 1			Monomer 2			Monomer 3		
Serial number	Charging capacity	Discharge capacity	Serial number	Charging capacity	Discharge capacity	Serial number	Charging capacity	Discharge capacity
1	1.472	1.388	1	1.624	1.395	1	1.535	1.402
2	1.406	1.294	2	1.532	1.396	2	1.486	1.401
3	1.387	1.390	3	1.508	1.395	3	1.460	1.404

lithium-ion batteries, spanning the entire lifecycle, from material development and cell design, through process optimization and system integration, to practical applications.

This section will systematically present the entirety of lithium-ion battery aging testing, including its fundamental concepts, aging mechanisms, testing methodologies, data analysis techniques, and extensive application scenarios.

1.2.5.1 Basic concepts and significance of aging tests

1.2.5.1.1 Definition of aging test

Aging tests refer to a series of experimental activities involving systematic measurement and recording of changes in battery performance parameters over time under simulated stress conditions that batteries may encounter during actual applications, or directly under real operating conditions, along with in-depth analysis of the underlying physical and chemical mechanisms.

1.2.5.1.2 The core objective of aging tests

Life span assessment and prediction: determining the service life of a battery under specific conditions and establishing models to forecast its long-term performance.

Reliability verification: Assessing the risk of battery failure under various stress conditions to ensure compliance with product specifications and safety requirements.

Mechanistic study: Identifying the primary aging mechanisms responsible for battery performance degradation, such as loss of active lithium, loss of active material, and an increase in interfacial impedance.

Optimized design: Providing data support and directional guidance for the optimization of battery materials, cell structures, manufacturing processes, and BMS.

Quality control: An essential method for screening and grading product quality consistency.

1.2.5.1.3 Classification of aging tests

Aging tests are primarily categorized into two major types based on the state of the battery during the testing process:

Cycling aging test: simulates the aging of a battery during charge/discharge cycling. During the test, the battery undergoes continuous charge/discharge cycles until its performance degrades to a predefined end-of-life condition.

Calendar aging test: simulates the aging of batteries under quiescent storage conditions. During the test, batteries are maintained at specific states of charge and temperatures without charging or discharging, with performance changes periodically monitored.

In practical applications, battery aging is a complex outcome resulting from the combined and mutually coupled effects of cycle aging and calendar aging.

1.2.5.2 Aging mechanisms of lithium-ion batteries

To understand aging tests, one must first comprehend the underlying aging mechanisms. Lithium-ion battery aging results from a series of complex physical and chemical side reactions occurring primarily at the electrode/electrolyte interface and within the bulk of the electrode materials.

1.2.5.2.1 Aging mechanisms on the negative electrode side

Evolution of the solid electrolyte interphase membrane:

Formation and Growth: During the initial charging and discharging cycles as well as subsequent usage, the electrolyte undergoes reduction at the negative electrode surface, forming a passivation layer known as the SEI film. An ideal SEI film is electronically insulating yet ionically conductive, effectively preventing further decomposition of the electrolyte. However, during actual aging processes, the SEI film continuously and slowly grows thicker. This irreversible growth consumes both the electrolyte and active lithium ions, leading to capacity fade. Additionally, the thickening SEI layer increases interfacial resistance to lithium-ion transport, resulting in elevated battery internal resistance and degraded power performance.

Reconstruction and decomposition: Under high temperature and high voltage conditions, the SEI layer may undergo decomposition and reconstruction, resulting in reduced stability, which further accelerates electrolyte decomposition and lithium consumption.

Lithium plating: under low temperature, high-rate charging, or excessively low anode potential conditions, the kinetics of lithium ion intercalation at the anode surface deteriorate, potentially causing some lithium ions to gain electrons and deposit on the anode surface in the form of metallic lithium instead of intercalating into the graphite layers. This phenomenon is termed "lithium plating."

Harm: Lithium plating directly consumes active lithium, resulting in capacity loss. The deposited metallic lithium is highly reactive and can react with the electrolyte to form a new SEI layer, further depleting both lithium and electrolyte. More seriously, the growth of dendritic lithium may penetrate the separator, causing internal short circuits and triggering severe safety issues.

Structural degradation of the negative electrode active material: During repeated lithium ion intercalation/deintercalation processes, graphite electrodes undergo volume expansion and contraction. Prolonged cycling stress may induce cracking and pulverization of graphite particles, disrupting the conductive network, increasing internal resistance, and causing partial active material to lose electrical contact with the current collector, leading to failure.

1.2.5.2.2 Aging mechanisms on the positive electrode side

Dissolution of transition metal ions: In particular, for cathode materials such as lithium manganese oxide and lithium-rich manganese-based materials, transition metal ions in the cathode dissolve into the electrolyte under high temperature and acidic conditions.

Harm: Dissolved metal ions can penetrate the separator and migrate to the surface of the negative electrode, destabilizing the solid electrolyte interphase layer and catalyzing its further growth, thereby accelerating capacity fading. Meanwhile, the structural integrity and capacity of the positive electrode material itself are also compromised due to the loss of metal ions.

Structural phase transformation of cathode materials: Under deep charge/discharge cycles or high voltage conditions, layered oxide cathodes undergo irreversible phase transformations. For example, a transition from a layered structure to spinel or rock-salt structures may occur. Such phase transformations obstruct lithium-ion migration pathways, resulting in reduced capacity and depressed voltage plateaus.

Formation of the cathode electrolyte interphase: Similar to the anode, at high voltages, the electrolyte undergoes oxidative decomposition on the cathode surface, leading to the formation of a cathode electrolyte interphase layer. The growth of the CEI layer similarly consumes active lithium and electrolyte, while increasing interfacial impedance.

Particle cracking and pulverization: During cycling, secondary spherical cathode materials also undergo volume changes, generating internal stresses that lead to the formation of microcracks within particles. These cracks expose fresh surfaces that react with the electrolyte, accelerating side reactions and disrupting the conductive pathways between particles.

1.2.5.2.3 Aging of the electrolyte and separator

Electrolyte consumption and decomposition: The continuous growth of the aforementioned SEI/CEI films and parasitic reactions involving metallic lithium continuously consume the electrolyte. Reduction in electrolyte quantity leads to decreased ionic conductivity and a significant increase in battery internal resistance.

Membrane aging and clogging: The separator may shrink or melt due to high temperatures, leading to internal short circuits. Additionally, metal ions dissolved from the positive electrode or fragments of electrode materials may block the pores of the separator, increasing resistance to ion transport.

In summary, the aging of lithium-ion batteries essentially involves a loss of active lithium inventory and a decline in the reversibility of active materials, accompanied by a general increase in interfacial impedance. Different stress conditions activate and exacerbate distinct dominant aging mechanisms. The purpose of aging tests is to design experiments that quantify these performance degradations and correlate them with specific aging mechanisms.

1.2.5.3 Implementation method and procedure of aging tests

A complete aging test cycle typically consists of four stages: pre-test preparation, test execution, performance monitoring, and post and test analysis. The specific procedure is illustrated in Fig. 1.3 below.

1.2.5.3.1 Preparation before testing

Sample Selection and Preprocessing: Select representative battery samples to ensure consistency in initial performance. Typically, all samples undergo multiple standardized charge/discharge cycles to stabilize their initial performance and to screen out defective units.

Initial performance characterization: Before initiating the aging test, each battery undergoes comprehensive initial performance testing, including:

Rated capacity test: Measure the battery's discharge capacity under standard conditions.

DC internal resistance test: The DC internal resistance at various SOC points is measured using pulse discharge or HPPC methods.

Electrochemical impedance spectroscopy testing: to obtain more detailed information regarding interfacial reactions and mass transfer processes.

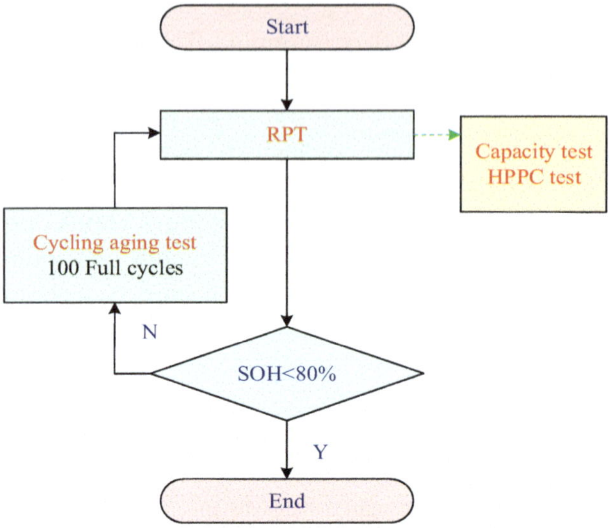

Figure 1.3 Flowchart of the aging test cycle.

Open-circuit voltage curve test: Record the correspondence between OCV and SOC.

Test plan design: Based on the test objectives, develop a detailed test plan that clearly defines stress conditions, test duration, and monitoring parameters.

1.2.5.3.2 Test execution and performance monitoring

This is the core execution phase of the aging test.

Cycling aging test program:

Stress factors: Charge/discharge rate (C-rate), Depth of discharge (DOD), cutoff voltage, and ambient temperature.

Typical protocol: Perform continuous charge-discharge cycles on the battery using a battery testing system within a constant-temperature chamber. For example, charge at a 1C rate via constant-current/constant-voltage mode and discharge at a constant current within the 20%–80% SOC range at 25°C.

Monitoring: Continuously record charging capacity, discharging capacity, charge/discharge energy, average voltage, cycle time, etc., for each cycle.

Calendar aging test plan:

Stress factors: storage temperature, state of charge.

Typical procedure: charge the battery to the target state of charge, then store it in a constant-temperature chamber.

Monitoring: Periodically remove the battery and perform performance characterization tests (such as capacity and internal resistance) under standard conditions, then return it to storage or recharge it to the original state of charge.

Actual operating condition simulation testing: To more accurately simulate application scenarios, real-world current/power profiles can be imported for testing, such as driving cycles for electric vehicles or frequency regulation commands from the power grid.

1.2.5.3.3 Post-test analysis

Post and test analysis is conducted when battery performance degrades to the end-of-life condition or when the predetermined test duration concludes.

Non-destructive analysis: Perform a comprehensive performance characterization of the aged batteries and compare the results with the initial data.

Degradation analysis: Disassemble batteries inside a glovebox and perform physical and chemical studies on components such as electrodes and separators to directly observe and validate degradation mechanisms. Commonly used techniques include:

Scanning electron microscopy: observation of surface morphology changes, cracks, and solid electrolyte interphase films in electrode materials.

X-ray diffraction: Analysis of changes in the crystal structure of anode and cathode materials.

X-ray photoelectron spectroscopy: analysis of the chemical composition of the SEI/CEI films.

Inductively coupled plasma spectrometry: detection of dissolved metal ion concentrations in the electrolyte.

1.2.5.4 Data analysis and modeling for aging tests

The vast amount of data obtained from aging tests needs to be scientifically analyzed in order to be transformed into valuable information and knowledge.

1.2.5.4.1 Performance degradation trajectory analysis

Capacity fade curve: Plotting the capacity retention ratio (current capacity/initial capacity) against the number of cycles or storage time. This serves as the most intuitive aging indicator.

Internal resistance growth curve: Plot the curve of internal resistance growth rate (current internal resistance/initial internal resistance) vs. the number of cycles or storage time.

Differential capacity analysis: The dQ/dV curve derived from charge/discharge profiles enables the identification of peak positions and intensities associated with electrode phase transitions, thereby allowing inference of structural degradation in anode and cathode materials, serving as a powerful tool for diagnosing aging mechanisms.

1.2.5.4.2 Empirical models and life prediction

Based on test data, empirical or semi-empirical models can be established to fit aging trajectories and predict long-term life.

Power-law model: Widely used for fitting cycle life. Capacity fade can be expressed as: where N represents the number of cycles, and a and b are fitting parameters. $Q_{loss} = a \times \hat{N}b$.

Arrhenius model: used to describe the influence of temperature on the aging rate. The relationship between the aging rate constant k and temperature T is given by: $K = A \times \exp(-Ea/RT)$, where Ea represents the activation energy and R is the gas constant. By conducting tests at different temperatures, Ea can be determined, allowing extrapolation to the aging rate under ambient conditions.

Square root model: commonly used to describe capacity loss due to SEI layer growth during calendar aging, which is proportional to the square root of time.

Although these models cannot reveal underlying mechanisms, they are computationally simple and exhibit good prediction accuracy under given stress conditions, and are widely applied in engineering practice.

1.2.5.4.3 Mechanistic model

As shown in Fig. 1.4, mechanistic models are based on electrochemical principles and aging reaction kinetics, aiming to describe the aging process from a fundamental physical and chemical perspective. These models are typically highly complex, involving multiple partial differential equations and coupled side reaction kinetic equations, yet they offer deeper mechanistic insights and broader predictive capabilities.

1.2.5.5 Application scenarios of aging tests

The application scope of aging tests is extremely extensive, covering nearly all fields related to lithium-ion batteries.

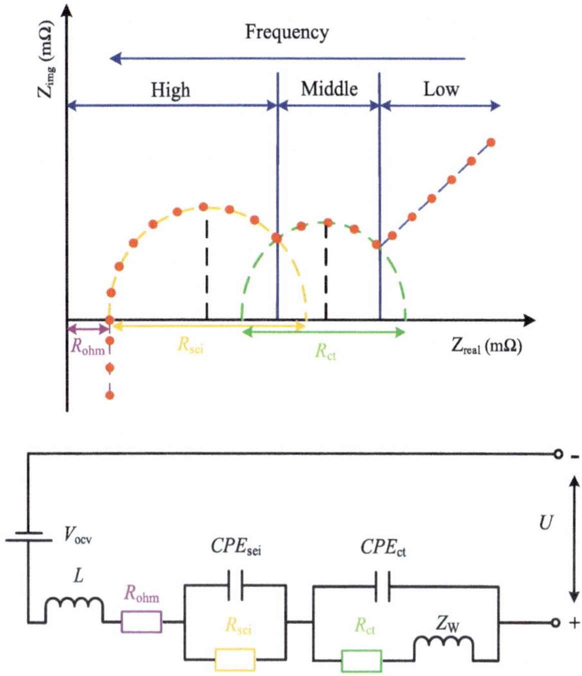

Figure 1.4 Schematic of the mechanism model.

1.2.5.5.1 Material development and screening

During the development phase of new materials, such as high-nickel cathodes, silicon-carbon anodes, and novel electrolytes, standardized aging tests enable rapid evaluation of cycle stability, high-temperature storage performance, and other critical characteristics, allowing direct comparison with existing material systems and providing essential feedback for material optimization.

1.2.5.5.2 Cell design and process optimization

Various cell design parameters, such as electrode thickness, compaction density, and N/P ratio, and manufacturing processes, including coating uniformity, drying temperature, and electrolyte filling volume, can all influence battery lifespan. By comparing the aging test results of different design and process samples, optimal solutions can be identified to enhance product performance.

1.2.5.5.3 Validation and optimization of battery management system strategies

One of the core functions of the BMS is to optimize battery usage in order to extend its lifespan. Fig. 1.5 provides a detailed illustration of the battery management system's strategy. Aging tests can provide data support for the development and validation of BMS strategies.

SOC window optimization: By testing cycle life under various DOD levels, it is demonstrated that shallow charging and discharging can significantly extend battery life, thereby guiding the daily SOC usage range setting in the BMS.

Charging strategy optimization: testing aging under various charging rates, particularly fast charging strategies, to assist in developing intelligent charging algorithms that meet fast-charging requirements while minimizing lithium plating and lifespan degradation.

Optimization of thermal management strategies: Clearly recognizing the significant impact of temperature on aging, reinforcing the importance of the thermal management system, and providing a basis for setting trigger thresholds for system cooling/heating.

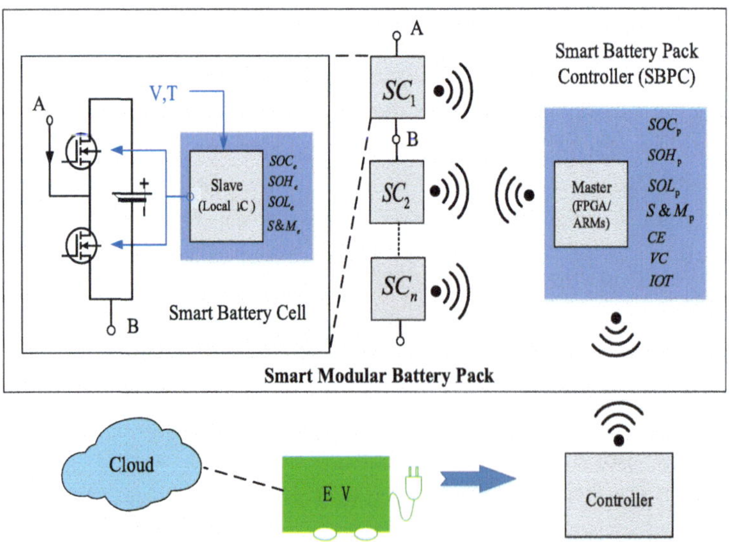

Figure 1.5 Block diagram of battery management system strategy.

1.2.5.5.4 Second-life assessment

When the capacity of a power battery degrades to approximately 80% of its initial value, it may no longer be suitable for use in electric vehicles, but still holds value for applications with lower requirements, such as energy storage systems or backup power supplies. Aging test data, particularly in-depth assessment of battery health status, serve as critical basis for determining whether a battery is suitable for second-life applications and how it should be sorted and grouped.

1.2.5.5.5 Quality assurance and reliability assessment

For battery manufacturers and end users, aging tests constitute a critical step in verifying whether battery products meet specified requirements for lifespan guarantee, reliability, and safety. They also serve as essential tools for consistency screening and quality grading during production.

1.2.5.5.6 Compliance testing with standards and regulations

Various countries and international organizations have established multiple testing standards for the lifespan and reliability of lithium-ion batteries. Aging tests are essential steps to meet these standard requirements and achieve market access.

1.2.5.6 Challenges in aging tests and future development trends

Although aging test technologies have become quite mature, they still face several challenges and continue to evolve with technological advancements.

1.2.5.6.1 Current challenges

Time and cost: Batteries can last several years or even a decade, but full-scale experimental testing takes a long time and is costly, making it difficult to meet the demands of rapid R&D and product iteration.

Representativeness and extrapolability of test results: In accelerated aging tests, although increasing stress levels can shorten testing duration, it may also alter the dominant degradation mechanisms, leading to discrepancies between predicted outcomes and real-world performance. Establishing reliable accelerated testing methodologies and models remains an ongoing research challenge.

Sample Consistency and Statistical Significance: Variability among individual batteries necessitates a sufficiently large sample size to achieve statistically significant conclusions, thereby further increasing the complexity and cost of testing.

Complexity of multi-stress coupling: In practical applications, batteries simultaneously endure electrical, thermal, and mechanical stresses. Designing and implementing testing protocols capable of accurately simulating the aging effects induced by such multi-stress coupling poses a significant challenge.

1.2.5.6.2 Future development trends

High-throughput automated testing: Integration of automated equipment and robotics enables parallel testing of hundreds or even thousands of battery samples simultaneously, significantly enhancing testing efficiency and data output.

Advanced in-situ/operando characterization techniques: Developing technologies capable of real-time monitoring of internal states during battery charging and discharging, such as in-situ XRD, in-situ SEM, and fiber-optic sensors, providing an unprecedented dynamic perspective for understanding degradation mechanisms.

Artificial Intelligence and Big Data Analytics: By leveraging machine learning and deep learning algorithms, hidden patterns and correlations are extracted from vast amounts of aging test data to establish more accurate and generalized lifetime prediction models, enabling early fault warning.

Multiscale modeling and simulation: By integrating multiscale models ranging from atomic and particulate levels to cell and module scales, virtual aging tests and lifespan predictions can be conducted on computers, reducing reliance on time-consuming and costly physical experiments and accelerating the design and development process.

Focus on extreme application scenarios: Develop specialized aging test protocols and evaluation systems for extreme application scenarios such as ultra-fast charging, extremely cold climates, and high altitudes.

1.3 Performance characteristics

1.3.1 Basic electrochemical properties

The electrochemical performance of lithium-ion batteries includes electromotive force, internal resistance, voltage, voltage characteristics, capacity, charge/discharge characteristics, cycle performance, self-discharge, and storage characteristics. This section primarily discusses the concepts of the electrochemical performance of lithium-ion batteries.

1.3.1.1 Electromotive force of the battery

The Electromotive force (EMF) of a battery is a physical quantity that measures the potential difference generated by the internal chemical reactions within the battery. It refers to the potential difference between the two electrodes in a closed circuit, corresponding to the direction of current flow from the positive to the negative electrode [3]. The magnitude of the battery's EMF depends on the ability of its internal chemical reactions and the characteristics of the electrode materials. Under ideal isothermal and isobaric conditions with thermodynamic reversibility, the decrease in Gibbs free energy of a system equals the maximum non-expansion work performed by the system. When only electrical work is involved, the relationship between the increase in Gibbs free energy and the reversible EMF of the battery can be expressed by Eqs. (1.16) and (1.17).

$$\Delta G_{T,p} = -nFE \tag{1.16}$$

$$E = -\frac{\Delta G_{T,p}}{nF} \tag{1.17}$$

In this Eqs. (1.16) and (1.17), E represents the reversible electromotive force of the battery; ΔG_{Tp} denotes the difference in Gibbs free energy of the redox reaction in the battery; n refers to the stoichiometric coefficient of electrons involved in the oxidation or reduction reaction; and F is the Faraday constant.

The maximum value ΔG_{Tp} of chemical energy converted into electrical energy in a battery corresponds to the maximum electromotive force of the battery. These values depend solely on the redox reaction system and are independent of the specific pathway by which the redox reaction proceeds. In practical batteries, the conversion of chemical energy into electrical energy typically occurs via thermodynamically irreversible processes; therefore, the actual electromotive force between the positive and negative electrodes of a battery must be less than this theoretical maximum.

1.3.1.2 Battery internal resistance

Inside a lithium-ion battery, lithium ions migrate from one electrode to the other. The collective resistance encountered by these ions during their movement constitutes the battery's internal resistance. Due to the effect of internal resistance, the terminal voltage of the battery is lower than its electromotive force and open-circuit voltage during discharge, and higher than both during charging. Current fundamentally arises from the directional movement of electric charge; as electrons move, they encounter resistance from the material itself and from magnetic fields. Internal resistance refers to the magnitude of opposition encountered by current flowing through the battery's interior during operation, serving as a quantitative measure. Internal resistance is one of the critical parameters of lithium-ion batteries, with its magnitude directly influencing key performance characteristics such as operating voltage, operating current, and capacity. The internal resistance of a lithium-ion battery is not constant; it varies during charging and discharging due to factors including temperature, charge/discharge rate, duration of charge/discharge cycles, electrolyte concentration, and changes in active material quality.

Battery internal resistance consists of ohmic resistance R_O and polarization resistance R_P. Ohmic resistance originates from the resistances of the electrolyte, separator, and contact resistances between various components. During the redox reactions inside the battery, an electric field is generated, causing dielectric materials to develop polarization charges due to the polarization effect. The polarization resistance refers to the opposition these polarization charges exert against current flow. During charging and discharging of lithium-ion batteries, internal resistance arises primarily from polarization reactions, which mainly include electrochemical polarization and concentration polarization. The intrinsic properties of active materials, electrode structure, battery manufacturing processes, and operating conditions all influence the magnitude of polarization resistance.

Among these factors, battery operating conditions are the primary determinants of polarization internal resistance. Total battery internal resistance equals the sum of ohmic resistance and polarization resistance, as expressed in Eq. (1.18).

$$R = R_o + R_p \tag{1.18}$$

The internal resistance characteristic of a battery refers to the resistance exhibited by the battery itself against current flow during operation. Different types of lithium-ion batteries display varying internal resistance characteristics. Even batteries of the same type may exhibit different internal resistance values due to variations in internal material composition, operating environment, and degree of aging. Therefore, studying the variation characteristics of battery internal resistance holds significant importance for modeling and estimation purposes.

Battery internal resistance generally comprises ohmic resistance and polarization resistance, with magnitudes typically on the order of micro-ohms or milli-ohms. Internal resistance is a crucial indicator for evaluating battery performance, and its characteristics are essential for SOC estimation, model development, and equivalent circuit modeling [4]. During battery cycle discharge and rest experiments, local voltage variation curves are illustrated in the figure below.

As shown in Fig. 1.6, when a battery is connected to a load and discharges or immediately after discharge ends, the battery terminal voltage experiences an abrupt drop or sudden rise. This phenomenon is attributed to the ohmic internal resistance inherent in lithium-ion batteries, causing significant voltage fluctuations during the initial stage of discharge or immediately after discharge termination. This is referred to as the internal resistance effect. In the first phase, the terminal voltage rapidly declines before gradually decreasing further, or upon termination of discharge, it slowly recovers.

This behavior results from polarization effects in lithium-ion batteries. After sufficient resting time, equilibrium is achieved within the battery, and polarization effects vanish [5]. Polarization effects generate a polarization internal resistance; battery manufacturing processes, internal structure, and operating conditions all influence this polarization resistance. However, discharge current and operating temperature exert particularly pronounced effects on polarization resistance due to their impact on lithium-ion mobility within the battery during operation.

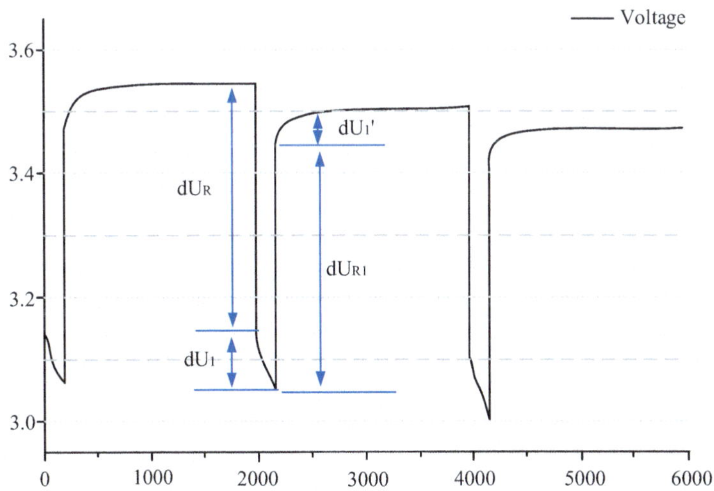

Figure 1.6 Voltage variation curve of the cyclic discharge and rest experiment.

Internal resistance is most sensitive to temperature, and its value can vary significantly under different temperature conditions. A major reason for the degradation of battery performance at low temperatures is excessively high internal resistance [6]. Generally, when lithium-ion batteries are used as power sources, from an external perspective, lower internal resistance is preferable. This is especially critical in high-power applications requiring low internal resistance.

1.3.1.3 Voltage characteristics

Terminal voltage refers to the potential difference between the positive and negative electrodes of a lithium-ion battery. Depending on the operating conditions of the circuit, it can be classified into Open-circuit voltage (OCV) and operating voltage [7]. Open-circuit voltage is the terminal voltage of a lithium-ion battery when no load is connected and no external power source is applied. Under normal conditions, due to the internal resistance of the battery, the operating voltage during discharge is lower than the open-circuit voltage, whereas during charging it is higher than the open-circuit voltage. The maximum voltage allowed at the end of charging is known as the charge cutoff voltage, while the minimum voltage permitted at the end of discharge is referred to as the discharge cutoff voltage. Exceeding these limits may cause irreversible damage to the battery; thus, cutoff voltages represent critical safety parameters.

Operating voltage refers to the potential difference between the positive and negative electrodes of a battery when it is in operation. Due to the internal resistance within the battery, the operating voltage is lower than the open-circuit voltage [8]. Similar to the classification of rated voltage, lithium-ion battery operating voltages are also categorized into rated voltage, theoretical voltage, open-circuit voltage, and loaded operating voltage based on their operational states. The rated voltage is directly specified by the battery manufacturer prior to shipment after production. The theoretical voltage represents the theoretically calculated potential difference between the positive and negative electrodes after internal electrochemical equilibrium has been established, denoted by E, with its calculation process as shown in the following equation.

$$E = \varphi_{+} - \varphi_{-} \tag{1.19}$$

In the above equation, φ_{+} is the electrode potential of the battery's positive terminal, and φ_{-} is the electrode potential of the battery's negative terminal. The open-circuit voltage can be directly measured by connecting a multimeter or voltmeter across the positive and negative terminals of the lithium-ion battery. The open-circuit voltage is the voltage used in the equivalent model simulation presented in this book. Since lithium-ion batteries are not ideal power sources and both the electrolyte and electrode materials possess internal resistance, the open-circuit voltage is slightly less than the electromotive force. However, when internal resistance is extremely small, the open-circuit voltage may be treated as equivalent to the electromotive force.

The operating voltage of a battery depends on its discharge regime. During constant-current discharge, the operating voltage continuously decreases until it reaches the permissible minimum voltage, at which point discharge terminates. This minimum voltage is determined based on considerations of battery safety and cycle life, and is referred to as the cut-off voltage for discharge, as illustrated in Fig. 1.7. The curve depicting the variation of battery operating voltage with discharge time is known as the discharge curve. In the discharge curve, the voltage range or duration during which the voltage remains relatively stable is termed the voltage plateau. Conversely, during charging, the voltage gradually increases, with corresponding concepts of charge termination voltage and charging voltage plateau.

Under actual operating conditions, the instantaneous voltage of lithium-ion batteries varies dynamically, during which overcharging and over-discharging are unavoidable and can cause damage to the batteries. Therefore, upper and lower voltage limits are set for lithium-ion batteries during practical use. These upper and lower cutoff voltages must be properly configured: setting the upper cutoff voltage too high may lead to safety hazards, while setting it too low may result in incomplete charging [9]. Similarly, setting the lower cutoff voltage too low may also pose safety risks; however, setting it too high will reduce the battery's energy utilization efficiency. Research indicates that under conditions of high discharge rate and low ambient temperature, the lower cutoff voltage for discharge can be set slightly lower when discharging lithium-ion batteries.

1.3.1.4 Current characteristics

There is no ideal power source in the world, whether a current source or a voltage source, all possess a certain internal resistance. Lithium-ion batteries, serving as automotive power sources, are no exception. When a battery continuously carries a large current, significant thermal energy is generated

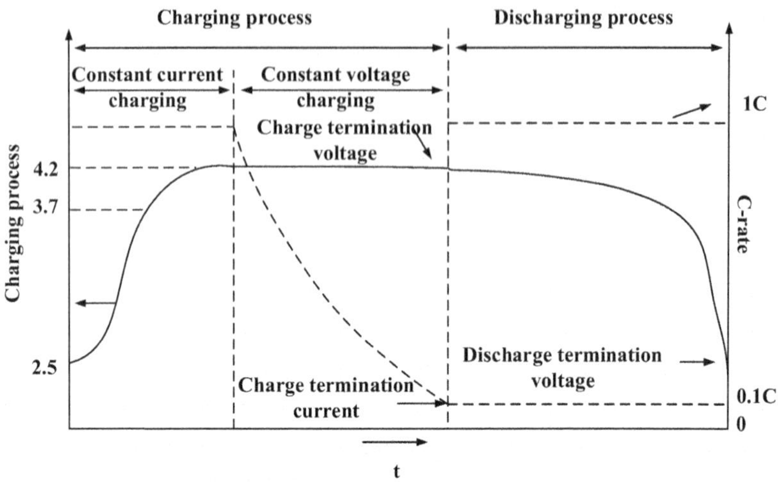

Figure 1.7 Typical charging and discharging curve of a lithium-ion battery.

due to the current passing through the resistance, as expressed by the following equation.

$$W = PT = I^2RT \tag{1.20}$$

This will cause the battery temperature to continuously rise, and the increasing temperature will further elevate the resistance, creating a vicious cycle. Once the battery management system detects abnormal temperature, it will take measures to initiate cooling [10]. Furthermore, prolonged high-current charging and discharging can lead to safety hazards; therefore, the current magnitude during lithium-ion battery charging and discharging must be continuously monitored.

1.3.1.5 Temperature characteristics

Battery temperature during real-time operation is an indispensable factor that cannot be overlooked. Battery activity is positively correlated with temperature: as temperature increases, the rate of lithium-ion exchange within the electrolyte accelerates, enhancing overall battery activity. This manifests as more complete energy output, increased available capacity, and improved operational efficiency. However, research has shown that prolonged exposure to high temperatures gradually degrades the structural stability of the positive electrode material's crystal lattice, significantly reducing battery safety and cycle life, thereby compromising product quality and causing it to fall below standard requirements. Conversely, under prolonged low-temperature conditions, the activity of both anode and cathode materials diminishes markedly; actual capacity drops significantly below rated capacity, charge/discharge efficiency declines, and overall utilization efficiency deteriorates.

Moreover, reduced material activity leads to diminished Li + exchange capability within the battery. The electrolyte's ability to transport Li + also deteriorates, resulting in Li + deposition at both electrodes and within the electrolyte, posing serious safety risks. Therefore, operating batteries at low temperatures with reduced charge/discharge current can mitigate Li + deposition and thus lower safety hazards [11]. In summary, sustained operation under either high or low temperatures adversely affects battery performance and significantly shortens battery lifespan.

1.3.1.6 Capacity characteristics

Battery capacity refers to the total amount of electrical charge that a lithium-ion battery can deliver when fully charged, typically denoted by the symbol Q and expressed in units of mAh (milliampere-hours) or Ah (ampere-hours). Battery capacity is a crucial indicator of battery quality and is closely related to the battery's operational duration [12]. In laboratory settings, lithium-ion batteries are generally discharged using either constant-current or constant-voltage methods. The battery's stored capacity is determined by measuring the total energy released during discharge or by calculating the product of discharge current and time, with the result expressed in Ampere-hours (Ah) or Milliampere-hours (mAh). Battery capacity can be classified into three categories: rated capacity, actual capacity, and theoretical capacity.

The theoretical capacity is the calculated value of the total charge of a lithium-ion battery, derived based on the electrochemical reactions occurring within the battery under the assumption that all lithium ions fully participate in the reaction, according to Faraday's law. Since the theoretical capacity represents an ideal value, it exceeds both the rated capacity and the actual capacity. The actual capacity is typically lower than the rated capacity due to degradation caused by battery usage [13]. The ideal capacity denotes the upper limit of electrical energy that a lithium-ion battery can deliver under theoretical conditions, corresponding to the complete and lossless conversion of stored chemical energy into electrical energy. In practical discharge processes, however, side reactions within the battery reduce efficiency, so that only a portion of the ideal capacity is actually achieved.

Rated capacity is determined by the battery manufacturer and directly marked on the battery's outer surface. It is a crucial indicator representing the amount of electric charge a battery can store, as well as a key metric for assessing battery operational duration. Rated capacity is also referred to as nominal capacity. According to regulations established by relevant national authorities, it denotes the minimum amount of charge that a battery is guaranteed to deliver under specified discharge conditions (such as temperature and discharge rate). The rated capacity corresponds to the Ah value specified by the manufacturer and constitutes one of the essential parameters of lithium-ion batteries.

The actual capacity of a lithium-ion battery refers to the maximum amount of charge it can deliver under specific discharge conditions immediately after purchase from a store. The actual capacity is typically calculated as the product of the discharge current and discharge time. This value does not necessarily correspond to the capacity of a brand-new battery; it may also represent the capacity of a used lithium-ion battery, in which case the actual capacity will inevitably be lower than that of a new battery. The actual capacity reflects the amount of electrical energy a battery can deliver during operation under real-world conditions. Numerous factors influence the actual capacity, including ambient temperature, the materials used in lithium-ion battery manufacturing, and the duration of usage. The charge delivered by a fully charged battery under certain conditions corresponds to the integral of discharge current over time, and this calculation process differs depending on whether it is conducted under constant current or alternating current conditions, as mentioned in Section 1.2.1.

1.3.1.7 Number of charge-discharge cycles

During the cycling charge and discharge process, lithium-ion batteries experience degradation in their electrolyte, positive and negative electrode materials, as well as external structure. The period from initial use to end-of-life is defined as the cycle life of the lithium-ion battery. This duration is determined by manufacturers through standardized charge/discharge testing under specified environmental conditions. Typically, the cycle life is defined as the number of charge/discharge cycles until the battery's actual capacity declines to 80% of its rated capacity.

1.3.1.8 Battery power of lithium batteries

The actual physical meaning of lithium-ion battery power is the amount of work done by the energy output generated through the internal chemical reactions of the battery per unit time, which corresponds to the quantity of energy released by the chemical reactions during that period. The unit is Watt (W). Depending on engineering applications, power can be further categorized into real power, instantaneous power, and others. The expression for lithium-ion battery power is given by the formula below, where E represents the electromotive force of the battery, R denotes the total internal resistance, and I is the average current per unit time.

$$P = UI = I(E - IR) = IE - I^2R \tag{1.21}$$

1.3.1.9 State of charge

State of Charge refers to the remaining capacity of a battery. The SOC value represents the ratio of the battery's remaining capacity under specific discharge conditions to its rated capacity. SOC is a relative quantity expressed as a percentage, with a range from 0% to 100%.

The state of charge of a lithium-ion battery reflects the remaining capacity of the battery at a given moment during its operation. Since the SOC cannot be directly measured, it must be estimated indirectly based on other observable external parameters of the battery, playing a crucial role in battery health state assessment. The SOC is defined as the ratio of the current remaining capacity to the fully charged capacity under conditions of full rest, as expressed in the following Eq. (1.22).

$$\mathrm{SOC}_t = \frac{Q_t}{Q_0} \times 100\% \tag{1.22}$$

In the equation, Q_t represents the remaining capacity of the battery at time t, and Q_0 denotes the rated capacity. The state of charge ranges between 0 and 1; an SOC value of one indicates a fully charged battery, while an SOC value of 0 indicates a completely discharged battery. The discharged capacity can be expressed as the integral of current over time. However, in practical operation, the actual charge that can be delivered by the battery is often lower than its nominal value due to internal resistance. Therefore, by incorporating Coulomb efficiency, also known as discharge efficiency, defined as the ratio of discharged capacity to charged capacity during the same cycle, the expression is corrected as shown in Eq. (1.23).

$$\mathrm{SOC}_t = \mathrm{SOC}_{t_0} - \frac{\int_{t_0}^{t} I(t)\eta dt}{Q_0} \tag{1.23}$$

where, SOC_{t_0} is the remaining capacity of the lithium-ion battery at time t_0; from time t_0 to the current moment, the battery discharges at a current $I_{(t)}$, with discharge direction defined as positive; $I_{(t)}$ represents the battery's operating current.

1.3.1.10 Depth of discharge

Depth of discharge (DOD) refers to the ratio of discharged capacity to the stored capacity (rated capacity). A smaller value indicates shallower discharge. The mathematical relationship between DOD and SOC is expressed as (1.24).

$$DOD = 100\% - SOC \tag{1.24}$$

1.3.1.11 Self-discharge rate

Self-discharge rate refers to the ratio of the amount of charge lost by a battery under no-load conditions to its rated capacity, used to indicate the rate at which battery capacity diminishes. It is primarily influenced by factors such as battery manufacturing processes, materials, and storage conditions. Typically, it is expressed as the percentage of capacity loss per unit time (per month or per year). Therefore, the self-discharge rate is equivalent to the leakage current rate, generally measured as the leakage current under open-circuit conditions. Since leakage currents are very small, the self-discharge rate of lithium-ion batteries is usually specified every month. The self-discharge rate can also be regarded as the battery's charge retention rate.

1.3.1.12 Coulomb effect

It has been demonstrated in capacity calibration experiments that during battery discharge, due to internal losses, the discharge capacity is always less than the charging capacity. Based on this observation, the concept of lithium-ion battery Coulombic efficiency was proposed. Coulombic efficiency is typically used to characterize the ability of a battery to deliver discharge capacity, defined as the ratio of discharge capacity at full charge to the corresponding charge capacity within the same cycle. It is generally a proper fraction less than 1.

The ratio of the charge capacity during the charging process to that during the discharging process. Generally, this is a proper fraction less than 1. Due to self-decomposition of the electrolyte, material aging, ambient temperature variations, and differing charge/discharge rates during operation, battery discharge efficiency is affected. In later simulation models, charge/discharge efficiency must be incorporated to compensate for and correct charge/discharge capacities. Combined with capacity calibration experiments, the Coulombic efficiency calculations for different charge/discharge processes are shown in Table 1.6.

As shown in Table 1.6, Coulomb efficiency is generally a proper fraction less than one under normal conditions. This occurs because the internal resistance inherent in the battery consumes part of the electrical energy during discharge, resulting in the total energy discharged being consistently less than the total energy charged. Experimental verification further indicates that as the discharge current increases, Coulomb efficiency decreases. This is because higher discharge currents intensify the electrochemical reactions within the battery, leading to increased internal resistance and greater self-consumption of energy, thereby reducing the amount of charge that can be discharged and consequently lowering Coulomb efficiency.

1.3.2 Dynamic response characteristics

During the operation of lithium-ion batteries, they are subjected to various pressures, which can be categorized into two main types based on their sources: internal stress and external stress. Internal stress arises from electrode material deformation and volume changes caused by lithium-ion insertion and extraction behaviors between the positive and negative electrodes, as well as various side reactions occurring during battery operation. External stress primarily results from mechanical constraints imposed during battery assembly, such as pressure generated by the casing during encapsulation, and rigid constraints applied during battery pack assembly to secure individual cells in place.

The influence of pressure on lithium-ion battery performance is dualistic. Appropriate pressure can enhance the contact tightness among various battery components, suppress lithium plating, and thereby improve cycle

Table 1.6 Coulomb effect.

Monomer 1		Monomer 2		Monomer 3	
Serial number	**Coulomb efficiency**	**Serial number**	**Coulomb efficiency**	**Serial number**	**Coulomb efficiency**
1	0.942	1	0.859	1	0.913
2	0.920	2	0.911	2	0.943
3	1.002	3	0.925	3	0.962

performance. However, excessive pressure may damage the structure of active materials and accelerate battery aging.

This section primarily discusses the causes of various stresses and their effects on lithium-ion battery performance.

1.3.2.1 External stress

The external stresses experienced by batteries mainly include two aspects: the external forces applied during battery manufacturing and mechanical stresses.

1.3.2.1.1 Force experienced during battery fabrication

The manufacturing process of lithium-ion batteries comprises three primary stages: (1) electrode fabrication, including material preparation, slurry mixing, coating, heating, drying, rolling, and cutting; (2) cell assembly, involving encapsulation, tab welding, housing installation, electrolyte filling, and sealing; and (3) module integration and system-level integration of subsystems such as the battery management system and cooling system. During several critical stages, including material preparation, electrode rolling, winding or stacking, encapsulation, and battery pack assembly, mechanical pressure is applied to the battery cells.

This pressure plays a crucial role in determining the final performance of lithium-ion batteries. The influence of pressure on battery performance is dualistic: appropriate pressure ensures tight contact among all components within the cell, prevents poor interfacial contact between electrodes and electrolyte, thereby improving lithium-ion deposition morphology and enhancing cycle stability. It also reduces internal resistance and polarization, thereby improving rate capability. Insufficient pressure may lead to delamination at the electrode-electrolyte interface after multiple charge/discharge cycles, resulting in rapid capacity fade and severely compromising battery lifespan. Excessive pressure can deteriorate interfacial integrity, cause a significant increase in internal resistance, accelerate capacity degradation, and in extreme cases may physically damage the cell structure leading to thermal runaway or loss of control.

1.3.2.1.2 Mechanical stress

The impact of mechanical pressure on lithium-ion battery performance is a significant research area. When subjected to external mechanical pressure, lithium-ion batteries undergo a series of complex changes in their internal structure and electrochemical properties. These alterations may directly affect the battery's capacity, cycle life, safety, and overall performance. Mechanical pressure can induce deformation and damage to the active materials within the battery, thereby impairing its capacity and energy density. Additionally, mechanical stress may alter the ion transport pathways inside the battery, reducing ionic conductivity and increasing internal resistance. This results in decreased efficiency during charging and discharging processes, ultimately degrading the battery's overall performance.

Mechanical pressure also significantly affects the safety of lithium-ion batteries. Excessive mechanical pressure may cause the internal separator to rupture, allowing direct contact between the positive and negative electrodes, thereby triggering a short circuit or even thermal runaway. This not only degrades battery performance but also poses serious safety risks.

Mechanical stress may also affect the cycle life of batteries. Under prolonged mechanical stress, the active materials and electrolyte within the battery may gradually degrade, leading to a progressive decline in battery capacity. This degradation compromises battery longevity and may accelerate the aging process. Mechanical stress can also induce the formation and fracture of the Solid electrolyte interphase (SEI) layer on the anode material, further impairing both cycle performance and safety.

One important aspect of mechanical pressure's influence on lithium metal battery performance is its effect on electrode material deformation. The lithium metal anode undergoes severe volume expansion, up to 79 times during charging, leading to morphological changes and dendrite growth. Such volume variations can cause particle fracture within the electrode, crack formation on the electrode surface, and poor electrical contact, thereby degrading battery cycle performance and safety. Mechanical pressure can induce deformation of $LiCoO_2$ particles in lithium-ion battery cathodes, consequently affecting their electrochemical performance. Cathode materials also experience deformation during lithium insertion and extraction processes, resulting in structural changes that reduce electrode conductivity and the utilization efficiency of active materials. Furthermore, mechanical pressure can indirectly influence battery performance by affecting electrolyte film thickness, alterations in electrode interface structure, and deformation of solid-state electrolytes.

In lithium metal batteries, the separator plays a critical role by establishing ionic conduction pathways between the cathode and anode, while simultaneously suppressing the formation of lithium dendrites and preventing internal short circuits. Due to the high reactivity of lithium metal, which readily reacts with numerous materials, the separator's structural integrity may be compromised through chemical degradation or mechanical deformation, significantly affecting battery performance. Mechanical stress induces strain in the polymer matrix of the separator, thereby impairing its ionic conductivity and electrical properties. Furthermore, mechanical pressure can alter the interfacial characteristics between the separator and both electrodes, disrupting electrochemical reactions within the cell.

Under typical operating conditions, lithium-ion batteries undergo repeated charge/discharge cycles; during these cycles, lithium ions repeatedly intercalate into and deintercalate from the lithium metal electrode, gradually leading to the formation of lithium dendrites. Once dendrites penetrate the separator, short circuits and battery failure may occur, potentially resulting in fire or even explosion in severe cases. The growth of lithium dendrites is rapid and difficult to control; their morphology and size evolve unpredictably over time due to continuous charging and discharging cycles. Additionally, electrolyte-induced corrosion further increases the risk of separator rupture.

Regarding mechanical deformation, although the separator experiences relatively uniform and minor compressive stress during battery rest periods, this does not imply that the separator remains immune to mechanical damage during charging and discharging cycles. Repeated charge/discharge cycles subject the battery to additional mechanical stress, further exacerbating deficiencies such as insufficient toughness, cracking, and deformation in the separator. The high reactivity of lithium metal can also adversely affect other performance aspects of the separator; upon contact with lithium-ion batteries, separators may exhibit various crack morphologies, including spiral cracking, radial cracking, or crack propagation along puncture grooves. Variations in environmental temperature and humidity at which the separator material is located may also influence its performance. Although such variations typically occur under non-extreme conditions, cyclic mechanical stress combined with fluctuating temperature and humidity increases the likelihood of mechanical degradation in separators. This can ultimately lead to premature failure of separators and subsequent battery malfunction, thereby impairing both battery performance and lifespan.

The electrolyte in lithium-ion batteries plays a critical role during charging and discharging processes. Mechanical pressure can influence the physical properties of the electrolyte, such as viscosity and dielectric constant, thereby affecting key performance parameters such as internal resistance and capacitance. Mechanical pressure may also induce decomposition or precipitation of certain electrolyte components, further degrading battery performance. The structural components of lithium-ion batteries, including the casing and sealing elements, protect the internal structure while simultaneously influencing battery performance. Mechanical pressure alters the stress distribution within these structural components, thereby impacting battery safety and cycle life. Additionally, mechanical pressure may lead to poor contact between structural components and internal battery structures, further compromising overall battery performance.

Mechanical pressure significantly influences the electrochemical reactions within batteries. Under mechanical stress, structural changes in active materials, deformation of electrode materials, and altered behavior of the electrolyte may occur, thereby affecting critical performance indicators such as capacity, cycle life, and safety.

At the same time, the impact of mechanical stress on battery performance is not unidirectional. During battery charging and discharging, volume changes in electrode materials generate internal stresses, which in turn interact with electrochemical reactions and jointly influence battery performance. When investigating the effects of mechanical pressure on lithium metal battery performance, it is essential to comprehensively consider the mutual interactions between mechanical stress and electrochemical reactions, as well as their dynamic evolution under varying conditions. Mechanical pressure is regarded as a potential solution that may suppress lithium dendrite formation by altering the structure of electrode materials. As lithium ions migrate from the cathode to the anode during charge and discharge cycles, they deposit on the lithium metal surface, leading to dendritic growth. Mechanical pressure can influence the flow and deposition behavior of lithium ions by modifying atomic arrangements and structural configurations within electrode materials.

Some studies indicate that appropriate mechanical pressure can promote the formation of more stable compounds on the lithium metal surface, thereby reducing dendrite formation. Furthermore, mechanical pressure can affect chemical stability by altering the morphology of electrode materials. Processing techniques such as rolling or stretching can produce electrode materials with more uniform grain sizes and denser crystalline structures. These structural modifications enhance both mechanical strength and chemical stability, enabling greater resistance to lithium dendrite propagation.

The impact of mechanical pressure on lithium metal battery performance, particularly cycle life and safety, is a double-edged sword. By employing appropriate strategies and methods to apply and regulate mechanical pressure, its beneficial effects can be maximized while potential risks are minimized. Future research will continue to explore effective ways to integrate and apply mechanical pressure in battery design, aiming to achieve lithium metal batteries with enhanced performance and improved safety.

1.3.2.2 Internal stress

During operation, the complex electrochemical-mechanical coupling processes are the primary cause of volume changes in lithium-ion batteries. During charging/discharging, lithium ions or atoms can intercalate/deintercalate into/from electrode materials via three distinct mechanisms: intercalation/deintercalation in graphite anodes and layered cathodes (such as LCO and Li[NixCoyMnz]O2, commonly known as NCM); alloying/dealloying in metallic anodes (e.g., Al and Si); and conversion reactions in novel anode materials (e.g., Co3O4 and Fe3O4). The continuous insertion and extraction of lithium ions between the cathode and anode inevitably induce structural changes in the electrode materials, which consist of both reversible and irreversible components.

The reversible portion refers to the material's ability to recover its initial shape after completion of electrochemical cycling. It is primarily caused by the repeated insertion and extraction of lithium ions between the positive and negative electrodes during battery operation.

The irreversible portion arises from various side reactions, such as active material dissolution, gas generation, formation of surface layers on electrodes (lithium plating, anode SEI film, cathode CEI film), and electrolyte decomposition. The primary causes of battery irreversibility are the growth of the cathode CEI film and anode SEI film, as well as gas generation from electrolyte decomposition. Studies have shown that the thickness of the cathode CEI film is on the order of several nanometers and does not induce noticeable swelling. Lithium plating tends to occur under low-temperature operation, whereas electrolyte decomposition predominantly occurs under high-temperature conditions or overcharging. Therefore, under normal operating conditions, the continuous thickening of the anode SEI film is the main contributor to irreversible deformation in batteries. Furthermore, due to external constraints, these deformations generate internal stresses within the battery. Such constraints mainly originate from non-uniform volume changes in electrode materials and mechanical limitations imposed by the battery casing.

The charge and discharge mechanism of lithium-ion batteries is illustrated in Fig. 1.8. A lithium-ion battery consists of a cathode material, an anode material, an electrolyte, a separator, and a binder. The cathode and anode serve as the primary storage sites for lithium ions, undergoing reversible delithiation and lithiation reactions during charging and discharging. The electrolyte facilitates the transport of lithium ions, which diffuse through the electrolyte during charge and discharge cycles. The separator physically isolates the cathode and anode materials. During the lithiation and delithiation processes at the electrodes, the insertion or extraction of lithium ions induces deformation in the electrode materials, resulting in Diffusion-induced stress (DIS). Excessive DIS may lead to fracture of electrode particles and other failure modes. Ultimately, these degradation mechanisms contribute to battery capacity fade, increased impedance, and shortened cycle life.

The operation of lithium-ion batteries involves the repeated insertion and extraction of lithium ions within electrode materials during electrochemical cycling. The mechanisms of insertion and extraction may vary depending on the type of active electrode material. For instance, in graphite or layered oxides such as $LiCoO_2$, lithium ion insertion and extraction occur via intercalation and deintercalation processes. In materials science, intercalation can be broadly defined as the reversible insertion of species between the crystallographic planes of a host lattice without disrupting the arrangement of these planes. In contrast, metallic anode materials such as Si, Sn, and Al undergo alloying and dealloying reactions that allow for lithium insertion and extraction through the formation of intermetallic

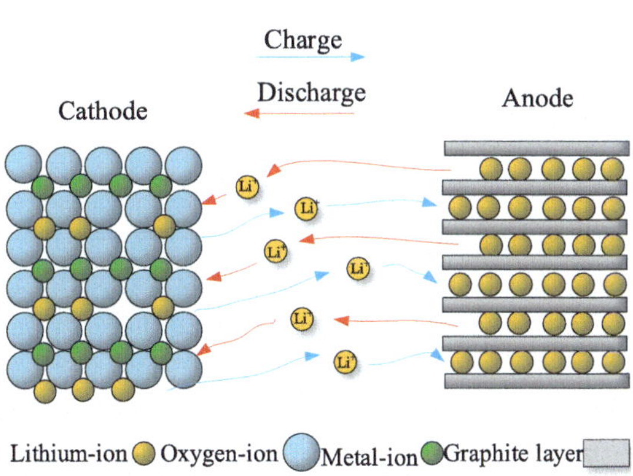

Figure 1.8 Schematic diagram of lithium battery operation.

compounds. In both cases, foreign lithium ions must be temporarily accommodated within the host lattice during repeated insertion and extraction cycles. The accommodation of these ions during electrochemical cycling alters the interplanar spacing within the host crystal structure, resulting in corresponding dimensional changes. Furthermore, such insertion and extraction processes can induce phase transitions and changes in crystal symmetry in many materials.

Stress induced by lithiation and delithiation during the electrochemical cycling of lithium-ion batteries arises from the insertion and extraction of lithium ions into and from the host lattice (electrode materials), resulting in significant stress generation within the electrode materials. The magnitude of this stress spans multiple orders of magnitude and depends on various factors, including the specific electrode material, electrode architecture, electrode composition, mechanisms of lithium insertion and extraction, electrochemical cycling rate, and the potential range under consideration. Furthermore, stress distribution within active electrode materials is often non-uniform. The primary mechanisms responsible for such stress generation are broadly categorized as illustrated in Fig. 1.9 and summarized as follows.

1.3.2.2.1 Physical constraints arising from dimensional changes in active materials lead to major structural alterations

When "foreign" lithium ions are inserted into or extracted from the bulk crystal lattice of active electrode materials, changes in lattice parameters or the formation of new crystalline phases may occur. However, the primary consequences of these phenomena are alterations in overall volume and morphology. In most cases, such volume and morphological changes are constrained by neighboring particles, substrate, current collector, or the available space within the battery (see Fig. 1.9a and b). These constraint effects represent the main source of stress development in many electrode materials.

1.3.2.2.2 Inter-particle contact

The most common electrode structure currently consists of active material particles, binder, and conductive additive. Due to the volume expansion of active particles during lithiation, particle contacts occur (see Fig. 1.9c), which may lead to significant stress at contact points.

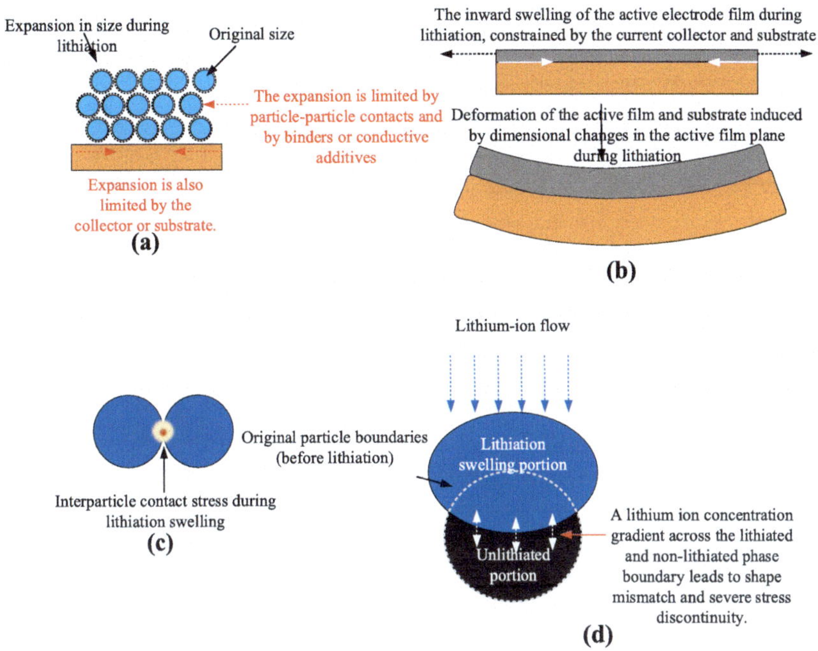

Figure 1.9 Primary causes of stress generation due to internal chemical cycling in lithium-ion batteries.

1.3.2.2.3 Mismatch between the crystal phase and the lithium concentration gradient

The changes occurring during lithium insertion and extraction are correlated with the lithium content. Furthermore, since lithiation and delithiation primarily proceed via diffusion of lithium ions within the active material, significant lithium-ion concentration gradients can develop within the active material. These lithium-ion concentration gradients depend on the lithium diffusion rate and current density, resulting in higher lithium ion concentrations at faster electrochemical cycling rates. Such concentration gradients lead to differing dimensional and volumetric changes as well as phase transformations from the battery surface to its core. Consequently, at any given moment, adjacent regions within the same active material may exhibit different crystal phases and distinct molar volumes. Additionally, these coexisting crystal phases may possess different elastic properties. The interactions and contacts between these regions generate stress.

It must be emphasized that the lithiation and delithiation occurring during electrochemical cycling are dynamic processes, thereby inducing repeated stress reversal. This repeated reversal further exacerbates the accumulation of damage caused by these stresses. This phenomenon can be analogized to more general fatigue mechanisms, which illustrate the increasing severity of stress effects due to cyclic loading.

Stress can directly induce structural disintegration of electrode materials. The extent of such disintegration depends on the electrode material and its structural characteristics. The degree of volume change during lithiation and delithiation, the mechanical properties of the material, and the degree of mechanical constraint collectively determine the severity of structural degradation. Extensive research has demonstrated that the electrochemical cycling rate and the extent of lithiation/delithiation also influence the degradation of electrode materials. Structural disintegration typically manifests as fracture and fragmentation of active electrode materials. In most cases, fractured segments lose contact with the electrolyte or other active materials, resulting in electrical isolation. Consequently, progressive fracture reduces the amount of active electrode material, thereby diminishing the capacity of lithium-ion batteries. Furthermore, stress generation can damage binders and reduce pore volume, compromising the mechanical integrity of electrodes. All these effects contribute to a reduction in battery capacity and a continuous decline in energy density with ongoing electrochemical cycling. Such stress-induced fracture and disintegration of electrode materials are considered one of the primary causes leading to ultimate failure in lithium-ion batteries. For certain electrode materials and structures, severe stress may result in structural fracture so profound that electrodes become non-functional after only one to a few electrochemical cycles.

In addition to capacity fade caused by electrical insulation resulting from structural fracture, fracture also generates new electrode surfaces. When these newly exposed surfaces come into contact with the electrolyte, further electrolyte reduction occurs, leading to the formation of a new SEI layer. Consequently, such stress-induced fractures result in irreversible lithium consumption, negatively affecting battery capacity. Furthermore, volume changes in active materials can also cause the existing SEI layer to fracture, triggering repeated SEI formation and reformation. Moreover, it is now widely recognized that SEI formation and deposition processes themselves induce significant stresses, further compromising the mechanical integrity of these layers. The resulting stresses and fractures within the SEI layer lead to continuous accumulation of the SEI after initial cycling, thereby causing ongoing irreversible lithium consumption within the system.

In addition, lithium intercalation and deintercalation can also lead to lithium plating. Lithium plating typically occurs at the anode, where a gray and white substance forms on the anode surface. Lithium plating refers to the deposition of lithium ions from the electrolyte onto the anode surface. The highly reactive metallic lithium deposited on the anode surface rapidly reacts with the electrolyte to form a SEI layer. As further lithium continues to deposit beneath the existing SEI layer, it eventually ruptures, prompting the formation of a new SEI layer. This process results in continuous consumption of electrolyte components, leading to a gradual decrease in electrolyte concentration. Subsequently, metallic lithium begins to grow perpendicularly to the electrode surface, forming lithium dendrites. The growth of lithium dendrites is considered one of the most detrimental side reactions. The dendritic morphology of lithium deposits provides a significantly increased surface area, enhancing contact between metallic lithium and the electrolyte and thereby accelerating consumption of both active lithium and electrolyte. Furthermore, due to their high hardness, these dendritic structures can fracture and damage the SEI layer. This exposes freshly deposited lithium on the anode surface to fresh electrolyte, intensifying side reactions and accelerating degradation of active materials within the battery, ultimately resulting in reduced capacity. Moreover, owing to their electronic insulating nature and brittleness, dendrite roots may fracture and detach from the electrode surface, forming freely dispersed dendritic fragments known as "dead" lithium. These "dead" lithium species neither participate in electrochemical reactions nor contribute to battery capacity. Their diffusion around the anode impedes ion transport and increases polarization and interfacial resistance, leading to additional voltage losses. Finally, continuous growth and partial dissolution of dendrites induce volume changes that further destabilize both electrode structure and charge/discharge performance. Dendrites growing at their tips may penetrate through the separator membrane, causing internal short circuits that result in battery failure or even explosion.

The delithiation and lithiation reactions occurring at the positive and negative electrodes of the battery also induce structural changes in the electrodes. In the absence of external structural constraints, the battery thickness will accordingly change. However, in energy storage battery modules, physical structural constraints exist externally on the battery, causing the original deformation to be converted into corresponding stress. Research has revealed that the positive and negative electrodes of lithium iron phosphate batteries exhibit different deformation characteristics. During battery charging, the volume of the lithium iron phosphate positive electrode shows an approximately linear decreasing trend, with thickness reduction reaching up to 7%. The graphite negative electrode also undergoes volume changes, displaying a non-monotonic variation pattern with a thickness increase of 8%. The combined effect results in a non-monotonic trend in overall battery thickness variation. Correspondingly, when mechanical constraints exist externally on the battery, surface stress during charging also exhibits a non-monotonic behavior, with an overall increasing trend. The discharge process is similar: surface stress likewise shows non-monotonic characteristics with an overall decreasing trend. The relationships among electrode thickness, battery stress, and SOC are illustrated in Fig. 1.10. This expansion generates external stress on the battery and requires greater space for accommodation, thereby affecting its normal operation.

1.3.3 Temperature dependence

The electrochemical performance of lithium-ion batteries is highly dependent on ambient temperature. Temperature not only influences the kinetics and thermodynamic equilibrium of electrochemical reactions within the battery, but also fundamentally determines the battery's energy release capability, cycle stability, and safety under various operating conditions by altering ion mobility, electrode interface charge transfer impedance, and

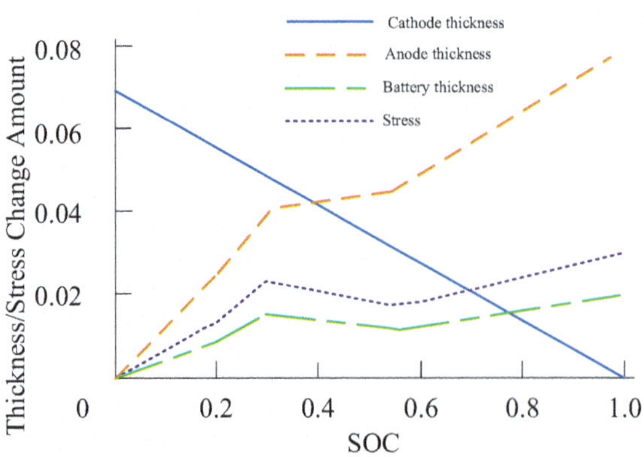

Figure 1.10 Correspondence among electrode thickness, battery stress, and battery SOC.

side reaction rates. Therefore, a thorough analysis of temperature dependence is crucial for establishing accurate battery models, designing effective thermal management systems, and enabling intelligent control in BMS. This paper systematically discusses the temperature dependence of lithium-ion batteries from five perspectives: open-circuit voltage characteristics, internal resistance characteristics, electrode kinetics, cycle life and rate capability, and impedance characteristics.

1.3.3.1 Temperature dependence of open-circuit voltage

The OCV of a battery under thermodynamic equilibrium is determined not only by the lithiation/delithiation potentials of the electrode materials but also closely related to the ambient temperature. The OCV essentially reflects the equilibrium potential of the electrode reaction, which depends on the change in Gibbs free energy associated with the reaction. According to fundamental thermodynamic relationships, the change in free energy is directly coupled with the entropy change of the reaction; thus, OCV exhibits a certain sensitivity to temperature. This temperature dependence is typically described by the Nernst equation, whose temperature coefficient is expressed as:

$$\frac{\partial U_{OCV}}{\partial T} = -\frac{\Delta S}{\mathrm{nF}} \tag{1.25}$$

In the above Eq. (1.25), U_{OCV} represents the open-circuit voltage, ΔS denotes the entropy change of the electrode reaction, n is the number of electrons transferred, and F is the Faraday constant. This equation indicates that the rate of change of OCV with respect to temperature is determined by the reaction entropy change. When $\Delta S < 0$, an increase in temperature leads to a decrease in OCV; conversely, when $\Delta S > 0$, a rise in temperature causes a slight increase in OCV. This phenomenon demonstrates that temperature variations not only affect the reversible reaction potential of batteries but also influence, to some extent, the availability of energy output. The figure below compares OCV and SOC curves at different temperatures: at low temperatures, the curves shift downward overall; at high temperature (45°C), they shift slightly upward; while at room temperature (25°C), the curve exhibits the most stability.

As shown in Fig. 1.11, under low-temperature conditions, the OCV curve shifts downward overall due to entropic effects. This implies that, at the same SOC, the battery's voltage level is relatively lower, resulting in a significant reduction in usable energy density. Meanwhile, ion diffusion

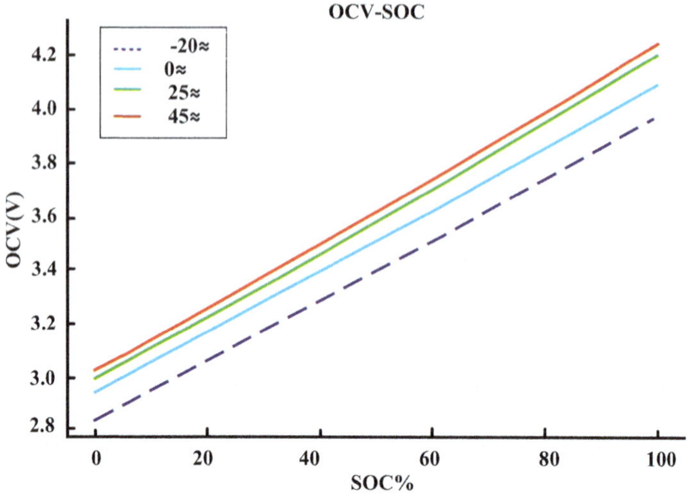

Figure 1.11 Comparison of OCV-SOC curves at different temperatures.

is restricted and polarization effects are exacerbated under low temperatures, further intensifying this downward trend and leading to a marked deterioration in the battery's actual discharge performance. Conversely, under high-temperature conditions, the OCV curve may exhibit a slight increase, manifesting as a marginal rise in voltage at the same SOC level, attributed to systems dominated by negative entropy changes. Although the elevated voltage level can temporarily enhance energy output, excessively high temperatures accelerate side reactions such as electrolyte decomposition and SEI layer degradation, ultimately shortening battery lifetime and introducing safety risks.

Therefore, the temperature dependence of OCV is not only a significant phenomenon in electrochemical theory but also holds practical importance in battery modeling and management. For algorithms estimating SOC based on the OCV and SOC curve, neglecting temperature correction for OCV will lead to substantial estimation errors. Particularly in electric vehicles and energy storage systems, where battery operating temperatures vary significantly, introducing a temperature compensation model is essential to ensure accurate estimation of both SOC and State of energy (SOE). This conclusion underscores the importance of temperature monitoring in BMS and provides a theoretical foundation for subsequent thermal management strategies.

1.3.3.2 Relationship between internal resistance and temperature

Internal resistance is one of the key parameters determining the power output capability and energy utilization efficiency of lithium-ion batteries, directly influencing the battery's terminal voltage, energy conversion efficiency, and heat generation level. Internal resistance is determined by multiple factors and is generally categorized into three main types: ohmic impedance, interfacial impedance, and diffusion impedance. Ohmic impedance primarily arises from the bulk electrical conductivity of current collectors, electrode materials, and electrolyte. Interfacial impedance is closely related to charge transfer processes at the electrode-electrolyte interface. Diffusion impedance reflects the difficulty of lithium-ion transport within both the electrolyte and solid-phase active materials. These three types of impedance exhibit significant differences under varying temperatures, resulting in a high sensitivity of internal resistance to temperature.

Under low-temperature conditions, the viscosity of the electrolyte significantly increases, leading to reduced lithium-ion mobility and consequently elevated ohmic impedance. Simultaneously, the activation energy effect of interfacial reactions is amplified, causing a slowdown in charge transfer processes and resulting in a rapid increase in interfacial impedance. Furthermore, the diffusion coefficients of lithium ions in graphite anodes and transition metal oxide cathodes sharply decrease with declining temperature, thereby increasing diffusion impedance. The cumulative effect of these three types of impedance causes a dramatic rise in battery internal resistance at low temperatures, manifesting as a pronounced voltage drop and insufficient dischargeable capacity. In contrast, under high-temperature conditions, the electrolyte's conductivity improves and ion mobility accelerates; meanwhile, the rate constants for interfacial reactions increase. As a result, both ohmic and interfacial impedances are significantly reduced, while diffusion impedance also decreases slightly, leading to an overall reduction in internal resistance.

This regularity can be formally described by the Arrhenius equation:

$$R(T) = R_0 \exp\left(\frac{E_a}{RT}\right) \tag{1.26}$$

In this Eq. (1.26), $R(T)$ represents the battery's internal resistance R_0 at a given temperature, A is a constant, Ea denotes the activation energy required for charge transfer or ion migration, R is the gas constant, and T is the absolute temperature. As indicated by this equation, as temperature increases, the denominator increases, causing the exponential term to decrease and thereby reducing the battery's internal resistance. Conversely, at low temperatures, the exponential term increases significantly, leading to a sharp rise in internal resistance. This formula not only elucidates the mechanism by which temperature governs battery internal resistance but also provides a mathematical expression essential for modeling. It holds significant importance in parameter identification and the development of equivalent circuit models.

From an application perspective, the temperature sensitivity of internal resistance directly affects battery performance under varying environmental conditions. Under low-temperature conditions, increased internal resistance during high-current discharge causes a rapid drop in terminal voltage, potentially triggering the protection circuit prematurely and preventing full discharge of the battery's energy. In electric vehicles operating during winter, this phenomenon manifests as a significant reduction in driving range. Conversely, at elevated temperatures, reduced internal resistance enhances power performance; however, accelerated side reactions and decreased electrolyte stability simultaneously increase capacity fade and safety risks. Therefore, the temperature and dependent behavior of internal resistance establishes a delicate balance among battery power performance, energy utilization efficiency, and lifetime evolution.

In experimental studies, the internal resistance at different temperatures is typically measured using EIS or pulse current testing methods, and analyzed in conjunction with capacity retention rates. Table 1.7 provides a quantitative comparison of the internal resistance variation rate and capacity retention rate across different temperatures. The data indicate that at −20°C, the battery's internal resistance increases by over 120% compared to room temperature, while the capacity retention rate drops to 65%. At 45°C, the internal resistance decreases by approximately 10% relative to room temperature, and the capacity remains around 95%. When the

Table 1.7 Internal resistance and capacity performance at different temperatures.

Temperature (°C)	Rate of change of internal resistance (%)	Capacity retention rate (%)
−20	+120	65
0	+40	85
25	Benchmark	100
45	−10	95
60	−25	80

temperature rises further to 60°C, the reduction in internal resistance becomes more pronounced; however, due to intensified side reactions, the capacity retention rate declines to 80%. These results not only validate the applicability of the Arrhenius relationship but also demonstrate that both excessively high and low temperatures lead to suboptimal battery performance.

Therefore, in BMS, a temperature compensation mechanism must be introduced to correct the influence of dynamic internal resistance variations on power prediction and life estimation. Meanwhile, an appropriate thermal management system should be configured at the battery pack level to maintain the battery operating temperature within an optimal range through heating and cooling strategies, thereby preventing power deficiency caused by excessively high internal resistance at low temperatures, as well as safety risks resulting from accelerated side reactions due to reduced internal resistance at high temperatures. It can be said that temperature's regulation of internal resistance constitutes a fundamental basis for battery operational safety and reliability, and represents one of the core scientific challenges in battery thermal management and performance optimization.

1.3.3.3 Temperature dependence of electrode kinetic rates

Electrode reaction kinetics is one of the key factors determining the reversibility and efficiency of lithium-ion battery charging and discharging processes. Its rate constant typically follows the Arrhenius relationship:

$$k(T) = k_0 \exp\left(-\frac{E_a}{RT}\right) \tag{1.27}$$

In this Eq. (1.27), k(T) denotes the temperature and dependent electrode reaction rate constant, k_0 represents the pre-exponential factor corresponding to the rate constant under ideal conditions with no energy barrier; *Ea* is the activation energy of the reaction, reflecting the energy barrier that must be overcome for the reaction to proceed; *R* is the gas constant; and *T* is the absolute temperature. This equation reveals an exponential relationship between temperature and reaction rate, indicating that an increase in temperature significantly reduces the restriction imposed by activation energy, thereby accelerating the insertion and extraction rates of lithium ions.

Under low-temperature conditions, the reaction rate constant significantly decreases due to the $\exp(-E_a/RT)k(T)$ sharp reduction in the relevant term, resulting in hindered charge transfer at the electrode interface and enhanced battery polarization. This is manifested as an increased voltage difference during charging and discharging and reduced capacity utilization. Meanwhile, the diffusion rate of lithium ions within the electrode slows down, making concentration polarization and lithium dendrite deposition more likely during high-rate charging and discharging, thereby posing safety risks.

Under ambient temperature conditions, the reaction rate constant is at an intermediate level, enabling a balance between kinetic reversibility and suppression of side reactions. Consequently, batteries typically exhibit optimal charge/discharge efficiency and cycle stability within the 20 to 30°C range. At this temperature range, the electrode reaction process approaches reversibility, the voltage plateau remains stable, and the capacity output approaches the theoretical value.

Under high-temperature conditions, the Arrhenius equation indicates that the rate constant will further increase, significantly accelerating electrode reaction kinetics and enabling the battery to maintain a stable voltage platform even under high-current charge and discharge conditions. However, this kinetic acceleration also entails significant side effects. As temperature rises, side reactions, such as electrolyte decomposition, SEI film rupture, and structural redox instability of electrode materials, proceed at an accelerated rate, leading to irreversible capacity loss and reduced cycle life. Particularly when operating above 50°C, the lattice stability of electrode materials deteriorates, potentially triggering severe side reactions such as oxygen release and thermal decomposition of the electrolyte, ultimately posing a risk of thermal runaway.

Thus, the relationship between electrode reaction rate and temperature exhibits a double-edged effect: moderate temperature elevation improves kinetic conditions and enhances power performance, whereas excessively high temperatures accelerate side reactions within the electrochemical system, compromising battery safety and lifespan. To achieve a balance between kinetic advantages and safety risks in practical applications, the BMS must continuously monitor temperature and regulate charge/discharge rates. In low-temperature environments, preheating is employed to reduce polarization; under high-temperature conditions, derating and cooling measures are implemented to suppress side reactions.

From a modeling perspective, Eq. (1.27) provides a theoretical foundation for temperature correction in battery electrode kinetics. The incorporation of rate constants $k(T)$ into both equivalent circuit models and electrochemical models enables accurate representation of how interfacial charge transfer processes vary with temperature, thereby enhancing the model's accuracy in predicting performance under diverse operating conditions. This characteristic is particularly critical in applications such as electric vehicles and energy storage power stations, where systems must contend with complex thermal environments arising from diurnal temperature variations, seasonal changes, and operational mode transitions.

1.3.3.4 Influence of temperature on cycle life and rate capability

The cycle life and rate capability of lithium-ion batteries are critical indicators determining their long-term stable operation in applications such as electric vehicles and energy storage systems, both of which are closely related to temperature. Temperature not only governs the reaction kinetics and internal resistance levels during individual charge/discharge cycles, but also directly influences the accumulation of side reactions over cycling, thereby significantly affecting the evolution of battery lifespan on a long timescale.

Under low-temperature conditions, the diffusion coefficient of lithium ions and the rate constant of electrode reactions both significantly decrease, resulting in reduced effective capacity of the battery. Due to severe polarization during high-rate charge and discharge processes, the battery terminal voltage rapidly declines, preventing full capacity release. Under these conditions, lithium plating is more likely to occur during high-current charging, potentially leading to the formation of lithium dendrites. This not only causes irreversible capacity loss but may also puncture the separator, triggering short circuits. Long-term cycling under low-temperature conditions leads to a steep decline in capacity fade curves and a substantial reduction in battery lifespan.

Within the ambient temperature range, the battery's electrochemical kinetics and transport properties are optimized, enabling a stable voltage platform under high-rate conditions while effectively suppressing the rapid occurrence of side reactions. Consequently, the cycle life at ambient temperature is typically the longest, with capacity retention remaining close to baseline levels after numerous cycles. For instance, experimental studies have shown that lithium iron phosphate batteries operated at 25°C maintain a capacity retention exceeding 90% after 1500 cycles, outperforming performance under low or high-temperature conditions.

Although electrode interface kinetics are improved under high-temperature conditions, resulting in significantly enhanced rate performance and the ability of the battery to maintain a high voltage platform even at 3C or higher rates, elevated temperature simultaneously accelerates parasitic reactions. These include electrolyte decomposition, degradation of the SEI layer, and structural instability of electrode materials. The cumulative effects of these parasitic reactions lead to accelerated capacity fading and a dramatic reduction in cycle life. Particularly when the temperature exceeds 50 to 60°C, these side reactions cease to accumulate gradually and may rapidly escalate into uncontrolled states, manifesting as abrupt capacity drop or even safety incidents.

In experiments and modeling, the relationship between cycle life and temperature is often fitted using an exponential relationship between degradation rate and temperature:

$$N_f(T) = N_0 \exp\left(-\frac{E_d}{RT}\right) \tag{1.28}$$

In this equation, $N_f(T)$ represents the cycle life at temperature T (expressed in number of cycles), N_0 is a constant, Ea is the apparent activation energy for life degradation, R is the gas constant, and T is the absolute temperature. This equation indicates that an increase in temperature significantly shortens the cycle life, which is highly consistent with the accelerated rate of side reactions.

Regarding rate performance, temperature significantly affects the battery's rate discharge curve. At low temperatures, the 1C discharge curve can still maintain a certain capacity; however, at 3C or higher rates, the terminal voltage drops rapidly, resulting in insufficient capacity release. At room temperature, the battery exhibits optimal rate capability, with a stable voltage plateau and nearly 100% capacity utilization. Under high-temperature conditions, rate performance further improves; however, this enhancement comes at the expense of reduced cycle life. Table 1.8 presents a comparative analysis of cycle life and rate performance under different temperatures, enabling a direct assessment of how temperature modulates the battery's power characteristics.

The data in the table clearly demonstrates the dual impact of temperature on battery cycle life and rate performance: at low temperatures, cycle life is significantly shortened and rate performance is poor, while at high temperatures, although rate performance improves, cycle life is severely reduced.

Table 1.8 Comparison of cycle life and rate performance at different temperatures.

Temperature (°C)	1C cycle life (times)	Capacity retention rate (%)	3C discharge capacity (%)	Engineering significance
−20	<300	65	40	High rate performance is extremely poor and unsuitable for low-temperature applications.
0	~600	85	70	Preheating measures are required
25	~1500	100	95	Best performance, most widely applied
45	~1000	95	98	Good power performance, but reduced lifespan
60	~500	80	90	Life expectancy drops sharply, with high safety risks

In engineering applications, this temperature dependence imposes higher requirements on battery management systems. Electric vehicles must employ battery heating measures during winter to ensure adequate capacity release and power output under low-temperature conditions; in summer or during high-load operations, efficient liquid cooling or air cooling systems are required to prevent rapid degradation of lifespan and increased safety risks under high temperatures. Energy storage systems similarly utilize thermal control systems to maintain an optimal operating environment around 25°C, thereby extending the battery's full lifecycle.

In summary, temperature exerts a significant bidirectional regulatory effect on the cycle life and rate performance of batteries. Moderate temperatures help maintain kinetic stability and reversible cycling, keeping battery performance within an optimal range; deviations from this optimal range result in varying degrees of degradation in both performance and lifespan. Therefore, appropriate thermal management and operational strategies are crucial for extending battery lifetime and ensuring high-power performance.

1.3.3.5 Temperature dependence of electrochemical impedance

EIS is a widely used technique for investigating the reaction kinetics and transport properties inside lithium-ion batteries. By applying a small-amplitude alternating current signal and measuring the battery's voltage response at various frequencies, Nyquist plots can be obtained, enabling the separation of individual impedance components such as ohmic resistance, charge transfer resistance, and diffusion resistance. The influence of temperature on these impedance elements is highly significant; thus, EIS serves as a powerful tool for elucidating temperature and dependent behaviors.

Under low-temperature conditions, the viscosity of the battery electrolyte increases and ionic conductivity decreases, resulting in a significant rise in ohmic impedance. More critically, the charge transfer process is limited by activation energy; at low temperatures, the rate constant decreases substantially, leading to a marked increase in charge transfer impedance, which manifests as an enlarged diameter of the high-frequency semicircle in the Nyquist plot. Furthermore, at low temperatures, the diffusion coefficient of lithium ions within the electrode solid phase declines, causing an elongation of the Warburg impedance linear segment in the low-frequency region and indicating severe concentration polarization. Consequently, the entire Nyquist plot shifts markedly toward the upper right, reflecting a substantial increase in impedance R_{ct}.

At room temperature, the electrolyte exhibits high conductivity, and the interfacial reaction rate and diffusion processes are in equilibrium. The Nyquist plot typically displays a small high-frequency semicircle and a moderate low-frequency diffusion tail, indicating that the battery's kinetic conditions are optimal within this temperature range, and the charge/discharge process approaches reversibility.

Under high-temperature conditions, both ohmic impedance and charge transfer impedance gradually decrease, and the semicircle in the Nyquist plot shrinks, indicating that the charge transfer process becomes easier. However, at excessively high temperatures, side reactions intensify, the SEI becomes unstable, electrolyte decomposition increases, and deposition of by-products accelerates. This leads to more complex diffusion characteristics in the low-frequency region, manifested as increased curvature of the curve. This indicates that although kinetic conditions are improved, the system's stability is significantly compromised.

The temperature dependence of electrochemical impedance can be described using an equivalent circuit model. The commonly used Randles circuit comprises solution resistance R_S, charge transfer resistance *Rct(T)*, double-layer capacitance C_{dl}(T), and Warburg impedance *Zw*(ω,T). The total impedance can be expressed as:

$$Z_{\text{total}} = R_s + \frac{1}{\frac{1}{R_{ct}(T)} + j\omega C_{dl}(T)} + Z_{\text{w}} \tag{1.29}$$

In Eq. (1.29), ω represents angular frequency, *Rs*(T) denotes the temperature and dependent solution resistance, *Rct(T)* is the charge transfer resistance, C_{dl}(T) is the double-layer capacitance, and *Zw*(ω,T) stands for Warburg impedance. Temperature affects these parameters by altering ion mobility, interfacial reaction rates, and diffusion coefficients. Experimental results indicate that the relationship between temperature and *Rct(T)* also follows an Arrhenius-type form:

$$R_0(T) = R_{ct,0} \exp\left(\frac{E_a}{RT}\right) \tag{1.30}$$

where *Rct,0* is a constant and *Ea* is the activation energy for charge transfer. This relationship indicates that the charge transfer resistance significantly increases at low temperatures, while it decreases at high temperatures, but safety consequently deteriorates.

This Fig. 1.12 presents the EIS Nyquist plots of a lithium-ion battery under conditions of −10, 25, and 45°C. The abscissa represents the real part of impedance Z'/Ω, reflecting solution resistance and charge transfer resistance, while the ordinate denotes the negative value of the imaginary part of impedance $-Z''/\Omega$, primarily indicating capacitive effects and diffusion characteristics. As shown in the figure, at low temperature, the entire curve shifts significantly to the right, and the diameter of the semicircle increases markedly, indicating elevated charge transfer resistance and restricted interfacial reaction kinetics. At ambient temperature, the curve is most compact, demonstrating optimal kinetic conditions. At elevated temperature, the high-frequency semicircle shrinks, indicating reduced charge transfer resistance; however, the low-frequency linear segment

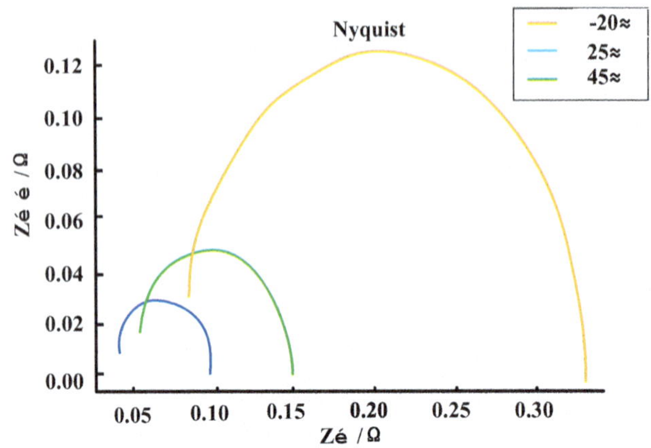

Figure 1.12 Comparison of Nyquist plots at different temperatures.

elongates, suggesting enhanced diffusion polarization and increased side reactions. These results provide a direct visualization of temperature's influence on battery impedance characteristics and offer experimental evidence for establishing temperature and dependent equivalent circuit models.

From an engineering perspective, EIS provides a window for analyzing battery operating states under varying temperatures. In low-temperature applications, high impedance limits battery power performance and necessitates heating measures to enhance ion migration and interfacial reaction rates. Conversely, in high-temperature applications, although impedance decreases, intensified side reactions require the BMS to implement effective thermal management and protective measures to prevent further temperature rise and potential thermal runaway. Furthermore, incorporating temperature and corrected impedance parameters into battery modeling significantly improves the accuracy of model predictions regarding voltage response and power performance under dynamic operating conditions.

In summary, the temperature dependence of electrochemical impedance not only reveals the operational mechanisms of batteries under various environmental conditions but also provides a scientific basis for optimizing battery management and thermal management strategies. Through experimental analysis of Nyquist plots and theoretical representation using equivalent circuit models, the influence of temperature on impedance characteristics can be comprehensively characterized, thereby playing a significant role in battery design, application, and safety control.

1.3.4 Aging behavior

The aging behavior of lithium-ion batteries refers to a complex process characterized by irreversible performance degradation over time and usage. This behavior directly determines the battery's lifespan, reliability, safety, and economic viability throughout its entire life cycle. A thorough understanding of aging phenomena is not only fundamental for battery state estimation, health management, and lifetime prediction but also serves as a critical basis for optimizing battery material systems, improving cell design, and establishing scientifically sound usage strategies. This section will systematically explore the macroscopic manifestations, underlying mechanisms, influencing factors, modeling approaches, and mitigation strategies of lithium-ion battery aging behavior.

1.3.4.1 Macroscopic manifestations and quantitative indicators of aging behavior

Battery aging ultimately manifests as changes in externally observable performance parameters. These macroscopic manifestations serve as direct indicators for assessing battery health status.

1.3.4.1.1 Capacity attenuation

This is the most closely monitored aging indicator. The available capacity of a battery continuously declines with increasing cycle counts or prolonged storage duration. When the capacity degrades to 80% of its rated capacity, this value being a conventional benchmark, though it may vary depending on specific application contexts, the battery is generally considered to have reached the end of its service life. Capacity fade directly reflects the loss of active materials within the battery and the reduction in the inventory of active lithium ions.

1.3.4.1.2 Internal resistance increases

The internal resistance of a battery, including ohmic resistance, electrochemical polarization resistance, and concentration polarization resistance, significantly increases with aging. An increase in internal resistance leads to the following consequences:

Variation in operating voltage: the voltage plateau decreases during discharge and increases during charging.

Power performance degradation: the battery's ability to output and accept power decreases, manifesting as weak acceleration and slower fast charging speeds.

Reduced energy efficiency: increased energy loss during charging and discharging processes (dissipated in the form of heat).

Thermal management burden is increasing: greater heat generation imposes higher demands on the thermal management system.

1.3.4.1.3 Decrease in energy and power efficiency

Energy efficiency refers to the ratio of discharge energy to charging energy. Due to increasing internal resistance and energy loss caused by side reactions, the energy efficiency of aged batteries continuously declines. Power efficiency reflects a battery's ability to maintain voltage stability during rapid charge and discharge cycles, which likewise deteriorates as internal resistance increases.

1.3.4.1.4 Increased self-discharge rate

A healthy battery slowly self-discharges during storage due to minor internal side reactions. As the battery ages, factors such as instability of the SEI layer and internal micro-short circuits lead to a significant increase in self-discharge rate, causing faster SOC loss during idle periods.

1.3.4.1.5 Thermal behavior changes

The heat generation characteristics and thermal stability of aged batteries undergo changes. Typically, an increased internal resistance leads to higher ohmic heating. Meanwhile, aged batteries may exhibit a lower onset temperature for thermal runaway, indicating deteriorated thermal stability and increased safety risks.

1.3.4.1.6 Dynamical performance degradation

The rate capability of batteries, namely their ability to charge and discharge at high rates, significantly degrades due to increased interfacial impedance and reduced lithium-ion diffusion coefficient. This limitation restricts the application of batteries in scenarios requiring high power output.

1.3.4.2 Intrinsic mechanisms of aging: from atoms to cells

Macroscopic aging manifestations originate from a series of complex physical and chemical changes occurring within the battery. These mechanisms primarily take place at the electrode-electrolyte interface and within the bulk of the electrode materials.

① Aging mechanisms on the negative electrode side

Evolution and growth of the solid electrolyte interphase membrane

Formation and Function: The SEI film forms on the surface of the graphite anode during the initial charging and discharging cycles of a battery, as a result of irreversible reduction reactions between the electrolyte and the anode. An ideal SEI film is dense and stable, acting as an electronic insulator while permitting lithium-ion conduction, thereby effectively preventing further decomposition of the electrolyte.

Aging growth: During prolonged cycling and storage, the SEI film undergoes continuous slow growth. The underlying growth mechanisms include: diffusion of electrolyte components through the SEI film to the graphite surface, where they undergo reduction reactions; dissolution and redeposition of the SEI film; and localized excessive thickening.

Consequence: The growth of the SEI layer irreversibly consumes the electrolyte and active lithium ions, leading to continuous capacity fading. Meanwhile, the thickened SEI layer increases the resistance to lithium-ion transport at the interface, representing one of the primary causes of elevated battery internal resistance.

② Lithium deposition (lithium plating)

Conditions for occurrence: Under low temperature, high-rate charging, or excessively low anode potential, the kinetics of lithium-ion intercalation into the graphite surface slow down. When the intercalation rate becomes lower than the migration rate of lithium ions from the electrolyte to the surface, lithium ions gain electrons at the anode surface and deposit in the form of metallic lithium.

Harm:

Reversible lithium loss: Some deposited metallic lithium becomes electrically disconnected from the host material and cannot be reutilized during discharge.

Irreversible lithium loss: the deposited metallic lithium is highly reactive and will react with the electrolyte, forming a new SEI layer, thereby permanently consuming active lithium and electrolyte.

Safety risk: The growth of dendritic lithium may penetrate the separator, causing internal short circuits and triggering thermal runaway.

③ Structure and failure of the negative electrode active material

Mechanical damage: Graphite undergoes approximately 10% volume expansion and contraction during lithium-ion intercalation and deintercalation. Prolonged stress cycling can induce microcracks and pulverization of graphite particles, disrupt the conductive network between particles, and cause some active material to lose electrical contact with the current collector, leading to failure.

Layered structure collapse: Over-discharge or prolonged exposure to low potential may cause the collapse of the graphite layered structure, resulting in loss of lithium storage capability. As shown in Fig. 1.13, the morphological comparison of the anode material before and after cycling under the electron microscope.

1.3.4.2.2 Aging mechanisms on the positive electrode side

① Dissolution of transition metal ions

Occurrence of materials: Particularly common in lithium manganese oxide and lithium-rich manganese-based cathode materials. Under high temperature, high voltage, and catalyzed by acidic impurities in the electrolyte (such as HF), transition metal ions (e.g., Mn^{2+}) from the cathode dissolve into the electrolyte.

Cascading degradation: dissolved metal ions migrate across the separator to the anode surface, catalyzing the decomposition and reformation of the SEI layer, thereby accelerating capacity fading and impedance increase. Meanwhile, the cathode material itself suffers capacity degradation due to lattice distortion.

② Structural phase transition of the cathode material

Mechanism: During deep charge/discharge cycles, particularly at high voltages, layered cathode materials undergo irreversible phase transformations, shifting from their active layered structure to electrochemically inactive spinel or rock-salt phases. As shown in Fig. 1.14, the morphological comparison of the cathode material before and after cycling under the electron microscope.

Consequence: Phase transformation blocks the diffusion pathways for lithium ions, resulting in capacity fading and a decrease in average operating voltage (voltage plateau degradation).

Formation of the positive electrode electrolyte interface.

Similar to the anode, at high voltages, the electrolyte undergoes oxidative decomposition at the surface of the cathode, forming a Cathode electrolyte interphase (CEI) layer. The growth of the CEI layer consumes active lithium and electrolyte while increasing interfacial impedance on the cathode side.

④ Particle cracking and mechanical failure

Cause: During the lithiation and delithiation processes, secondary spherical cathode materials also undergo lattice volume changes. Anisotropic stresses between different crystallites can induce microcracks within the particles.

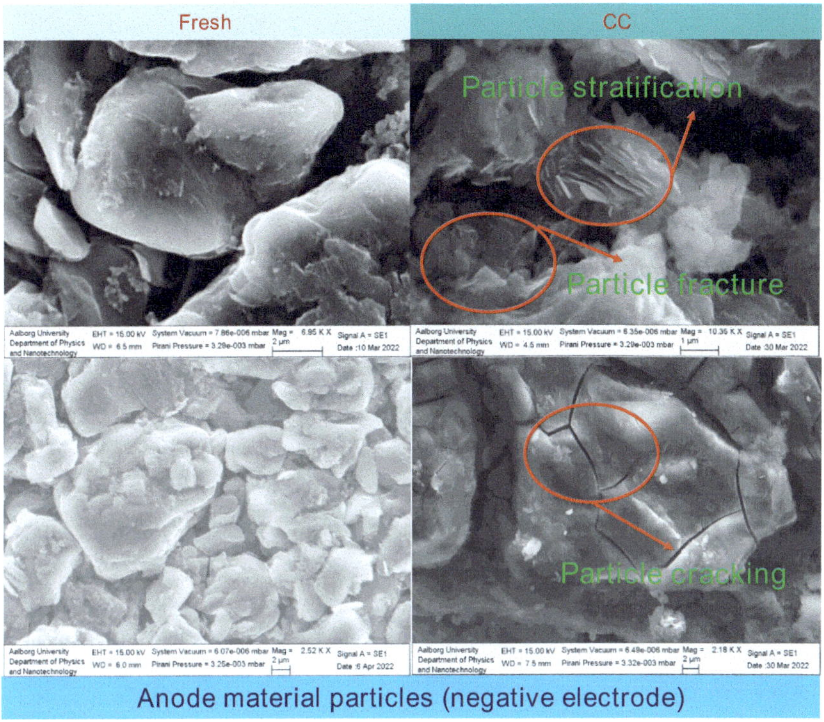

Figure 1.13 Comparison of SEI morphology on the negative electrode material before and after cycling.

Impact: Cracks expose new surfaces that react with the electrolyte, accelerating side reactions. Cracks can also cause partial separation of active materials from the conductive agent, leading to "electrochemical deactivation."

1.3.4.2.3 Aging of the electrolyte and the separator

Electrolyte consumption and decomposition: Continuous growth of the SEI/CEI films and lithium plating side reactions continuously consume the electrolyte. A reduction in electrolyte quantity directly leads to decreased ionic conductivity and a significant increase in battery internal resistance, with performance degradation being particularly severe under low-temperature conditions.

Membrane aging: The separator may shrink or melt due to high temperatures, leading to internal short circuits. Additionally, metal ions dissolved

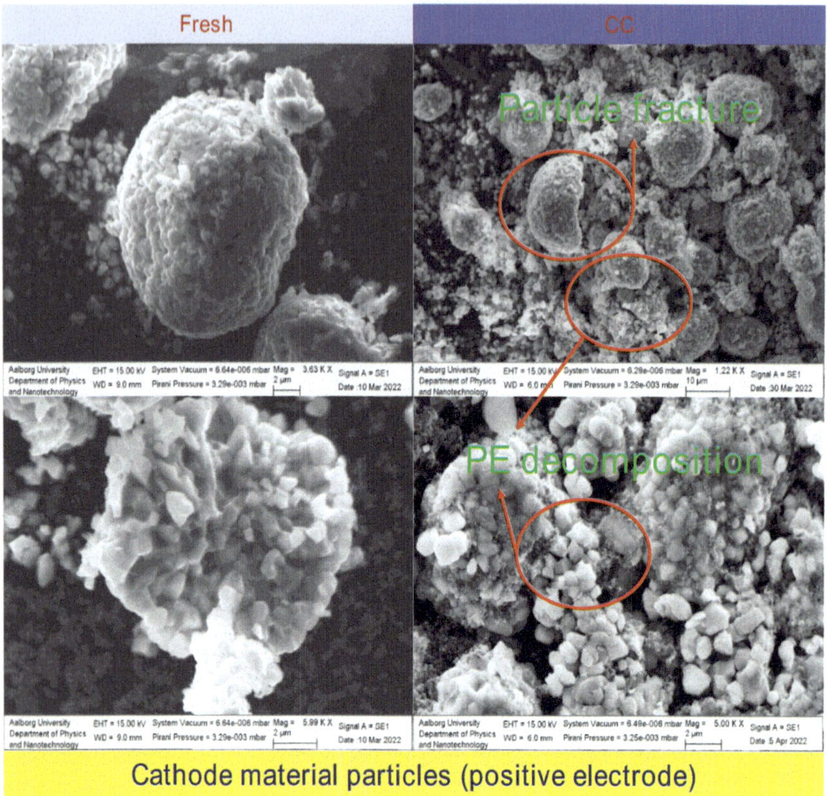

Figure 1.14 Comparison of SEI morphology on the positive electrode material before and after cycling.

from the positive electrode or electrode debris may block the pores of the separator, increasing resistance to ion transport.

Binder degradation: The binder in the electrode may degrade or lose its adhesive properties under prolonged high-temperature or high-potential conditions, leading to the detachment of active materials from the current collector and exacerbating capacity fading.

1.3.4.2.4 Coupling and mutual interactions of aging mechanisms

It should be emphasized that the aforementioned degradation mechanisms do not occur in isolation, but rather form a complex, interconnected network that exacerbates one another. For example, Mn^{2+} ions dissolved from the cathode migrate to the anode and catalyze the growth of the SEI. Consumption of the electrolyte intensifies local SOC non-uniformity, thereby promoting lithium plating. Loss of active material and depletion of lithium inventory are intertwined processes that collectively contribute to capacity fade.

1.3.4.3 Key factors influencing aging behavior

The aging trajectory and rate of batteries strongly depend on their operating conditions and usage history.

1.3.4.3.1 Electrical factors

① State of charge:

High SOC: accelerates structural degradation on the cathode side (phase transitions, oxygen release) and electrolyte oxidation; simultaneously increases the potential at the anode, elevating the risk of lithium plating. Long-term storage at high SOC is a primary stressor contributing to calendar aging.

Low SOC: May lead to dissolution of the anode copper current collector, causing catastrophic failure.

SOC window/cycling depth: Shallow charging and discharging can significantly extend cycle life by reducing the amplitude of volume changes and mechanical stress in the electrode materials.

② Current and C-rate:

High-rate charging: exacerbates battery polarization and easily causes the negative electrode potential to enter the lithium plating region, representing the primary degradation mechanism in fast charging applications.

High-rate discharge: leads to greater voltage polarization and a drop in the voltage plateau. Although it does not directly consume lithium, it generates more heat, indirectly accelerating aging.

Voltage window: Overcharging irreversibly damages the cathode structure and accelerates electrolyte oxidation; over-discharging causes dissolution of the anode copper current collector and decomposition of the SEI layer. Strictly limiting the voltage range is critical to ensuring long cycle life.

1.3.4.3.2 Temperature factor

High temperature: acts as a "catalyst" for all chemical side reactions. It significantly accelerates processes such as the growth of SEI/CEI films, transition metal dissolution, and electrolyte decomposition, serving as the primary accelerating factor for both calendar aging and cycle aging.

Low temperature: primarily affects kinetic processes. The migration and diffusion rates of lithium ions in the electrolyte and electrode materials slow down, leading to increased polarization, particularly causing lithium plating during charging, which results in severe irreversible capacity loss.

Temperature cycling/variation: Temperature non-uniformity within the battery pack can lead to uneven current distribution, causing certain areas to age more rapidly and exacerbating overall inconsistency.

1.3.4.3.3 Temporal factors

Calendar aging continues to occur over time, even during periods of non-use, with its rate determined by storage temperature and SOC. Cycle aging is related to the accumulated number of cycles, energy throughput, and time.

1.3.4.3.4 Mechanical stress

External mechanical stresses such as vibration, impact, and compression may cause electrode structural damage, internal short circuits, or failure of component connections, directly leading to performance degradation or safety issues.

1.3.4.4 Modeling and simulation of aging behavior

Establishing an accurate aging model is crucial for predicting battery lifetime and optimizing usage strategies.

1.3.4.4.1 Empirical and semi-empirical models

Principle: Based on test data, capacity fade or internal resistance growth trajectories are fitted using mathematical formulas.

Common models:

Power-law model: used for fatigue life fitting. $Q_{loss} = a \times \hat{N}b$

Arrhenius model: used to describe the influence of temperature on the aging rate. $K = A \times \exp(-Ea/RT)$

Square root model: used to describe capacity loss controlled by the SEI layer during calendar aging. $Q_{loss} \propto \sqrt{t}$

Advantages: simple formulation, low computational cost, and good prediction accuracy within a specific stress range.

Drawbacks: lack of physical significance, poor extrapolation capability, and inability to reveal underlying mechanisms.

1.3.4.4.2 Mechanistic model

Principle: Described by coupled partial differential equations based on electrochemistry and aging reaction kinetics, capturing the lithium ion concentration, potential distribution, and the rates of side reactions (such as SEI growth and lithium plating).

Representative models include the extended single-particle model, pseudo-two-dimensional models, etc.

Advantages: clear physical interpretation, capable of deeply revealing the aging process, strong predictive capability, and suitable for complex operating conditions.

Drawbacks: The model is complex, with a large number of parameters and high computational cost.

1.3.4.4.3 Data-driven models

Principle: Utilizing machine learning, deep learning, and other algorithms to learn degradation patterns from massive amounts of battery operational data and aging test data.

Methods: Gaussian process regression, support vector machines, neural networks, etc.

Advantages: No explicit physical equations required; capable of handling high-dimensional and nonlinear problems, suitable for direct learning from BMS data.

Drawbacks: reliance on a large volume of high-quality data, poor model interpretability, and uncertainty in extrapolation capability.

1.3.4.5 Mitigation and management strategies for aging behavior

By understanding the aging mechanisms and influencing factors, various strategies can be employed to retard battery aging and extend its service life.

1.3.4.5.1 Innovation at the material level

Cathode materials: Develop high-stability cathode materials through surface coating and bulk doping to suppress phase transitions and metal dissolution.

Anode materials: When using silicon-carbon composite materials, optimize binders and conductive networks to buffer volume expansion effects; develop artificial SEI technology to pre-form stable interfaces.

Electrolyte: Developing novel additives to form more stable and dense SEI/CEI films; employing high-concentration or localized high-concentration electrolytes to widen the electrochemical window.

Separator: Ceramic-coated separators are employed to enhance thermal stability and resistance to metal ion penetration.

1.3.4.5.2 Optimization of cell design and manufacturing processes

Design: Optimize electrode thickness, porosity, and N/P ratio (negative-to-positive electrode capacity ratio) to reduce lithium plating risk at the source.

Process: Ensure uniform coating and thorough drying of the electrode sheets, minimize moisture and impurities, and enhance manufacturing consistency.

1.3.4.5.3 Battery management at the system level

Optimize thermal management: ensure the battery operates within its optimal temperature range, avoiding abuse under high or low temperature conditions. Efficient liquid cooling or direct cooling systems are critical.

Intelligent charging strategy:

Avoid long-term storage at full charge.

Pulse charging and reduced current slow charging strategies are employed at the end of fast charging to minimize polarization and prevent lithium plating.

Dynamically adjust the charging curve according to battery temperature and health status.

Reasonable SOC usage window: Set daily upper and lower limits for SOC in the BMS, and promote the "shallow charging and shallow discharging" principle.

Balanced management: By employing active or passive balancing techniques, inconsistencies among individual cells within a battery pack are reduced, preventing individual cells from overcharging or overdischarging, thereby enhancing the overall lifespan of the battery pack.

1.3.4.5.4 Second-life applications

When the capacity of a power battery degrades to below 80%, its performance under vehicle-mounted conditions no longer meets requirements. However, it still retains sufficient remaining life and capacity for lower-demand applications such as grid energy storage, backup power for base stations, and low-speed electric vehicles. Through precise state-of-health assessment and sorting/reconfiguration, second-life utilization of batteries can be achieved, a highly effective strategy from a full lifecycle perspective to alleviate resource pressure and reduce carbon emissions.

1.3.5 Safety boundary characteristics

Lithium-ion batteries, as high-energy-density energy storage devices, possess a highly complex electrochemical system whose operational state is subject to the coupled effects of numerous factors. Under prolonged cycling and high-power applications, the absence of effective monitoring and constraint mechanisms may lead to performance degradation or even severe safety incidents due to uncontrolled reactions. The so-called safety boundaries refer to the permissible limits of battery operation in terms of voltage, current, temperature, and power. These boundary conditions are constrained by the thermodynamics and kinetics of electrochemical reactions, as well as by material stability, cell design parameters, and external operating environments. When battery operation exceeds these limits, a series of side reactions may occur internally, including SEI film degradation, electrolyte decomposition, lithium dendrite deposition, transition metal ion dissolution, and electrode structural collapse. These side reactions not only result in reduced capacity and energy efficiency but also accumulate risks that may ultimately trigger thermal runaway. Therefore, in-depth research into the characteristics of battery safety boundaries holds significant theoretical and engineering importance for ensuring reliable operation of battery systems, extending their lifespan, and optimizing management strategies.

First, voltage boundaries represent the most fundamental and critical operational constraints for batteries. Taking commercial NMC lithium-ion batteries as an example, the operating voltage of a single cell is typically confined within approximately 2.5 to 4.2 V. When the voltage falls below the lower limit Umin, the reversible delithiation of lithium from the graphite anode is restricted, leading to the breakdown of the SEI film, interface instability, and potentially the formation of metallic lithium deposition. The deposited metallic lithium may further evolve into dendrites that penetrate the separator, causing internal micro-short circuits or even catastrophic failure. Conversely, when the voltage exceeds the upper limit Umax, irreversible phase transitions occur in the layered structure of the cathode, high-valent transition metal ions dissolve into the electrolyte, and intense electrolyte decomposition ensues, generating gas and releasing substantial heat, thereby inducing a rapid increase in internal pressure and temperature. Voltage boundary conditions are generally expressed as:

$$U_{\min} \leq U \leq U_{\max} \tag{1.31}$$

In Eq. (1.31), U represents the single cell voltage, while U_{min} and U_{max} denote the lower and upper voltage limits, respectively. It is noteworthy that the voltage boundaries vary significantly across different battery systems. As shown in Table 1.9, the upper voltage limit for lithium iron phosphate batteries is approximately 3.6 V, whereas high-nickel ternary batteries may extend beyond 4.3 V. This directly determines the trade-off between energy density and safety.

Secondly, the current boundary reflects the rate limitations of the battery during charging and discharging. Its expression is:

$$|I| \leq C \cdot \alpha \tag{1.32}$$

where I represents the charge and discharge current, C denotes the battery's rated capacity, and α is the rate coefficient. In practical applications, the charge and discharge rate of electric vehicle batteries is typically limited between 1 and 3C. As shown in Table 1.10, the internal resistance of lithium-ion batteries significantly increases with rising charge/discharge rate, from 0.5 to 3C, while the capacity retention rate continuously declines, and the risk level rises from low to high.

Excessive charging current significantly increases the overpotential at the electrode surface, causing the electrode reaction to deviate from reversible conditions and promoting lithium dendrite deposition. Excessive discharge current leads to a rapid drop in terminal voltage, insufficient capacity output, and substantial I^2R heat accumulation, resulting in rapid temperature rise that may trigger overcurrent protection or directly damage electrode materials. Particularly under low-temperature conditions, even

Table 1.9 Comparison of voltage boundaries among different battery systems.

Battery system	Lower limit voltage (V)	Upper limit voltage (V)	Peculiarity
Lithium iron phosphate	2.5	3.6	High safety, low energy density
Ternary lithium	2.5	4.2	Balancing energy density and safety
High nickel ternary	2.7	4.3–4.35	High energy density, poor thermal stability
Lithium cobalt oxide	2.5	4.2–4.3	High voltage stability, but high cost.

Table 1.10 Battery performance and risk under different current multiples.

Magnification (C)	Internal resistance change (%)	Capacity retention rate (%)	Risk level
0.5	Benchmark	100	Low
1	+10	95	Low
2	+30	85	Middle
3	+60	70	High

when the current remains within the rated range, polarization can cause the effective current density to exceed safe thresholds, thereby exacerbating failure risks.

The temperature boundary is one of the most critical parameters for battery safety. The normal operating temperature range of batteries is generally between −20 and 60°C, Table 1.11 shows that lithium-ion batteries perform optimally at 25°C.

$$T_{\min} \leq T \leq T_{\max} \tag{1.33}$$

When the temperature falls below the lower limit, the lithium-ion transference number decreases, the diffusion coefficient declines, and the interfacial reaction rate constant significantly reduces, resulting in increased internal resistance and a sharp drop in capacity utilization. Meanwhile, lithium plating is prone to occur during high-rate charging, posing a risk of short circuits. When the temperature exceeds the upper limit, electrolyte decomposition and gas evolution are intensified, the SEI becomes unstable, the cathode lattice structure collapses, and side reaction rates increase exponentially. Substantial experimental evidence indicates that when the temperature surpasses 80°C, battery cells often enter thermal runaway, characterized by intense exothermic reactions, gas expansion, and potential fire or explosion.

Table 1.11 Battery performance at different temperatures.

Temperature (°C)	Capacity retention rate (%)	Cycle life (times)	Security risks
−20	65	300	High
0	85	600	Middle
25	100	1500	Low
45	95	1000	Middle
60	80	500	High

The power boundary is jointly determined by voltage and current, constraining the input and output limits of the battery in high-power applications:

$$|P| \leq U \cdot I_{\max} \tag{1.34}$$

where P represents battery power and $I_{\max}$ denotes the allowable maximum current. Power boundary constraints limit extreme operating conditions of the battery during rapid acceleration of electric vehicles and high-power energy storage charge/discharge cycles. Exceeding this boundary would result in substantial Joule heating and polarization heat generation, leading to localized hotspots and cell imbalance, ultimately reducing battery lifespan and increasing safety risks.

The risk of thermal runaway significantly increases when the battery operates near or beyond these boundaries. The thermal runaway process can be described by an energy balance equation:

$$C_{th}\frac{dT}{dt} = Q_{gen} - Q_{loss} \tag{1.35}$$

In this equation, C_{th} represents the thermal capacity of the battery, Q_{gen} denotes the heat generated during battery operation, including ohmic heat, polarization heat, and side reaction heat, and Q_{loss} stands for the heat dissipation. When Q_{gen} exceeds Q_{loss}, the battery temperature will continuously rise. Once it surpasses a critical threshold, a series of cascading reactions will be triggered, such as decomposition of the SEI film, violent reactions of the electrolyte, and oxygen release from electrode materials, ultimately leading to thermal runaway and safety incidents.

To verify the reasonableness of safety boundaries, researchers commonly employ various experimental methods, including overcharge testing, overdischarge testing, nail penetration and compression tests, and high-temperature cycling experiments, as shown in Table 1.12. Overcharge testing reveals the mechanisms underlying side reactions induced under high-voltage conditions; overdischarge testing demonstrates that low-voltage operation leads to the degradation of the SEI layer and irreversible capacity loss. Nail penetration and compression tests simulate the risk of short circuits triggered by external mechanical damage, while high-temperature cycling experiments illustrate the progressive deterioration of battery life due to cumulative side reactions. Typical experimental results show that under these extreme conditions, batteries exhibit abrupt voltage drop, rapid temperature increase, gas evolution, and even fire or explosion.

Table 1.12 Safety boundary experiment and performance.

Experimental methods	Trigger conditions	Main manifestations	Risk level
Overcharged	Voltage >4.25 V	Electrolyte decomposition, gas release	High
Over-release	Voltage <2.0 V	SEI film destruction, capacity loss	Middle
Acupuncture	Internal short circuit	Rapid temperature rise, fire hazard	Extremely high
Extrusion	External mechanical damage	Internal short circuit, thermal runaway	High
High temperature cycling	$T > 60°C$	Material degradation, shortened lifespan	High

To comprehensively characterize these safety constraints, the concept of a Safe operating area (SOA), defined as the region within which a battery can stably operate under specified voltage, current, and temperature ranges, has been proposed.

To more comprehensively characterize the safety of batteries under various operating conditions, researchers have proposed the concept of the SOA. The SOA refers to a permissible operational range defined by simultaneously considering multiple critical parameters, such as voltage, current, and temperature during battery operation. Within this range, the battery maintains relative thermodynamic and kinetic stability, enabling reversible energy conversion and ensuring long-term cycling safety. The introduction of the SOA concept not only overcomes the limitations of single-boundary metrics in fully capturing the operational limits of batteries but also provides a multidimensional, unified framework for battery safety evaluation and engineering applications. Within this region, batteries operate under controlled electrochemical reactions with suppressed side reactions; however, exceeding these boundaries may lead to hazardous conditions such as voltage overshoot, current overload, or thermal imbalance, resulting in rapid performance degradation or even thermal runaway. Therefore, the SOA concept holds significant importance for uncovering the comprehensive safety behavior of batteries, guiding control strategies in BMS, and establishing industry-wide safety standards.

In practical engineering applications, the BMS is regarded as the "guardian of safety" during battery operation. Its core function is to ensure that batteries consistently operate within a safe working range and remain distant from potential risks through real-time monitoring, assessment, and control. The BMS typically acquires key operational parameters of the battery, such as individual cell voltages, total current, and cell temperatures via sensors and monitoring circuits. These data are then dynamically compared against pre-defined safety boundary conditions to determine whether the battery is approaching or has exceeded allowable operational limits. If the monitoring results indicate that the system is nearing these boundaries, the BMS promptly initiates integrated software and hardware intervention measures. For example, it may adjust control algorithms to reduce charge/discharge current rates or activate forced air cooling, liquid cooling, or heating systems within the thermal management system to maintain temperature within acceptable ranges.

Additionally, the BMS issues graded alarm signals based on risk levels to alert users or higher-level systems for further action. When battery operation significantly exceeds safety thresholds, the BMS triggers its highest-level protective actions, such as forcibly disconnecting circuits, deactivating relays, or isolating battery modules to prevent catastrophic events such as thermal runaway. Thus, the role of the BMS extends beyond passive monitoring; it serves as a critical bridge that transforms theoretical safety boundaries into executable engineering control logic and hardware-based protection mechanisms. By enabling real-time data processing and decision-making, it converts complex electrochemical constraints into clear operational commands, thereby ensuring stable and safe battery system performance throughout its entire lifecycle.

In summary, the safety boundary characteristics reveal the operational limits of batteries in terms of voltage, current, temperature, and power. These limit conditions not only reflect electrochemical principles but also directly constitute the engineering for practical battery applications. Through theoretical formulas, experimental validation, and model-based analysis, battery safety boundaries can be quantitatively described. Furthermore, real-time monitoring and control by the BMS enable these boundaries to be translated into practical protective measures, ensuring safe and stable battery operation throughout its entire lifecycle. This research provides fundamental support for understanding battery failure mechanisms and thermal runaway risks, offers theoretical foundations for designing thermal management and protection strategies, and holds significant implications for enhancing safety in applications such as electric vehicles and energy storage power stations.

1.4 Application scenarios

1.4.1 Electric vehicles

With the rapid development of society and economy, new-energy vehicles are gradually replacing conventional gasoline-powered vehicles and have become a popular mode of transportation. New-energy vehicles utilize lithium-ion batteries as their power source, which exhibit excellent physicochemical properties such as high energy density, no memory effect, low self-discharge rate, and favorable form factors.

Lithium-ion batteries are widely applied in new energy vehicles primarily due to their high energy density and power density. Modern lithium-ion batteries can store substantial amounts of energy within a relatively compact and lightweight package. This enables electric vehicles to achieve reasonable driving ranges, which is one of the primary concerns for consumers when considering electric mobility. Power density is another critical characteristic of lithium-ion batteries. It determines how quickly the battery can release energy, directly influencing a vehicle's acceleration performance and overall dynamic capability. The high power density of lithium-ion batteries allows electric vehicles to accelerate rapidly. This not only enhances the driving experience but also ensures that electric vehicles can effectively handle a wide range of driving conditions.

Second, the service life of lithium-ion batteries is also a key concern for electric vehicle manufacturers and consumers. Well-designed lithium-ion batteries can endure thousands of charge/discharge cycles before their capacity degrades to a level that significantly affects the vehicle's driving range. This extended lifespan helps reduce the total cost of ownership for electric vehicles, as battery replacements are not required frequently. Furthermore, advancements in BMS play a crucial role in prolonging battery life. A BMS monitors and controls various parameters, such as voltage, current, and temperature, of individual cells within a battery pack, ensuring that lithium-ion batteries operate under optimal conditions and preventing issues such as overcharging or overheating.

Simultaneously, the development of charging infrastructure is closely linked to the widespread adoption of electric vehicles, and lithium-ion batteries are compatible with various charging speeds. Level 1 charging uses standard household outlets and is the slowest but most accessible method. Level 2 charging requires dedicated charging stations and offers faster charging speeds. Moreover, lithium-ion batteries can withstand fast charging when properly managed without significant degradation, thereby reducing downtime for electric vehicle owners.

Moreover, the environmental impact of lithium-ion battery-powered electric vehicles is a highly significant topic. Although the production of lithium-ion batteries does pose environmental challenges, primarily due to the extraction and processing of raw materials such as lithium, cobalt, and nickel, the overall lifecycle emissions of electric vehicles are substantially lower than those of internal combustion engine vehicles. As power grids increasingly rely on renewable energy sources such as solar and wind, the carbon footprint of electric vehicles will continue to decline, making them an increasingly sustainable transportation option.

In vehicle design, the use of lithium-ion batteries offers greater flexibility. Battery packs can be engineered into various shapes and sizes to accommodate the vehicle's structure, thereby optimizing interior space and overall weight distribution. Proper weight distribution enhances vehicle handling and stability, further improving the driving experience. Moreover, since battery packs are typically mounted at the bottom of the vehicle, the center of gravity is lowered, significantly contributing to driving stability, particularly during cornering.

From an economic perspective, the development of electric vehicles has also driven growth in related industries, with the lithium-ion battery industry serving as a central component. Battery production, research and development, and other stages have created substantial employment opportunities while simultaneously fostering technological innovation. As battery technology continues to advance, production costs have gradually declined, making electric vehicles more price-competitive and further accelerating their market penetration.

Moreover, the intelligent development of electric vehicles is closely linked to lithium-ion batteries. Advanced battery management systems not only regulate fundamental battery parameters but also integrate with the vehicle's intelligent systems to achieve more efficient energy utilization. For instance, by analyzing driving habits and road conditions, these systems can intelligently adjust the battery's output power, optimize vehicle energy consumption, and further extend driving range. At the same time, real-time transmission of battery status information to the vehicle's central control system enables drivers to continuously monitor battery health and remaining charge levels, thereby enhancing both convenience and safety during vehicle operation.

In the future, as lithium-ion battery technology continues to advance, the development and application of novel lithium-ion battery technologies, such as solid-state batteries will further enhance the performance of electric vehicles. Driving ranges will be extended, charging speeds will increase, and battery lifespans will be significantly prolonged. These improvements will enable electric vehicles to play an increasingly vital role in transportation and become one of the predominant choices for future mobility.

1.4.2 Energy storage systems

With the continuous advancement of new energy systems, energy storage systems play a crucial role in modern energy frameworks, effectively addressing the spatiotemporal mismatch between energy supply and demand. Lithium-ion batteries, owing to their numerous advantages, have become one of the core technologies in energy storage systems.

From the perspective of the power grid, lithium-ion battery energy storage systems can effectively stabilize grid load. During periods of low electricity demand, such as nighttime, power generation may exceed consumption. At such times, excess electrical energy can be stored using lithium-ion battery energy storage systems. During peak demand periods, such as daytime office hours, this stored energy can then be released back into the grid, thereby reducing the difference between peak and valley loads, enhancing both the operational efficiency and stability of the grid. This is of significant importance for ensuring reliable power supply and delaying investments in grid upgrades and modernization.

Currently, the intermittency and volatility of renewable energy represent one of the primary challenges for their large-scale integration into power grids. Lithium-ion battery energy storage systems offer an effective solution to this issue. For instance, solar power generation depends on sunlight, while wind power relies on wind availability; consequently, their output power fluctuates significantly with changing natural conditions. Lithium-ion battery energy storage systems can store excess electrical energy when renewable generation exceeds grid demand, and release stored energy when generation falls short of demand. This enables a smoothed output of renewable energy, increases the share of renewables in the energy mix, and promotes the transition toward cleaner, low-carbon energy systems.

On the user side, lithium-ion battery energy storage systems also have widespread applications. For industrial and commercial users, installing lithium-ion battery energy storage systems enables demand management. By charging during periods of low electricity prices and discharging during peak pricing periods, users can significantly reduce their electricity costs. Additionally, in the event of grid failures or power outages, these energy storage systems can serve as backup power sources, ensuring the continuous operation of critical equipment and preventing production losses due to power interruptions. For residential users, home-based lithium-ion battery energy storage systems can be integrated with solar photovoltaic systems to store excess electricity generated during daytime hours for use at night, thereby increasing household energy self-sufficiency. Furthermore, such systems can provide emergency power during grid outages.

The performance advantages of lithium-ion battery energy storage systems form the foundation for their widespread application. First, they exhibit rapid response times, enabling charge and discharge switching within milliseconds or seconds, allowing quick adaptation to fluctuations in grid power demand. Second, they demonstrate high energy efficiency, with charge and discharge efficiencies typically ranging from 85% to 95%, thereby minimizing energy losses during storage and release. Additionally, lithium-ion batteries possess a long cycle life; high-quality units can endure thousands of charge/discharge cycles, resulting in low replacement costs over prolonged system operation.

From the perspective of technological development, the integration technologies of lithium-ion battery energy storage systems continue to advance. Currently, through advanced control strategies and energy management systems, multiple battery packs can be collaboratively controlled, enabling efficient operation of large-scale energy storage systems. Meanwhile, the integration of energy storage systems with smart grids is becoming increasingly deep. Energy storage systems can now serve as essential components of smart grids, participating in various functions such as grid dispatching and demand response, thereby further enhancing the level of grid intelligence.

In terms of safety in energy storage systems, lithium-ion batteries continue to undergo improvements. Battery management systems not only regulate charging and discharging but also enforce strict control over battery temperature and voltage balancing, thereby preventing safety issues such as thermal runaway. Furthermore, novel battery materials and structural designs are enhancing the safety performance of lithium-ion batteries, for instance, by employing ceramic separators and flame-retardant electrolytes to reduce the risk of safety incidents.

In the future, as energy transition continues to advance, the demand for energy storage systems will keep growing. Lithium-ion battery energy storage systems will further develop toward higher energy density, longer lifespan, lower cost, and enhanced safety. Meanwhile, the application scenarios of energy storage systems will continue to expand. Beyond existing applications on the grid side and user side, they are expected to play increasingly important roles in microgrids and electric vehicle charging stations, becoming key enabling technologies for building new-generation power systems.

1.4.3 Portable electronic devices

With the continuous development of society and economy, portable electronic devices such as smartphones, laptops, tablets, and smartwatches have become indispensable components of daily life and work. Lithium-ion batteries, with their outstanding performance, have provided reliable power support for these devices, significantly promoting the development and widespread adoption of portable electronics.

Smartphones are among the most widely used portable electronic devices employing lithium-ion batteries. Modern smartphones are increasingly powerful, integrating high-definition displays, high-performance processors, multiple sensors, and a wide range of applications, all of which require substantial electrical energy to operate. The high energy density of lithium-ion batteries enables smartphones to store sufficient charge within a compact form factor, meeting users' power demands for 1 day or even longer. Additionally, the low self-discharge rate of lithium-ion batteries represents a significant advantage in their application to smartphones and similar devices. Even when the device is in standby mode, the battery's energy loss is minimal, ensuring that sufficient charge remains available after prolonged periods of inactivity.

Laptops also demand high performance from batteries, requiring them to support users in prolonged work, study, or entertainment sessions without external power sources. The high energy density and extended cycle life of lithium-ion batteries meet these requirements effectively. A laptop equipped with a high-quality lithium-ion battery can sustain several hours of continuous operation, such as document processing or video playback, after a full charge, enabling users to work or study freely in locations without power access, such as coffee shops, airplanes, or outdoor settings, thereby enhancing flexibility in both work and learning. Furthermore, lithium-ion batteries are lightweight; their reduced weight contributes to lowering the overall weight of the laptop, improving portability and making it easier for users to carry.

Tablets integrate certain functionalities of smartphones and laptops, requiring both adequate battery life and a slim, lightweight form factor. The characteristics of lithium-ion batteries precisely meet the demands of tablets. They provide sufficient power to support prolonged activities such as reading, gaming, and video watching, without making the device excessively bulky, thus ensuring good ergonomics and portability.

Wearable devices such as smartwatches demand high energy density and compact volume from batteries due to their small form factor. Lithium-ion batteries are capable of storing sufficient electrical energy within an extremely limited space, enabling smartwatches to support functions including time display, health monitoring (e.g., heart rate tracking, step counting), and notification alerts, while maintaining operational endurance of several days or even up to a week, thereby minimizing the inconvenience of frequent charging for users.

From the perspective of technological development, in response to the ever-increasing performance demands of portable electronic devices, lithium-ion battery technology continues to advance. Battery manufacturers are constantly researching new battery materials and structural designs to enhance energy density. For instance, replacing conventional graphite anode materials with silicon-based anode materials holds the potential to significantly increase battery energy density, thereby further extending the operational range of portable electronic devices.

At the same time, the development of fast charging technology is closely related to lithium-ion batteries. Currently, many portable electronic devices support fast charging functions, which is made possible by the excellent performance of lithium-ion batteries in fast charging. By optimizing the battery's chemical system and charging protocols, lithium-ion batteries can store large amounts of electrical energy in a short period of time, significantly reducing charging duration and enhancing user experience.

With regard to battery safety, lithium-ion batteries have adopted a series of safety measures tailored to the characteristics of portable electronic devices. The battery management system precisely controls the charging and discharging processes, preventing occurrences such as overcharging, over-discharging, and excessive current. Furthermore, the battery's casing and internal structural design have been enhanced for greater robustness, enabling them to withstand external forces to a certain extent and thereby reducing the likelihood of safety incidents.

In the future, as portable electronic devices evolve toward being lighter, thinner, smarter, and more powerful, the demands placed on lithium-ion batteries will continue to rise. Lithium-ion batteries will undergo ongoing improvements in energy density, charging speed, cycle life, and safety to meet the growing power requirements of portable electronic devices, thereby driving sustained development in the portable electronics industry.

1.4.4 Aerospace and military

With the ongoing complex evolution of the global industrial landscape and the rapid, healthy development of China's national economy, the demand for non-renewable traditional energy sources continues to rise, placing traditional energy reserves at an increasingly severe risk of depletion. Therefore, in order to meet the demands of society and economic development, it has become imperative to identify next-generation energy sources capable of addressing the issue of non-renewability inherent in conventional energy. Lithium-ion batteries, owing to their numerous advantages, including high cycle life, low mass, long storage duration, high efficiency, cleanliness, and environmental friendliness, have emerged as a primary power source in fields such as aerospace and defense, demonstrating exceptional superiority.

Aerospace and military applications have consistently served as frontiers for advanced technological development, demanding energy systems of near-exacting standards. These systems are often required to operate reliably under the most extreme environmental conditions imaginable, ranging from the vacuum and intense radiation of outer space, to the low temperature and low pressure at high altitudes, to the severe mechanical shock, wide temperature fluctuations, and electromagnetic interference encountered in combat zones. Consequently, battery systems powering such equipment must not only meet basic energy supply requirements but also exhibit a comprehensive set of advanced characteristics, including high specific energy, lightweight design, extended cycle life, superior safety performance, and robust environmental adaptability. In this context, lithium-ion batteries have progressively replaced conventional Nickel-metal hydride (Ni-MH) and Silver-zinc (Ag–Zn) batteries due to their revolutionary advantages, emerging as the preferred energy solution in these critical domains.

Compared to traditional battery technologies, lithium-ion batteries offer significantly higher energy density, commercial products currently achieve 200 to 300 Wh/kg, while laboratory prototypes have surpassed 500 Wh/kg. This translates into substantial reductions in weight and volume for equivalent energy output, providing decisive advantages in reducing launch mass for spacecraft, extending flight duration of Unmanned aerial vehicles (UAVs), or minimizing individual soldier load. Furthermore, lithium-ion batteries exhibit extremely low self-discharge rates, typically below 2% to 3% per month, enabling long-term charge retention essential for military equipment requiring prolonged standby or long-duration missions. Additionally, through specialized electrolyte formulations and system engineering designs, lithium-ion batteries demonstrate an exceptionally broad operational temperature range, from extreme cold at −40°C to high temperatures up to +60°C—making them highly suitable for deployment across satellites, UAVs, underwater vehicles, and various portable military systems. As such, they have become one of the core enabling technologies underpinning modern defense capabilities and space exploration endeavors.

To meet these high-end applications, lithium-ion batteries must satisfy a series of stringent and diverse key performance requirements:

First, high specific energy and lightweight design are fundamental objectives, directly determining the flight duration, range of aircraft, or the effective payload capacity of satellites. In aerospace applications, reducing battery weight by just 1 kilogram can save tens of thousands to hundreds of thousands of dollars in launch costs; in individual soldier systems, reduced battery weight enables carrying more ammunition, supplies, or sensors, directly enhancing combat effectiveness.

Second, wide-temperature adaptability (from −40 to +60°C) is a basic prerequisite ensuring reliable battery operation under extreme conditions, such as the frigid shadows and intense solar heating experienced in space orbits (with temperature differentials reaching hundreds of degrees Celsius), or the harsh climates on Earth's battlefield, ranging from polar cold to desert heat [14].

Third, high reliability and safety are critical life-support factors encompassing radiation resistance, explosion prevention, shock resistance, and vibration tolerance. High-energy particle fluxes in space accelerate battery degradation; meanwhile, military applications demand that batteries avoid catastrophic failures, such as explosions or leaks under ballistic impacts or extreme overloads. Finally, long cycle life and rapid response capability are essential for mission economics and emergency responsiveness. For example, power systems for geostationary orbit communication satellites must operate reliably in orbit for over 15 years without maintenance; meanwhile, directed-energy weapons such as electromagnetic railguns and laser systems require battery packs capable of delivering massive pulsed power within seconds or even milliseconds. These requirements vary significantly depending on mission profiles and operational environments, as summarized in Table 1.13 below, which clearly illustrates performance priorities and differentiated demands across various application scenarios.

Lithium-ion batteries applied in the aviation field are required to possess advantages such as compact size, high capacity, and high power density. As a critical component of aircraft emergency and auxiliary power systems, aviation lithium-ion batteries can provide emergency power to essential equipment such as navigation, communication, and flight control systems when the primary aircraft power source fails, thereby preventing aviation safety incidents. Currently, aviation lithium-ion batteries have largely replaced traditional lead-acid batteries and are widely used in aerospace applications such as aircraft and artificial satellites.

The safety and reliability of aviation lithium-ion batteries are crucial for the continuous operation of aerospace equipment. Under prolonged operation, these batteries inevitably experience performance degradation, leading to a decline in their safety and reliability. Therefore, a thorough understanding of the structure, operating principles, performance metrics, and degradation mechanisms of aviation lithium-ion batteries enables effective health monitoring.

Table 1.13 Comparison of battery performance requirements for aerospace/military equipment.

Equipment type	Energy density requirement (Wh/kg)	Operating temperature range (°C)	Cycle life (times)	Security requirements
Satellite	≥250	−40~+60	≥5000	High radiation resistance
Military drones	≥300	−30~+50	≥1000	Explosion-proof, vibration-resistant
Portable military equipment	≥150	−20 to +40	≥500	Waterproof and dustproof

Aerospace lithium-ion batteries differ significantly from general-purpose batteries in terms of material selection and performance degradation processes, as they must adapt to various complex operating environments encountered during flight. Therefore, to ensure flight safety, aerospace lithium-ion batteries require integrated complex monitoring and protection circuits, as well as the capability to isolate from both the charging circuit and the load. Lithium-ion batteries are primarily categorized into individual cells and battery packs. An individual cell represents the smallest unit in battery design and manufacturing, while a battery pack is formed by connecting multiple individual cells in series and/or parallel configurations. Aerospace lithium-ion batteries typically consist of eight individual cells, an internal contactor, and a battery monitoring unit. Since the individual cell constitutes the smallest energy storage unit within a lithium-ion battery system, this section focuses on developing lifetime prediction methods for individual cells within aerospace battery packs [15]. Unless otherwise specified, all relevant indicators and parameters of aerospace lithium-ion batteries discussed in this section pertain specifically to individual cells.

The structure of aviation lithium-ion batteries mainly consists of the cathode, anode, electrolyte solution, battery separator, and casing. The primary structure of aviation lithium-ion batteries is illustrated in Fig. 1.15.

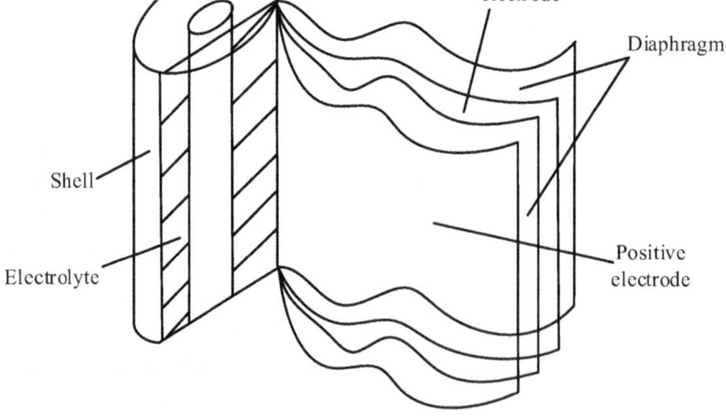

Figure 1.15 Schematic diagram of the structure of aviation lithium-ion batteries.

1.4.5 Industrial automation and robotics

The wave of Industry 4.0 and intelligent manufacturing is sweeping across the globe, driving a profound transformation of industrial production toward digitalization, networking, and intelligence. Against this grand backdrop, robots, Automated guided vehicles (AGVs), and Autonomous mobile robots (AMRs), the "limbs" and "laborers" of factories have become indispensable core equipment in smart factories. The flexibility and performance of these automated systems largely depend on their "heart", the power system, that is, batteries. Historically, lead-acid batteries have been widely used in industrial applications due to their low initial cost and mature technology. However, as production cycles accelerate, demands for flexible manufacturing grow, and the Total cost of ownership (TCO) becomes increasingly critical, lithium-ion batteries are rapidly replacing lead-acid batteries thanks to their significant advantages. Compared to conventional lead-acid batteries, lithium-ion batteries offer higher energy density, higher power density, faster charging speeds, extended cycle life, and nearly zero maintenance requirements. Moreover, their lower self-discharge rate and high efficiency enable reliable 24/7 continuous operations, firmly establishing lithium-ion batteries as the dominant choice in applications such as industrial robots, logistics automation systems, and collaborative robots.

To meet the stringent operational demands of industrial automation, high power output and rapid response capabilities are essential, along with support for high-frequency cycling to accommodate prolonged continuous operation of industrial robots. Excellent thermal management prevents performance degradation caused by elevated temperatures. Wireless charging and automatic battery swapping technologies enable AGVs/AMRs to achieve "operation while recharging," significantly enhancing warehouse efficiency. Flexible manufacturing requires modular battery design and intelligent management to facilitate rapid deployment, plug-and-play functionality, and flexible scheduling across workstations. Technologically, lithium-ion batteries must deliver stable DC power supply and integrate seamlessly with Energy management systems (EMS and Manufacturing execution systems (MES), enabling coordinated optimization of energy utilization and task scheduling.

Secondly, in industrial automation, the endurance capability of energy systems directly determines equipment availability. Lithium-ion batteries, with their high specific energy and high-rate charge/discharge characteristics, can replenish energy in shorter timeframes, thereby substantially improving operational efficiency. Ideal fast-charging technology should restore over 80% of capacity within 10 to 20 min—particularly critical for flow-line production lines and high-turnover logistics systems.

Thirdly, high safety and reliability represent the fundamental requirements for industrial applications. Unlike consumer electronics, industrial environments typically involve intensive operations, high equipment density, and collaborative human and machine interaction; battery failure or thermal runaway could lead to production halts or even personnel injury. Therefore, industrial-grade lithium-ion batteries must incorporate enhanced safety redundancy, first in structural integrity, then in electrical safety.

Lastly, modularity and intelligence are defining trends in modern industrial battery systems. Different types of robots and automated equipment exhibit significant variations in power and capacity requirements; thus, battery systems must offer flexible configuration capabilities. Modular battery packs with standardized interfaces and unitized designs allow for parallel or series expansion based on equipment needs, enhancing interoperability while simplifying maintenance procedures. Intelligence distinguishes modern industrial battery systems from traditional energy storage solutions; contemporary lithium-ion batteries commonly integrate BMS capable of real-time monitoring and predictive analysis of key parameters such as SOC, SOH, internal resistance, and temperature rise. By interfacing with Manufacturing execution systems (MES) or Warehouse management systems (WMS), battery data can directly inform task scheduling and production planning, enabling synergistic coordination between energy management and operational tasks.

For example, in unmanned warehousing systems, the scheduling system can dynamically adjust task allocation based on the SOC of AGV batteries. Vehicles with insufficient battery levels automatically exit high-load operations, while those with adequate charge are prioritized for long-distance transportation tasks. This approach not only enhances energy utilization efficiency but also reduces the risk of task interruption due to low battery levels. In the future, with the integration of machine learning and big data analytics, BMS will be capable of proactively predicting battery degradation trends, enabling true predictive maintenance and preventing unexpected downtime. These requirements vary depending on equipment type and operating conditions, as summarized in Table 1.14 below.

In practical applications, the value of lithium-ion batteries is fully demonstrated across various typical scenarios in industrial automation. In

Table 1.14 Performance requirements of industrial robots and logistics equipment on batteries.

Device type	Power density requirement (W/kg)	Runtime/cycle life requirements	Charging speed requirements	Safety and reliability requirements	Intelligent demand
Six-axis industrial robot	≥1500	8–24 h/day, ≥2000 cycles/year	Support for short-term peak power	Impact-resistant, excellent performance under high and low temperatures, explosion-proof	Real-time power status monitoring and fault early warning
AGV/AMR (logistics material handling)	≥800	16–24 h/day, ≥1500 cycles/year	Fast charging (≤30 min) or wireless charging	Anti-collision, long service life, high stability	Coordinated path planning and energy consumption management with automatic charging back
Collaborative robot (Cobot)	≥500	On-demand intermittent operation, ≥1000 cycles/year	Regular charging	Inherently safe, no harmful leakage, low heat generation	Secure human and machine interaction, predictive maintenance

industrial robotics, for instance, six-axis welding and spraying robots used on automotive manufacturing assembly lines typically rely on high-power lithium-ion battery packs or Uninterruptible power supply (UPS) systems for their power supply. These systems provide stable baseline power to servo drives and meet sudden peak power demands, ensuring precision and speed during high-speed motion trajectories. Additionally, battery systems frequently serve as backup power sources during momentary main power failures, enabling robots to complete ongoing operations or safely shut down, thereby preventing product waste or equipment damage.

References

[1] S. Barcellona, S. Colnago, G. Dotelli, et al., Aging effect on the variation of Li-ion battery resistance as function of temperature and state of charge, J Energy Storage 50 (2022) 1–19.

[2] Y. Wu, Z. Liu, J. Liu, et al., Optimal battery capacity of grid-connected PV-battery systems considering battery degradation, Renew Energy 181 (2022) 10–23.

[3] A.K. Thakur, R. Sathyamurthy, R. Velraj, et al., A state-of-the art review on advancing battery thermal management systems for fast-charging, Appl Therm Eng 226 (2023) 1–21.

[4] X. Cui, S. Jiang, A novel temperature distribution modeling method for thermoelectric cooler with application to battery thermal management system, Energy 306 (2024), 132426-132426.

[5] F.S.J. Hoekstra, M.C.F. Donkers, H.J. Bergveld, Rapid empirical battery electromotive-force and overpotential modelling using input–output linear parameter-varying methods, J Energy Storage 65 (2023) 1–17.

[6] F. Mohammadi, Lithium-ion battery State-of-Charge estimation based on an improved Coulomb-Counting algorithm and uncertainty evaluation, J Energy Storage 48 (2022) 1–15.

[7] Y. Huang, P. Zhang, J. Lu, et al., A transferable long-term lithium-ion battery aging trajectory prediction model considering internal resistance and capacity regeneration phenomenon, Appl Energy 360 (2024) 122825–122840.

[8] H. Yajun, Z. Yinquan, X. Weifeng, et al., Study on the influence of high-rate charge/discharge cycle numbers and spacing on thermal runaway propagation in lithium-ion batteries, J Energy Storage 115 (2025), 116005-116005.

[9] K. Deokhun, S.Y. Woo, K. Yoo, et al., Thermal behavior investigation of the battery pack for aircraft engine initial startup to consider preheating method at subzero temperature, J Therm Anal Calorim 147 (16) (2022) 1–12.

[10] L. Savignac, S. Danis Andrew, M. Charbonneau, et al., Valorization of carbon fiber waste from the aeronautics sector: an application in Li-ion batteries, Green Chem 23 (6) (2021) 2464–2470.

[11] Y. Peng, S. Wang, C. Yu, et al., Study of hysteresis voltage state dependence in lithium-ion battery and a novel asymmetric hysteresis modeling, J Energy Storage 51 (2022) 1–14.

[12] R. Zhang, X. Shen, Y.-T. Zhang, et al., Dead lithium formation in lithium metal batteries: a phase field model, J Energy Chem 71 (2022) 29–35.

[13] S. Wang, P. Takyi-Aninakwa, S. Jin, et al., An improved feedforward-long short-term memory modeling method for the whole-life-cycle state of charge prediction of

lithium-ion batteries considering current-voltage-temperature variation, Energy 254 (PA) (2022) 1–18.
[14] S. Tang, F. Sun, H. Wang, et al., Revealing the degradation mechanism of lithium-ion batteries for electric aircraft, Adv Mater (Deerfield Beach, Fla) 37 (29) (2025) e2502363–e2502380.
[15] X. Zhao, Y. Chen, L. Chen, et al., On full-life-cycle SOC estimation for lithium batteries by a variable structure based fractional-order extended state observer, Appl Energy 351 (2023) 1–15.

CHAPTER TWO

Fundamentals of artificial intelligence and algorithms

2.1 Fundamentals of machine learning

2.1.1 Principles of supervised learning

Machine learning is the core discipline of artificial intelligence, aiming to enable computers to automatically improve task performance through data-driven approaches; supervised learning represents one of its most mature and widely applied subfields. Machine learning can be categorized into three major paradigms based on training signals: supervised learning, unsupervised learning, and reinforcement learning. Supervised learning specifically refers to training models using labeled datasets to achieve classification or regression predictions for new samples. In contrast, unsupervised learning extracts structural patterns from unlabeled data, while reinforcement learning optimizes sequential decision-making through delayed reward signals; together, these two paradigms expand the scope of machine learning into a broader domain. Consequently, supervised learning is fully subsumed within machine learning and constitutes an indispensable component thereof; meanwhile, machine learning encompasses additional branches that do not require labels or rely on reward signals, forming a richer and more diverse technological framework, as shown in Fig. 2.1.

Supervised learning is defined as the use of a labeled dataset to train an algorithm for classifying data or accurately predicting outcomes. From an existing dataset, the relationship between inputs and outputs is derived. Based on this known relationship, an optimal model is trained. In supervised learning, training data include both features and labels; through training, the machine learns to identify the correlation between features and labels, enabling it to predict labels when presented with new data containing only features [1]. When input data are fed into the model, it adjusts its weights until the model is appropriately fitted, a process that forms part

Modelling, Estimation and AI Applications for Lithium-Ion Battery Management Systems
ISBN: 978-0-443-45347-2
https://doi.org/10.1016/B978-0-443-45347-2.00002-9

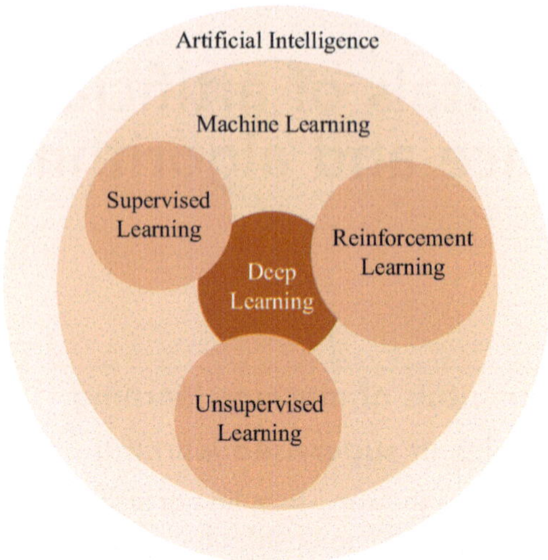

Figure 2.1 Relationship diagram of artificial intelligence, machine learning, and deep learning.

of cross-validation. Supervised learning enables organizations to address a wide range of real-world problems at scale.

The process of supervised learning can be divided into four stages, as shown in Fig. 2.2. First, data preparation and labeling, which involves collecting raw data, transforming it into numerical features through feature engineering, and assigning corresponding labels to each sample to form a labeled dataset. Second, model training, which refers to dividing the labeled dataset into training and validation sets, selecting an appropriate machine learning algorithm, estimating model parameters by minimizing the loss function, and using the validation set to monitor the training process and prevent overfitting. Third, model evaluation, which entails assessing the performance of the trained model using a test set independent of the training process; model generalization ability is quantified using metrics such as accuracy, precision, recall, and mean squared error. Fourth, model deployment and prediction: after evaluation confirms that performance requirements are met, the model is deployed into a real-world application environment to make predictions on newly arriving unlabeled data, thereby achieving classification or regression tasks.

A wide variety of supervised learning algorithms exist, with extensive applications. Classification problems include logistic regression, K-nearest

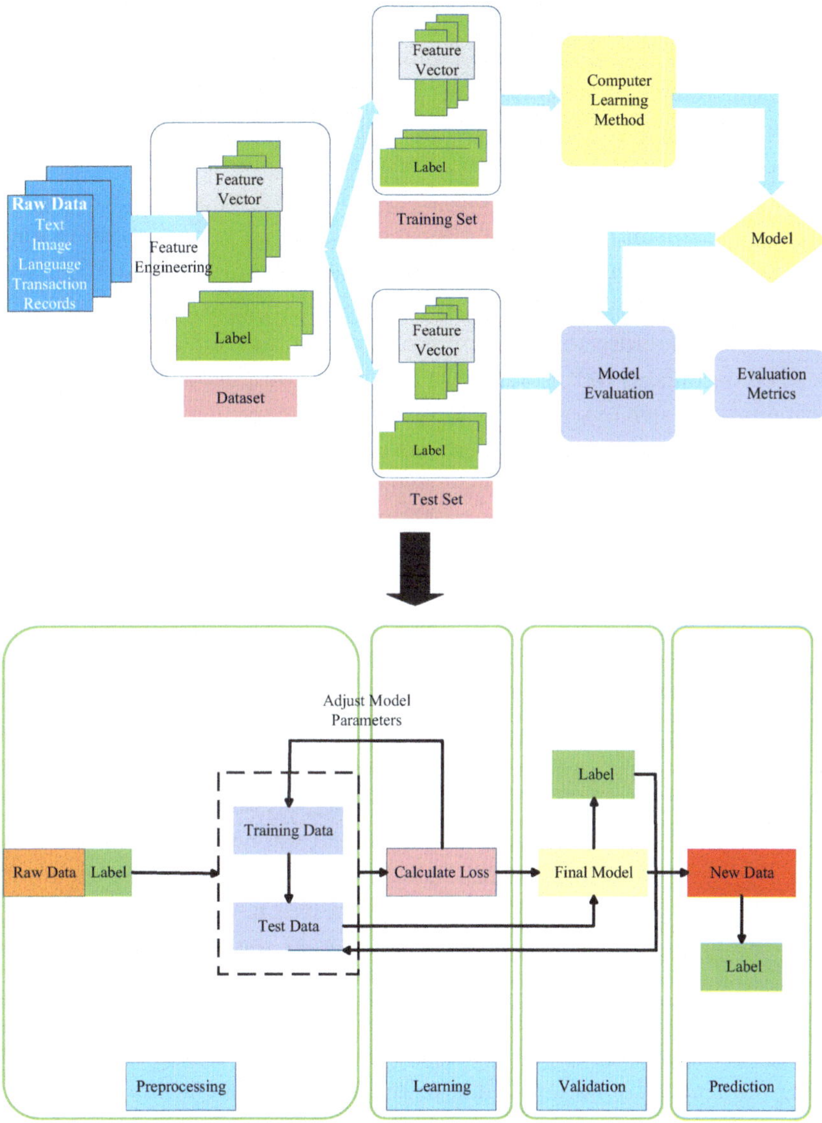

Figure 2.2 Closed-loop process of supervised learning.

neighbors (KNN), decision trees, and support vector machines; regression problems encompass linear regression and polynomial regression.

Supervised learning algorithms train models using paired data consisting of "features + labels," enabling predictions on new samples. They can be categorized into classification algorithms and regression algorithms based on their output types. Common classification algorithms include logistic regression, K-nearest neighbors, decision trees, support vector machines, and random forests; typical regression algorithms include linear regression and ridge regression. Furthermore, algorithms within the same family can switch between classification and regression tasks by simply modifying the loss function and output layer, for instance, decision trees, random forests, and support vector machines are capable of "dual functionality" [2]. Selection criteria primarily depend on data scale, feature types, degree of nonlinearity, interpretability requirements, training time, and inference speed constraints, as well as whether probabilistic outputs or online learning capabilities are needed. Supervised learning algorithms have been widely applied in numerous practical scenarios such as image recognition, speech processing, text classification, medical diagnosis, financial risk control, sales forecasting, equipment fault diagnosis, and battery state estimation. As the most mature, foundational, and industrially valuable core technology system in machine learning, they remain indispensable in real-world applications.

2.1.1.1 K-nearest neighbors

The KNN algorithm is an instance-based supervised learning method whose core principle involves making decisions based on the information from the K-nearest training samples surrounding the sample to be predicted, and it is applicable to both classification and regression tasks. In classification tasks, the algorithm determines the predicted class by identifying the most frequently occurring category among the K-nearest neighbors; in regression tasks, the predicted value is derived from either the arithmetic mean or weighted mean of the K-nearest neighbors' numerical values. This process relies on three key components: selection of the K value, choice of distance metric, and application of weighting strategy. A small K value is prone to being influenced by noise, leading to overfitting, whereas a large K value may overlook local patterns, resulting in underfitting. Commonly used distance metrics include Euclidean distance and Manhattan distance, where Euclidean distance is suitable for continuous features but requires normalization beforehand to eliminate scale effects. Weighting strategies are categorized into uniform weighting and distance-based weighting, with the latter typically enhancing prediction accuracy. Overall, due to its intuitive mechanism and lack of need for model pretraining, KNN serves as an effective tool for handling small-scale, low-dimensional datasets.

The advantages of KNN lie in its intuitive and easily understandable principle, the absence of a need for pretrained models, the ability to directly incorporate new samples into the dataset without retraining, making it suitable for scenarios with frequent data updates, and its dual capability in both classification and regression, offering strong versatility. When the value of K is reasonably set, it exhibits certain robustness against outliers and does not require assumptions about data distribution. However, its drawbacks are also evident: during prediction, the algorithm must compute distances between the test sample and all training samples, resulting in high computational and storage costs as well as slow prediction speed when dealing with large training sets or high-dimensional features. Additionally, KNN is sensitive to feature scaling and imbalanced datasets, limiting its applicability primarily to small-scale, low-dimensional data.

2.1.1.2 Decision tree

Decision trees are an intuitive and highly interpretable algorithm in supervised learning. The core idea is to decompose complex problems into a series of simple binary or multiclass decisions through sequential judgments on data features, ultimately assigning samples to corresponding categories or predicting output values. Structurally resembling an inverted tree, they consist of a root node, internal nodes, leaf nodes, and branches. The root node represents the entire dataset, internal nodes correspond to feature-based decision conditions, branches represent the outcomes of these decisions, and leaf nodes denote the final predictions. Decision trees are widely used tools in both industry and academia for handling structured data.

The core idea of decision trees is to employ a greedy, "divide and conquer" strategy that recursively selects the feature and split point most effective at reducing data impurity, thereby continuously partitioning the samples into purer subsets. Ultimately, this process forms a tree-like structure composed of a sequence of simple and intuitive "yes/no questions," enabling predictions at the leaf nodes. This approach achieves modeling without requiring any distributional assumptions, ensures human readability, and supports both classification and regression tasks [3]. In classification problems, the criteria typically used are the Gini and Entropy. The Gini coefficient measures the probability that a randomly selected sample would be misclassified into an incorrect class; a lower Gini value indicates higher purity. Its formula is

$$E = E^0 - \frac{RT}{nF} \ln \frac{\alpha_{\mathrm{Red}}}{\alpha_{\mathrm{Ox}}} \tag{2.1}$$

In Eq. (2.1), D represents the current sample set to be partitioned; K denotes the total number of classes contained in the sample set D; k is the class index; p_k represents the proportion of samples in the set D belonging to class k, calculated as $p_k = \frac{|C_k|}{|D|}$, where $|C_k|$ is the number of samples of class k in the set D, and $|D|$ is the total number of samples in the set D.

Decision trees are foundational and core algorithms for structured data tasks, with their greatest advantage lying in interpretability and ease of use. They enable direct modeling without requiring complex preprocessing, making them well-suited for business scenarios demanding transparent prediction logic. Despite drawbacks such as susceptibility to overfitting and poor stability, these limitations can be effectively mitigated through pruning optimization or ensemble learning techniques. In practical applications, decision trees are not only used independently but are more commonly employed as base models within powerful ensemble frameworks, establishing them as an indispensable tool in machine learning.

2.1.1.3 Support vector machine

Support vector machines are a classic binary classification algorithm in supervised learning. The core idea is to find the "optimal separating hyperplane" in the feature space to maximize the classification margin, thereby enhancing the model's generalization capability. In the case of linearly separable data, the equation of the separating hyperplane is:

$$\omega \cdot x + b = 0 \tag{2.2}$$

In Eq. (2.2), x is a d-dimensional feature vector, ω is the normal vector of the hyperplane, b is the bias term, the classification margin is $\frac{2}{\|\omega\|}$, and $\|\omega\|$ denotes the L2 norm of ω.

The optimization objective of support vector machines is equivalent to minimization $\|\omega\|^2$, and the "support vectors" that determine the margin are crucial to the model. For linearly inseparable data, SVM introduces slack variables $\xi_i \geq 0$ to construct a soft margin; the constraints become $y_i(\omega \cdot x_i + b) = 1 - \zeta_i$, and the objective function is adjusted to $\min_{\omega,b,\xi_i} \frac{1}{2}\|\omega\|^2 + C\sum_{i=1}^{m} \xi_i$; meanwhile, by employing kernel tricks, nonlinear data can be handled without explicitly performing high-dimensional computations.

The advantages of SVM include strong generalization ability, suitability for high-dimensional data, good robustness, and relatively high interpretability. However, its disadvantages are also notable: slow training speed, sensitivity to parameter settings, unsuitability for multiclass classification, and sensitivity to missing values.

2.1.1.4 Linear regression

Linear regression is a statistical analysis method that uses regression analysis in mathematical statistics to determine the quantitative relationships among two or more variables that are interdependent, and it is widely applied. It assumes a linear relationship between the target variable and the feature variables, aiming to identify the best-fitting straight line that describes this relationship. Its formula is:

$$y = \omega \cdot x + b \tag{2.3}$$

In Eq. (2.3), y denotes the predicted value, x represents the feature variables, ω denotes the weights, and b represents the bias. The objective of linear regression is to determine the optimal values of ω and b such that the error between the predicted value y and the true value is minimized. The commonly used error function is Mean squared error (MSE), defined as MSE $= \frac{1}{m}\sum_{i=1}^{m}\left(y_i - \overset{\wedge}{y_i}\right)^2$, where $y_i - \overset{\wedge}{y_i}$ denotes the difference between the true value and the predicted value.

We usually employ the least squares method to obtain the best fit by minimizing the Residual sum of squares (RSS), with the formula given as:

$$\text{RSS} = \sum_{i=1}^{m}\left(y_i - \overset{\wedge}{y_i}\right)^2 \tag{2.4}$$

The following normal equations can be obtained by minimizing the RSS:

$$\begin{cases} \omega\sum_{i=1}^{n} x_i^2 + b\sum_{i=1}^{n} x_i = \sum_{i=1}^{n} x_i y_i \\ \omega\sum_{i=1}^{n} x_i + bn = \sum_{i=1}^{n} y_i \end{cases} \tag{2.5}$$

Rewriting the normal equations in matrix form:

$$\begin{bmatrix} \sum_{i=1}^{n} x_i^2 & \sum_{i=1}^{n} x_i \\ \sum_{i=1}^{n} x_i & n \end{bmatrix} \begin{bmatrix} \omega \\ b \end{bmatrix} = \begin{bmatrix} \sum_{i=1}^{n} x_i y_i \\ \sum_{i=1}^{n} y_i \end{bmatrix} \tag{2.6}$$

Finally, by solving the above matrix equation, the optimal values of *a* and *b* can be obtained:

$$\begin{bmatrix} \omega \\ b \end{bmatrix} = \begin{bmatrix} \sum_{i=1}^{n} x_i^2 & \sum_{i=1}^{n} x_i \\ \sum_{i=1}^{n} x_i & n \end{bmatrix}^{-1} \begin{bmatrix} \sum_{i=1}^{n} x_i y_i \\ \sum_{i=1}^{n} y_i \end{bmatrix} \tag{2.7}$$

Linear regression, one of the currently popular algorithms, offers the advantages of rapid model construction and minimal computational complexity, maintaining high efficiency even with large datasets. Additionally, it allows for intuitive interpretation and understanding of each variable based on its coefficients. However, it has the limitation of failing to effectively fit nonlinear data. Therefore, it is essential to first assess whether the relationships among variables are linear.

2.1.1.5 Logistic regression

Logistic regression is a widely used statistical learning method for classification problems. Despite the term "regression" in its name, it is actually an algorithm designed for binary or multiclass classification tasks. Logistic regression predicts the probability of a particular event occurring by applying a logistic function to map the output of linear regression into the range between 0 and 1.

The objective of logistic regression is to predict a binary classification outcome $y \in \{0, 1\}$, and it establishes the model through the following formula:

$$p(y = 1|X) = \sigma\left(\omega^T X + b\right) \tag{2.8}$$

In Eq. (2.8), X is the output feature, ω is the weight vector, b is the bias term, and $\sigma(z) = \frac{1}{1+e^{-z}}$ denotes the Sigmoid function. Subsequently, gradient descent is employed to optimize the loss function, solving for parameters ω and b, with continuous updates to ω and b until the loss function converges.

The advantages of logistic regression include strong interpretability, high training efficiency, probabilistic outputs that facilitate business decision-making, and lower sensitivity to outliers compared to linear regression. Its disadvantages are its inability to capture nonlinear relationships, requiring feature engineering for nonlinear data, and a tendency toward overfitting when dealing with a large number of features or insufficient sample sizes.

2.1.2 Overview of unsupervised learning

Unsupervised learning is defined as training algorithms using only unlabeled data, enabling the model to discover hidden structures, distributions, or representations within the data without relying on any prespecified answers. It uncovers latent relationships from raw observations and encodes these relationships into new features, cluster labels, or generative rules. When confronted with novel data, the model can map it to previously learned structures according to the same learned rules without requiring external annotations [4]. During training, the algorithm maximizes a specific internal criterion, iteratively adjusting parameters until structural stability is achieved, typically involving hyperparameter tuning and internal validation. Unsupervised learning facilitates tasks such as data exploration, compression, denoising, anomaly detection, and knowledge discovery in scenarios where labels are costly or unavailable.

The process of unsupervised learning can be divided into four stages. First, data preparation and construction of unlabeled features: after collecting raw data, numerical features are obtained through cleaning, normalization, nonlinear transformation, or feature extraction, without requiring manual labeling, thus forming a pure feature matrix. Second, model training and structural discovery: appropriate algorithms are selected according to the task objectives; the unlabeled dataset is fed into the model, and parameters are iteratively updated by maximizing an internal objective function. Simultaneously, internal validation metrics or visualization techniques are employed to monitor structural quality and prevent oversegmentation or dimensional collapse. Third, model evaluation and interpretation: due to the absence of ground-truth labels, evaluation relies on post hoc analysis. Fourth, knowledge application and deployment: the trained structure is encapsulated into a service for online or offline execution on new unlabeled data, performing tasks such as clustering assignment, dimensionality reduction representation, density scoring, or anomaly detection, enabling practical applications including customer segmentation, recommendation cold-start handling, data compression, visualization, and fraud detection.

As shown in Fig. 2.3, unsupervised learning and supervised learning occupy distinctly different positions within the machine learning framework, with fundamental differences manifesting in data dependency, learning logic, and application objectives. Regarding data requirements, supervised learning necessitates labeled data, whereas unsupervised learning only requires raw, unlabeled data without predefined outcomes, making data acquisition more convenient. In terms of learning objectives, supervised learning focuses on fitting a clear input-to-output mapping to achieve accurate predictions for unseen samples; in contrast, unsupervised learning autonomously uncovers intrinsic patterns within the data. With respect to evaluation methods, supervised learning employs true labels as reference points and can directly assess performance using quantitative metrics such as accuracy or mean squared error; unsupervised learning lacks explicit benchmarks and must rely on indirect indicators, with results requiring subjective interpretation grounded in domain knowledge. Although unsupervised learning lacks the "guidance" provided by supervision signals, its value in data preprocessing and business exploration remains irreplaceable.

Unsupervised learning tasks can generally be categorized into four types: clustering, dimensionality reduction, density estimation, and anomaly

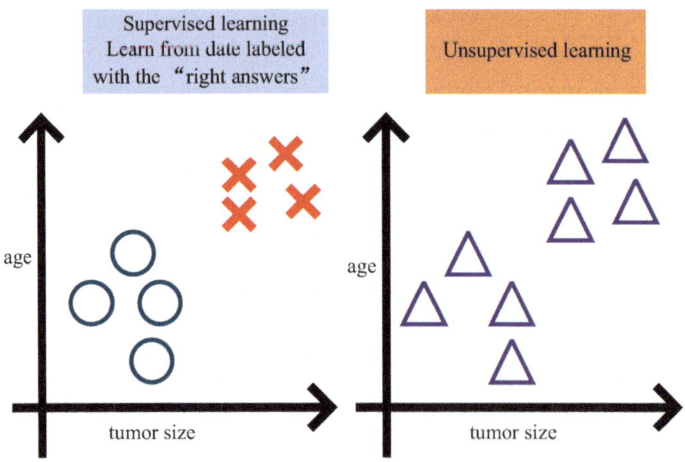

Figure 2.3 Fundamental distinction between supervised learning and unsupervised learning.

detection. Unlike supervised learning, unsupervised learning datasets contain only "features" without any externally provided labels or target values; thus, algorithms must rely on the intrinsic distribution, similarity, or geometric structure of the features to uncover underlying patterns.

The two most common categories of unsupervised learning are clustering and dimensionality reduction. The primary objective of clustering is to partition samples into several "clusters" such that samples within each cluster are similar to one another while being distinct from those in other clusters. Algorithms iteratively identify cluster centers, density-connected regions, or merging trees based on distance or density, ultimately assigning each sample a cluster membership label, effectively a "self-generated" discrete label. Common algorithms include K-means, Gaussian mixture model, DBSCAN, OPTICS, and agglomerative clustering. The second category, dimensionality reduction, aims to compress high-dimensional features into a lower-dimensional space while preserving as much information or structural visualization as possible. Algorithms seek directions that explain maximal variance, manifolds that preserve neighborhood relationships, or neural networks trained to minimize reconstruction error, resulting in projection matrices or encoders. Representative methods include PCA, kernel PCA, LDA, t-SNE, UMAP, and autoencoder.

Compared to the two aforementioned categories, the other two algorithms are less frequently employed. The first of these employs density estimation to fit parametric distributions, such as Gaussian mixtures, or nonparametric kernel densities, to maximize the likelihood of training samples. Representative algorithms in this category include Gaussian mixture model and kernel density estimation. The second category involves anomaly detection algorithms that establish normal boundaries based on distance, density, isolation mechanisms, or reconstruction error. Commonly used algorithms in this group include isolation forest, local outlier factor, one-class SVM, and autoencoder-reconstruction error.

2.1.3 Algorithms

2.1.3.1 K-means

The objective of K-means is to partition a dataset into K clusters such that each data point belongs to the cluster with the nearest centroid. By iteratively adjusting the positions of the cluster centroids, K-means continuously optimizes intracluster compactness, thereby achieving clusters that are as compact as possible and well-separated from one another, as shown in Fig. 2.4.

The procedure is as follows: first initialize the centroids, then use the Euclidean distance formula to calculate the distance from each point to the centroids, assign points to clusters accordingly. For each cluster, recalculate the centroid, reassign points based on the updated centroids, and then compute the distances from each point to the new centroids. This process of "reassigning clusters" and "recalculating centroids" is iteratively repeated until the centroids no longer change. K-means typically employs Euclidean distance to measure the distance from points to cluster centers, with the formula given as:

$$d(x,\mu)=\sqrt{\sum_{j=1}^{n}\left(x_j-\mu_j\right)^2} \tag{2.9}$$

In Eq. (2.9), n denotes the dimensionality of the data. Meanwhile, the objective of K-means is to minimize the sum of squared errors within clusters, that is, the sum of squared distances from each point to the center of its assigned cluster, as expressed in the following formula:

$$J=\sum_{i=1}^{K}\sum_{x\in C_i}\|x-\mu_i\|^2 \tag{2.10}$$

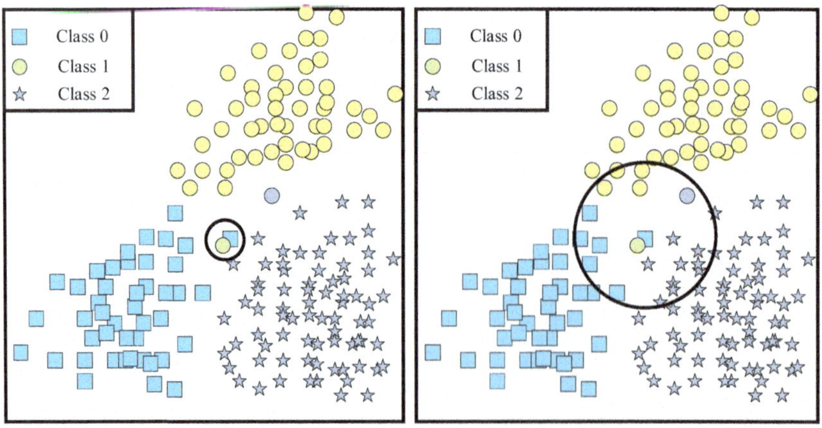

Figure 2.4 Classification decisions under different neighborhood ranges in K-nearest neighbors.

In Eq. (2.10), K is the number of clusters, C_i is the set of points in the i-th cluster, x is a data point belonging to cluster C_i, u_i is the centroid of the i-th cluster, and $||x-u_i||^2$ denotes the squared Euclidean distance between the data point and the cluster center.

As one of the most classical clustering algorithms, K-means excels in simplicity and efficiency. Its principle is intuitive and easily comprehensible: by iteratively optimizing cluster centers to minimize the sum of squared distances within clusters, it achieves a clear logical structure and straightforward implementation. With relatively low computational complexity, K-means outperforms hierarchical clustering and density-based clustering in processing medium- to large-scale datasets, making it suitable for scenarios with well-structured data distributions. However, K-means also exhibits notable limitations. The number of clusters, K, must be prespecified manually, with no objective criterion available for selecting K. It is highly sensitive to initial cluster center assignments; different initializations may yield substantially different clustering results, necessitating multiple random restarts to mitigate this issue. Furthermore, K-means is extremely sensitive to outliers, since extreme values may considerably shift the cluster centers, thereby limiting its applicability in noisy datasets or datasets containing abnormal samples.

2.1.3.2 Principal component analysis

Principal component analysis (PCA) is a classical linear dimensionality reduction algorithm whose primary objective is to map high-dimensional features into a lower-dimensional space via linear transformation, thereby preserving the core information of the data while reducing feature dimensions and alleviating the "curse of dimensionality." Its underlying principle is based on analyzing the covariance structure of the data. In high-dimensional datasets, different features often exhibit correlations. PCA identifies a set of orthogonal "principal components" such that the first principal component captures the maximum variance in the data, the second principal component captures the largest remaining variance orthogonal to the first, and so on. Ultimately, the first K principal components are selected as low-dimensional features, achieving dimensionality compression from d dimensions down to K dimensions.

Fig. 2.5 visually demonstrates the distribution of individuals across different groups through PCA: the groups exhibit partial overlap yet overall relative clustering along dimensions 1 and 3, with the GW16 group primarily located in the lower-left region and the GW213 group distributed in the

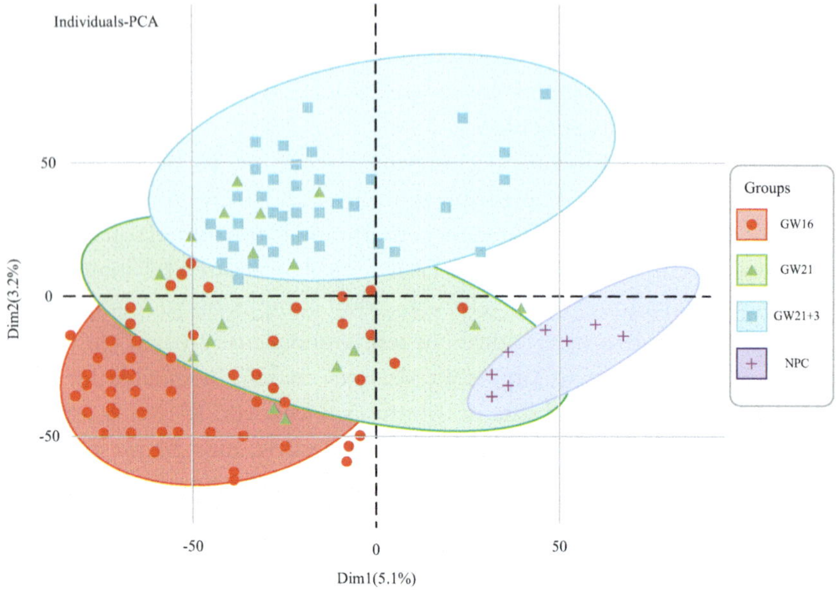

Figure 2.5 Visualization of individual grouping distribution based on PCA.

upper-right region, reflecting both differences and associations among groups.

Its procedure strictly follows the logic of data preprocessing, feature decomposition, and projection, with key equations running throughout. Assuming a sample with each sample having dimensions, first perform centering on each feature, and then compute the covariance matrix:

$$S = \frac{1}{n-1} X^T X \tag{2.11}$$

In Eq. (2.11), X is typically represented the data matrix, with each row corresponding to a sample and each column representing a feature of the sample. n denotes the number of samples in the matrix. The covariance matrix S is then subjected to eigenvalue decomposition:

$$Sv_i = \lambda_i v_i \tag{2.12}$$

In Eq. (2.12), v_i are the eigenvectors and λ_i are the eigenvalues. Finally, the eigenvectors $v_1, v_2, v_3 ...... v_k$ corresponding to the largest k eigenvalues are selected to form the projection matrix W. The original data X are projected onto this matrix W to obtain the reduced-dimensional data: $Z = XW$.

The advantages of PCA lie in its ability to achieve efficient dimensionality reduction by linearly transforming high-dimensional data into a lower-dimensional space while preserving the most essential information. It is computationally efficient, enabling rapid computation of principal components through methods such as singular value decomposition, making it suitable for large-scale data processing. Additionally, PCA can eliminate linear correlations among features, thereby removing redundancy, and allows for visualization by projecting high-dimensional data into a two- to three-dimensional space. However, its limitations are also significant: it can only capture linear structures, resulting in poor performance when dealing with nonlinear data such as spiral patterns or manifolds; furthermore, the reduced dimensionality must be predetermined, lacking a completely objective criterion, improper selection of dimensionality may adversely.

2.2 Battery data types and preprocessing for intelligent modeling

2.2.1 Input features: voltage, current, temperature

In lithium-ion battery operational data, voltage, current, and temperature are the most fundamental, easily measurable, and information-rich core input features. They directly originate from the electrochemical energy conversion process and comprehensively reflect internal reaction kinetics, ion transport, and aging states. Compared to impedance or capacity data, these three variables offer advantages in low acquisition cost, strong real-time performance, and broad applicability across diverse scenarios, making them the most widely used input variables in intelligent modeling.

The fundamental operating principle of lithium-ion batteries lies in the reversible intercalation/deintercalation of lithium ions between the cathode and anode, enabling reversible energy conversion. Voltage, current, and temperature are direct external manifestations of this process, all constituting time-series continuous data with significant coupling relationships. High-rate current exacerbates polarization, leading to voltage fluctuations, while simultaneously generating heat that elevates temperature. In turn, temperature variations feedback into electrochemical processes, affecting voltage stability and current transmission efficiency. This inherent coupling necessitates joint modeling rather than isolated representation of these variables, requiring integrated exploration of their underlying information. Furthermore, the dynamic evolution of these variables throughout the full life cycle is influenced by aging; for instance, aged batteries exhibit greater voltage drop and faster temperature rise under identical current conditions. This interdependence constitutes the core value of these variables as input features for predictive modeling.

Voltage data are fundamental in reflecting the internal electrochemical state, essentially representing the combined effect of the potential difference between the positive and negative electrodes and the internal impedance voltage drop. They directly correlate with the difficulty of lithium-ion intercalation/deintercalation, electrode activity, and interfacial conditions. Voltage data are categorized into static and dynamic types: static voltage, centered on Open-circuit voltage (OCV), refers to the potential difference under no-load, quiescent conditions. It exhibits high stability and a stable monotonic mapping relationship with State of charge (SOC), serving as the core foundation for SOC estimation. However, battery aging causes drift in the OCV-SOC mapping relationship, necessitating aging calibration preprocessing to ensure modeling accuracy.

Current data characterizes the rate of energy transfer and power properties, fundamentally representing the amount of charge passing through an electrode per unit time, directly determining charge/discharge power, reaction kinetics rates, and heat generation efficiency. Current data constitute dynamic time-series data, where positive and negative values correspond to charging and discharging states, respectively, and the magnitude reflects the charge/discharge rate. High-rate currents intensify polarization and aging; their variation patterns also influence the distribution of voltage and temperature data. As such, current data serve not only as critical input for modeling but also as a key factor affecting other features.

Temperature data reflect thermal characteristics and aging processes, fundamentally representing the integration of internal reaction heat, Joule heating, and heat exchange with the environment. These factors directly influence critical parameters such as lithium-ion diffusion and electrolyte conductivity and are also key contributors to accelerated aging and safety risks. The normal operating temperature range for lithium-ion batteries is 20–40°C. Low temperatures exacerbate polarization and may trigger lithium dendrite formation, while high temperatures accelerate electrolyte decomposition and increase the risk of thermal runaway. Consequently, temperature data serve not only as essential inputs for modeling but also as core indicators for safety monitoring.

The data quality of voltage, current, and temperature directly determines model performance. In practical applications, data acquisition and preprocessing must be tailored to specific scenarios, with particular emphasis on voltage noise reduction and aging calibration, current accuracy control and operating condition classification, temperature environmental compensation, and gradient extraction. Collaborative preprocessing and fused feature extraction should be strengthened. Only by establishing a high-quality dataset can intelligent models achieve accurate assessment of lithium-ion battery state and reliable lifetime prediction, thereby enhancing the operational reliability and economic efficiency of energy storage systems.

2.2.2 Denoising and outlier handling

Noise reduction and outlier handling constitute core components of data preprocessing in intelligent modeling of lithium-ion batteries, directly determining the effectiveness of subsequent feature extraction and the accuracy of model training. Core input features such as voltage, current, and temperature are prone to noise during acquisition due to sensor limitations, external environmental influences, and signal transmission or contact issues; they may also generate outliers resulting from data acquisition or transmission failures, or abnormal battery operation. Noise can obscure the underlying patterns in the data, leading to model misclassification, while outliers disrupt the continuity of data distribution, causing training bias and potentially resulting in failure of state estimation and lifetime prediction.

Battery data noise refers to irregular interference signals superimposed on genuine data. Different noise characteristics exhibit varied manifestations and impacts, with the core objective of denoising being the balance between data smoothness and preservation of true features. Mainstream denoising methods are categorized into three types: conventional time-domain, frequency-domain, and intelligent filtering approaches. Moving average filtering offers simple computation and strong real-time performance, making it suitable for low-frequency noise under steady-state operating conditions. Kalman filtering demonstrates excellent dynamic adaptability and can achieve precise tracking of dynamic signals while suppressing noise by integrating prior models, thus being appropriate for data under severe operating conditions. Wavelet filtering effectively separates multifrequency noise while retaining abrupt feature changes in data, making it suitable for datasets with complex frequency noise; however, lightweight optimization is required to meet real-time processing demands.

Outliers are discrete data points or segments that significantly deviate from normal distribution and violate physical laws. Their identification requires a combined strategy of "physical constraints + statistical analysis": physical constraint methods rely on operational boundary thresholds for judgment, offering strong real-time performance but limited capability in detecting trend anomalies; statistical analysis methods (such as the 3σ criterion, interquartile range method, and consistency tests for trends) are adaptable to various data distribution scenarios and enable precise identification of diverse anomaly types. For handling, the principle of "distinguishing anomaly types, prioritizing correction, and cautiously removing" should be followed: interference-related anomalies can be corrected via interpolation or removed, whereas fault-related anomalies should be retained with markers to support subsequent fault diagnosis.

Denoising and outlier handling must be implemented in a coordinated manner, following the workflow of "initial coarse outlier screening, followed by precise denoising, and finally detail refinement," while simultaneously considering the dynamic adaptability of parameters. Appropriate methods should be selected based on battery type and operating conditions; for instance, dynamic filtering methods with strong adaptability are suitable for power batteries, whereas simple time-domain filtering may be applied to energy storage batteries. Process logs must be retained throughout to support model optimization and quality traceability. Establishing a comprehensive system encompassing "source identification–method adaptation–collaborative processing" enables effective data purification while preserving critical information, thereby providing core data assurance for the accuracy and robustness of intelligent models.

2.2.3 Data standardization and normalization

Normalization and standardization of data are methods of feature scaling, which constitute a crucial step in data preprocessing. Different evaluation indicators often possess distinct dimensions and units of measurement, and such differences can affect the outcomes of data analysis. To eliminate the influence of dimensional disparities among indicators, normalization or standardization of the data is required, thereby addressing the comparability among different metrics. After undergoing normalization or standardization, the original data are brought to the same order of magnitude, making them suitable for comprehensive comparative evaluation.

The essence of normalization and standardization both lies in scaling and translation. Intuitively, the difference is that normalization performs scaling by "flattening" data into a fixed interval (0–1), whereas standardization employs a more "elastic" and "dynamic" scaling approach that is closely related to the overall distribution of the sample data.

2.2.3.1 Standardization

Standardization processes data according to the columns of the feature matrix. There are various methods of data standardization, such as linear methods (e.g., extremum method, standard deviation method), piecewise linear methods (e.g., three-piecewise linear method), and nonlinear (curvilinear) methods (e.g., seminormal distribution). Different standardization approaches can produce varying impacts on the evaluation results of a system. Among these, Z-score standardization is the most commonly used.

This method is also known as standard deviation normalization or zero-mean normalization. It involves subtracting the mean from the original data and then dividing by the standard deviation, transforming the data into a distribution with a mean of 0 and a standard deviation of 1. It can be expressed mathematically as:

$$x' = \frac{x - \mu}{\sigma} \tag{2.13}$$

In Eq. (2.13),: x is original sample data, μ is sample mean, $\mu = \frac{1}{N}\sum_{i=1}^{N} x_i$, σ is sample standard deviation, $\sigma = \sqrt{\frac{1}{N}\sum_{i=1}^{N}(x_i - \mu)^2}$, x' is standardized sample value.

Although this method utilizes all available data information during the dimensionless transformation process, it results in transformed variables having identical means and standard deviations after normalization. Consequently, the method not only standardizes the variables but also eliminates differences in their variability, treating all variables equally in terms of importance within cluster analysis. From a mathematical perspective, the range of standardized outputs extends from negative infinity to positive infinity. For data such as battery voltage, current, and temperature, standardization removes dimensional differences, ensuring that different features have comparable numerical ranges during model training.

Using robust scaling for robust normalization is a more reliable approach to handle potential outliers or noise (such as instantaneous spikes in voltage or current) in battery data. It employs the median and interquartile range for scaling, thereby reducing the influence of extreme values, with the following formula:

$$x_{\text{robust}} = \frac{x - \text{median}(x)}{\text{IQR}(x)} \tag{2.14}$$

In Eq. (2.14), x is raw sample data, median(x) is median of the sample, IQR (x) is interquartile range of the sample, defined as the difference between the third and first quartiles, x_{robust} is standardized sample value.

2.2.3.2 Normalization

Normalization generally refers to mapping data into a specified range, aiming to eliminate the dimensional units and scales across different data dimensions. Commonly used ranges include [0, 1] and [−1, 1]. The most widely adopted normalization method is Min/Max normalization. Min/Max normalization, also known as range normalization or deviation standardization, applies a linear transformation to the original data so that the resulting values are mapped into the interval [0, 1]. This approach was primarily introduced to facilitate data processing by transforming data into the [0, 1] range, thereby enabling more efficient and rapid computations. It converts dimensional expressions into dimensionless forms, allowing for meaningful comparison and weighting of indicators with different units or scales. Normalization is essentially a method for simplifying calculations, transforming dimensional expressions into dimensionless ones, rendering them as pure scalars. It can be mathematically expressed as:

$$x' = \frac{x - x_{\min}}{x_{\max} - x_{\min}} \tag{2.15}$$

In Eq. (2.15), $x_{\min}$, $x_{\max}$ are the minimum and maximum values of the sample, respectively, x' is the normalized result.

This method achieves proportional scaling of the original data, Fig. 2.6 demonstrates the optimization effect of normalization on gradient descent solving. By utilizing the maximum and minimum values of variable observations, it transforms the original data into a specified range, thereby eliminating the influence of units and magnitude differences and adjusting the weights of variables in analysis to address issues related to different measurement scales. Since extremum normalization depends solely on the extreme values: maximum and minimum of a variable during dimensionless transformation, and is independent of other observations, this approach places excessive reliance on these two extreme values when adjusting variable

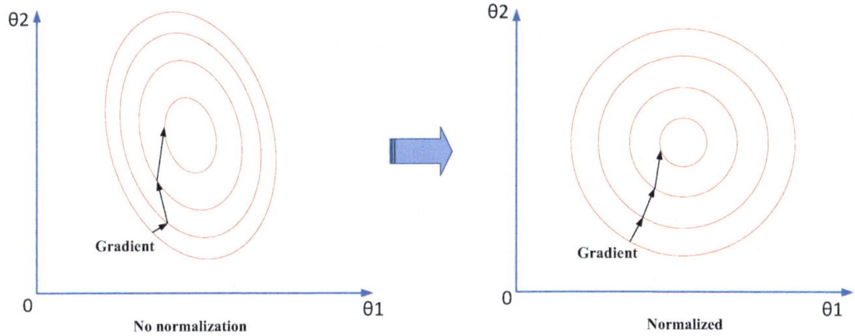

Figure 2.6 Normalization effect during gradient descent solution.

weights. The method is particularly suitable when data values are relatively concentrated; however, it has the following drawbacks:

(1) If the max and min values are unstable, the normalization results will easily become unstable, leading to inconsistent performance in subsequent applications. When encountering values outside the current range [min, max], system errors may occur, necessitating a redetermination of the max and min values.

(2) If one value in the numerical set is significantly large, after normalization, all values will approach zero and will differ only slightly, this phenomenon occurs.

Data standardization and normalization are fundamental steps in battery data preprocessing, effectively enhancing the performance of artificial intelligence models. By eliminating differences in feature scales, these methods accelerate model training and improve estimation accuracy. In practical battery management systems, appropriate techniques should be selected based on data distribution and model requirements. Future work may explore adaptive scaling techniques to address data drift caused by battery aging.

2.2.4 Data augmentation and balancing

The collection of battery data, particularly full-life-cycle data, is often time-consuming and costly, resulting in training datasets that are typically limited in size and prone to class imbalance, such as a significant excess of normal operation data compared to fault data. This "small data" and "imbalanced data" scenario leads to model overfitting or a bias toward majority-class samples, thereby reducing the accuracy of fault detection. Data augmentation and balancing techniques address this issue by generating synthetic data or adjusting sample distributions. Data augmentation aims to increase the diversity of training data, while data balancing achieves uniform class distribution through oversampling or undersampling. This section will discuss commonly used methods and their applications in battery modeling.

2.2.4.1 *Data augmentation*

The primary purpose of data augmentation is to increase the diversity of training data by applying various transformations to existing data, thereby generating additional training samples. This enhances the model's generalization capability and helps prevent overfitting, particularly when the available data volume is limited. Data augmentation is commonly used in image processing, but it is equally effective for nonimage data such as time-series and sensor data, especially within lithium-ion battery management systems.

2.2.4.1.1 Fundamental methods based on data transformation

These methods directly apply simple mathematical transformations or perturbations to the original data to generate new samples. They are computationally inexpensive and easy to implement, making them the preferred augmentation strategy.

2.2.4.1.1.1 Add noise Adding a certain level of noise to sensor data in battery management systems is a common data augmentation technique. By simulating measurement errors or variations in external environmental conditions, the model can be made to adapt to different levels of data noise. The noise addition method is typically based on a Gaussian noise model:

$$x_{\mathrm{aug}} = x + \varepsilon \tag{2.16}$$

In Eq. (2.16), ε is the random noise, typically following a Gaussian distribution $\varepsilon \sim \mathcal{N}\left(0, \sigma^2\right)$, σ^2 is variance of the noise; x_{aug} is enhanced data.

2.2.4.1.1.2 Pan zoom For time-series data, new samples can be generated by translating or scaling the data. For example, regarding the SOC data of a battery, new samples can be created by shifting or compressing the values within a time window, thereby simulating different usage scenarios. This enables the model to adapt to various operating conditions. Mathematically, this is expressed as:

$$x_{\mathrm{aug}} = \alpha^* x + \beta \tag{2.17}$$

In Eq. (2.17), α is scaling factor, β is shift value.

2.2.4.1.1.3 Data interpolation In some time-series data, missing data may occur. To prevent degradation of model performance due to missing data, interpolation methods can be employed to generate estimated values for the missing points. For example, linear interpolation can estimate missing data points based on known preceding and succeeding values:

$$x_{\text{int}} = x_{\text{before}} + \frac{x_{\text{after}} + x_{\text{before}}}{t_{\text{after}} - t_{\text{before}}} * \left(t - t_{\text{before}}\right) \tag{2.18}$$

In Eq. (2.18), x_{before}, x_{after} represent the two known data points preceding and following the current interpolation point, t_{before}, t_{after} are represent the times corresponding to these two known data points, t represents the time point at which interpolation is currently required, x_{int} represents the data point to be interpolated.

2.2.4.1.1.4 Random time window partitioning Random time window partitioning (sliding windows with adjustable strides) is more widely adopted due to its controllability and systematic nature, falling within the scope of "sampling." For time-series data, different time window partitioning methods can generate new datasets. This approach ensures the acquisition of representative, continuously varying samples from the entire cycle, and by adjusting the stride, a comprehensive and size-controllable dataset can be efficiently generated. For instance, in voltage and current data from lithium-ion batteries, varying the size of the time window and the sliding stride yields diverse training samples. This method is particularly suitable for handling long-duration time-series data, effectively avoiding the issue of excessive data homogeneity within training sets.

2.2.4.1.2 Generative methods based on deep learning

When data are extremely scarce or when highly complex and realistic new data need to be generated, deep learning–based methods demonstrate significant advantages.

Generative adversarial networks (GANs) consist of a Generator (G) and a Discriminator (D), as shown in Fig. 2.7. The generator attempts to generate data from random noise that is sufficiently "convincing" to be mistaken for real data, while the discriminator strives to distinguish between real data and generated data. Through adversarial training, both components coevolve and ultimately enable the generator to produce highly realistic data. Mathematically, this is expressed as:

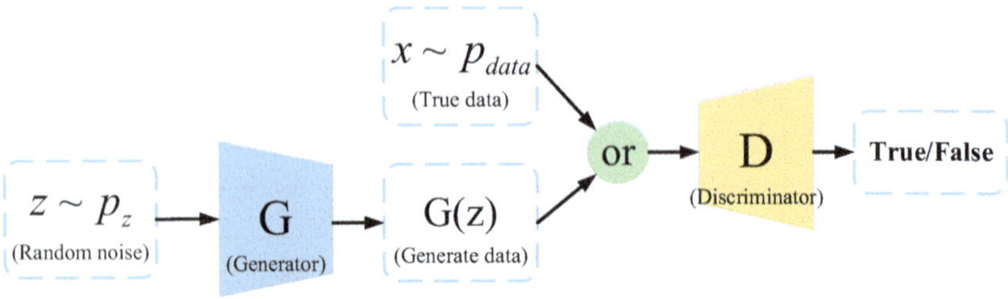

Figure 2.7 Main architecture of GAN.

$$\min_{G}\max_{D} V(D, G) = E_{x\sim P_{\text{data}}(x)} \log D(x) + E_{z\sim P_z(z)}[\log(1 - D(G(z)))] \tag{2.19}$$

In Eq. (2.19), z represents the random noise vector; $\min_{G}\max_{D}$ represents the minimax optimization objective. describes the adversarial game between the generator and discriminator; $V(D,G)$ represents the value function, composed of two terms corresponding to the discriminator's judgments on two types of data; $P_{\text{data}}(x)$ denotes the probability distribution of real data; $P_z(z)$ denotes the prior probability distribution of noise; $G(z)$ denotes the synthetic data generated by the generator; $D(\cdot)$ denotes the probability that the discriminator assigns to a given data sample being real.

Its primary application in battery data lies in generating battery fault data that are difficult to collect under normal operating conditions, thereby addressing the severe imbalance of fault data; it also generates charge/discharge curves under various aging levels for training more accurate SOH estimation models.

2.2.4.1.2.1 Data balancing In lithium-ion battery management systems, data imbalance is a prevalent issue, particularly when tasks involve multiple battery states such as discharging, charging, and fault conditions. Due to the lower frequency of occurrence of certain states compared to others, the dataset becomes biased toward the majority classes during training, resulting in poor learning performance for minority classes. This data imbalance can lead to suboptimal model performance in practical applications. The primary impacts of data imbalance include overfitting and low recall rates. Overfitting refers to the tendency of models to excessively fit the majority class samples under imbalanced conditions, thereby impairing their ability to correctly identify minority classes. This may manifest as high overall accuracy but poor predictive performance for minority class instances. Low recall rates indicate that the model typically fails to detect a significant portion of minority class instances, meaning that even with high overall accuracy, numerous false negatives or false positives may occur when handling rare events. For example, in battery management systems, fault diagnosis constitutes a minority class problem; data imbalance can result in low identification rates for fault states.

Data balancing methods aim to adjust the class distribution within the training set to construct a more balanced dataset, thereby enhancing the model's ability to recognize minority classes.

2.2.4.1.2.2 Oversampling The core idea of oversampling is to increase the number of minority class samples so that it approaches the number of majority class samples, thereby balancing the dataset by either replicating minority class samples or generating new synthetic ones. SMOTE (Synthetic minority oversampling technique) is a commonly used method that generates new minority class samples through interpolation in the feature space. SMOTE effectively increases the number of minority class samples by selecting existing minority samples and generating new ones based on their neighboring samples, as shown in Fig. 2.8. The mathematical formulation is as follows:

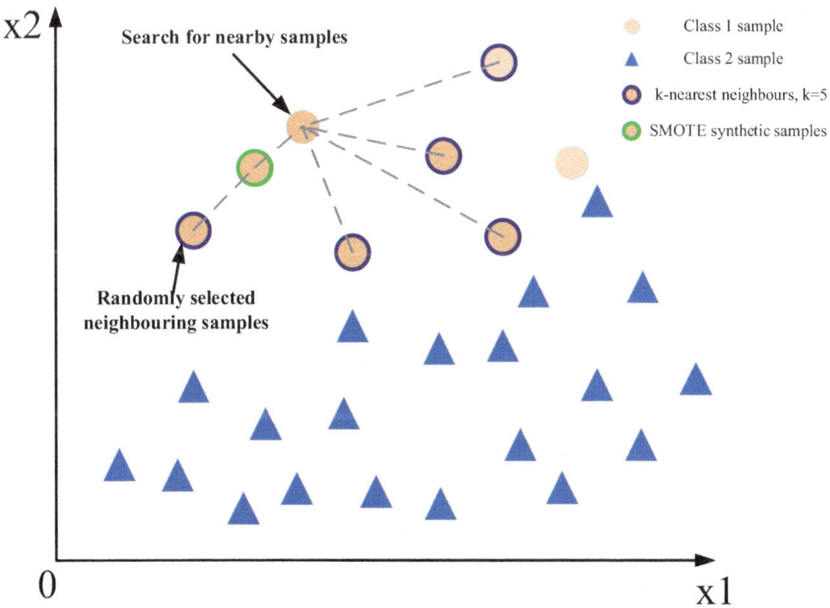

Figure 2.8 Schematic diagram of SMOTE principle.

$$x_{\text{new}} = x_i + \lambda^* (x_j - x_i) \tag{2.20}$$

In Eq. (2.20), x_i are minority class samples, x_j are randomly selected neighbors, λ is random numbers in the range [0,1], x_{new} are synthesized new samples.

2.2.4.1.2.3 Undersampling The core idea of undersampling is to reduce the number of majority class samples so that it approaches the number of minority class samples, thereby balancing the dataset by decreasing the majority class sample size. This approach often results in the loss of some useful information from the majority class; however, it remains a viable option in scenarios with extremely imbalanced data. The key challenge in undersampling lies in preserving data diversity as much as possible while reducing sample size.

2.2.4.1.2.4 Ensemble learning By aggregating multiple models, ensemble learning can reduce errors that may arise from individual models, smooth out anomalous predictions, and thereby yield more comprehensive and reliable conclusions. For SOC estimation, random forest or XGBoost can be employed with historical sequences of voltage, current, and temperature as inputs for regression prediction. Ensemble models effectively mitigate the impact of measurement noise and variations across different operating conditions. For SOH estimation, bagging or boosting methods can be applied to estimate capacity degradation and internal resistance growth based on features extracted from charge/discharge curves (e.g., changes in peak values of incremental capacity curves, constant-current charging time), providing robust estimations. For Remaining useful life (RUL) estimation, gradient boosting (e.g., XGBoost) can leverage early-cycle data to predict the battery's remaining service life. Boosting models are capable of capturing complex nonlinear patterns in the capacity degradation trajectory.

For applications requiring high reliability and high precision, such as lithium-ion Battery management systems (BMS), ensemble learning offers a powerful and practical technical approach. By leveraging the "wisdom of crowds," it significantly enhances the performance of AI models in critical tasks such as state estimation and fault diagnosis, making it an indispensable tool for developing the next generation of intelligent and reliable BMS. In practice, random forests and XGBoost are typically preferred baseline models and efficient solutions.

Data augmentation and balancing are effective approaches to addressing the issue of imbalanced battery data. Through oversampling, undersampling, or generating synthetic data, these techniques can enhance the robustness and accuracy of machine learning models. In battery management systems, such methods contribute to improved fault detection and state estimation performance. Future research directions include developing battery-specific augmentation algorithms and real-time balancing strategies.

2.2.5 Time-series alignment and windowing

In intelligent modeling of battery data, the preprocessing stage is critical, with "time-series alignment and windowing" serving as its core step. This process bridges raw temporal data with subsequent algorithms, aiming to achieve two primary objectives: first, harmonizing the temporal resolution across multisource data; second, generating structured and model-friendly input samples. During battery operation, time-series data such as voltage, current, and temperature are continuously generated. However, in practical data acquisition, differences in device sampling frequencies or dynamic changes in battery operating conditions often result in misaligned timestamps and varying sequence lengths. Direct utilization of such raw data disrupts the intrinsic temporal relationships among features, thereby degrading model performance. Therefore, time alignment and windowing are essential to reconstruct a consistent temporal data structure, providing a reliable and effective foundation for downstream modeling.

2.2.5.1 Time-series alignment: Reestablishing temporal consistency of multisource data

The battery system involves multiple critical parameters, including voltage, current, and temperature. Voltage reflects the internal potential difference within the battery, current indicates the rate of charge flow, and temperature is associated with thermal variations. These parameters must be recorded synchronously over time to accurately analyze the battery's state. In practical data acquisition, different sensors operate at varying sampling frequencies. Moreover, the startup times of individual sensors exhibit random discrepancies, causing temporal misalignment in the collected data. This temporal misalignment adversely affects the analysis of battery characteristics. It becomes difficult to precisely observe how current variations induce temperature responses, thereby impairing our understanding of battery behavior.

Complex scenarios rely on Dynamic time warping (DTW). DTW enables optimal alignment of temporally similar yet length-different sequences by flexibly stretching or compressing the time axis. Event-triggered alignment is also applicable to battery-specific operating conditions; we can designate the "fully charged event" as a critical reference point, setting the moment when battery voltage reaches 4.2 V as time zero, and then resynchronizing subsequent discharge data relative to this zero point. This comparative approach facilitates the straightforward extraction of differences in discharge characteristics following full charge across different cycles, and such data can further support battery cycle life analysis.

2.2.5.2 Time-series windowing: Decomposing long time series into local feature units

The aligned sequential data remains a continuous long sequence, whereas intelligent models such as deep learning typically require "sampled" inputs, similar to how image recognition involves dividing an entire image into multiple small patches (local regions), sequential data must also be segmented into "windowed" sub-sequences, enabling the model to focus on characteristic patterns within localized time intervals.

The essence of windowing lies in the design of window types and parameters:

1) Fixed window: Segment the sequence using a fixed duration or a fixed number of data points, suitable for stages where the battery state is relatively stable.

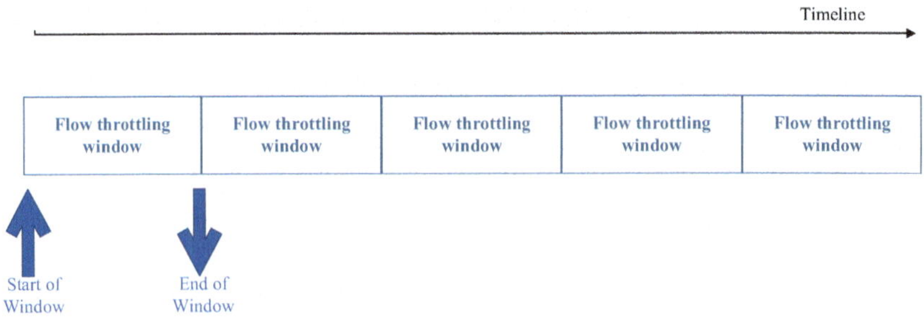

2) Sliding window: The window slides over the long sequence with a fixed step size, and adjacent windows have overlapping regions, which improves data utilization and captures the continuity of temporal patterns.

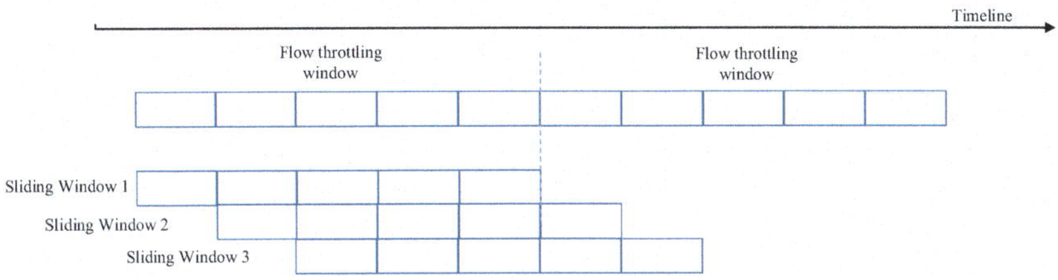

3) Adaptive window: Dynamically adjust the window boundaries based on abrupt events in battery state to ensure each window contains a complete "event-response" cycle. The window length and step size must be jointly determined by balancing the battery's dynamic characteristics and modeling objectives, preserving the full charge/discharge features of each cycle while using overlapping steps to reflect the gradual trend of capacity differences between cycles.

2.2.5.3 *The supporting value of alignment and windowing for intelligent modeling*

After being processed through "time-series alignment + windowing," the battery data exhibit temporal consistency across multiple features and form locally interpretable sample units, providing critical support for subsequent intelligent modeling.

From the perspective of model adaptability:

Convolutional neural networks (CNNs) can extract "local temporal patterns" from windowed sequential data, paralleling the logic of extracting "local textures" in images. Features such as the "distribution of fluctuation frequencies" within a window, "changes in differences between adjacent sampling points," or the "phase difference of synchronous fluctuations" between current and temperature can serve as "temporal texture" features for CNNs, enabling the identification of battery anomalies such as lithium plating or separator damage.

Recurrent neural networks (RNNs) or Long short-term memory networks (LSTMs) can learn the "temporal dependencies among multiple features within a window" based on aligned window sequences, such as how the current's "rise-steady-fall" process drives the voltage's "platform formation-maintenance-collapse," and how temperature evolves from "slow rise" to "rapid increase" during this process. These dependencies are crucial for deciphering the electrochemical–thermal coupling mechanisms in batteries and serve as the fundamental basis for LSTMs to "remember long-term state variations" when predicting battery remaining useful life.

The transformer model can utilize the windowed data to capture long-range temporal dependencies through its self-attention mechanism, thereby quantifying the cumulative effect of "capacity degradation," offering a global perspective for long-term life prediction.

In practical applications, alignment resolves the "temporal synchronization" issue among multisource data, laying the foundation for feature association analysis; windowing achieves the decomposition of "long-term sequences into sample units," providing intelligent models with interpretable and easily learnable input formats. Together, these two mechanisms transform battery time-series data from a "raw and fragmented" state into a "structured and high-value" modeling resource, serving as a fundamental prerequisite for enhancing the accuracy and reliability of battery-oriented intelligent algorithms.

2.3 Fundamentals of deep learning

2.3.1 Artificial neural network architectures

Artificial neural networks (ANN) are inspired by the information transmission mechanisms of biological neural systems. By simulating the interconnection and collaborative operation of numerous "artificial neurons," ANNs enable the learning and representation of complex nonlinear relationships, forming the foundational core of deep learning technologies. In the field of intelligent battery modeling, ANNs can effectively uncover latent correlations among multisource data such as voltage, current, and temperature, providing robust model support for tasks including battery state assessment and lifetime prediction.

2.3.1.1 Artificial neuron: the fundamental computational unit of a network

The artificial neuron is the fundamental functional unit of an NN, whose structure mimics the "reception–integration–transmission" process of biological neurons. A typical artificial neuron comprises the following core components:

Input: This corresponds to the dendrites of biological neurons, receiving signals from other neurons or external data. In battery application scenarios, inputs may include monitored features such as battery voltage, current, temperature, or intermediate features transmitted by other neurons.

Weight: This represents the strength of connections between simulated biological neurons' synapses, used to measure the importance of input signals. In battery health state modeling, the weight assigned to current signals influencing "lithium plating risk" may be higher than that for temperature signals, with the network automatically adjusting these weights through training.

Summation and bias: The weighted sum of all input signals is computed and then a bias term is added, yielding the neuron's "net input." The bias provides additional adjustable capacity, enabling the neuron to fit data more flexibly.

Activation function: This performs a nonlinear transformation on the net input to generate the neuron's output. Common activation functions include sigmoid (which compresses outputs into the range of 0–1, suitable for binary classification tasks) and ReLU (rectified linear unit, a linear rectifier function that alleviates the vanishing gradient problem and accelerates network training). Through activation functions, artificial neurons acquire nonlinear characteristics, enabling the entire network to learn complex nonlinear relationships.

Output: This corresponds to the axon of a biological neuron, transmitting the processed signal to the next layer of neurons or serving as the final output of the network.

2.3.1.2 Hierarchical structure of neural networks

ANNs organize a large number of artificial neurons in the form of "layers," as shown in Fig. 2.9. Typical hierarchical structures include the following.

Input layer: This directly receives raw data, with the number of neurons matching the dimensionality of the data features. For instance, if the input features are battery voltage, current, temperature, and cycle count, the input layer comprises four neurons, each corresponding to the numerical value of one feature.

Hidden layer: Located between the input layer and the output layer, it constitutes the core region of the network for feature extraction and transformation. The hidden layer may consist of multiple layers (when the number of hidden layers is ≥ 2, the network is referred to as a "deep neural network"). Neurons in each layer are connected to the previous layer via weights, progressively transforming input features into more abstract and discriminative representations.

Output layer: This outputs the final result of the network, with the number of neurons determined by the task objective. For a battery

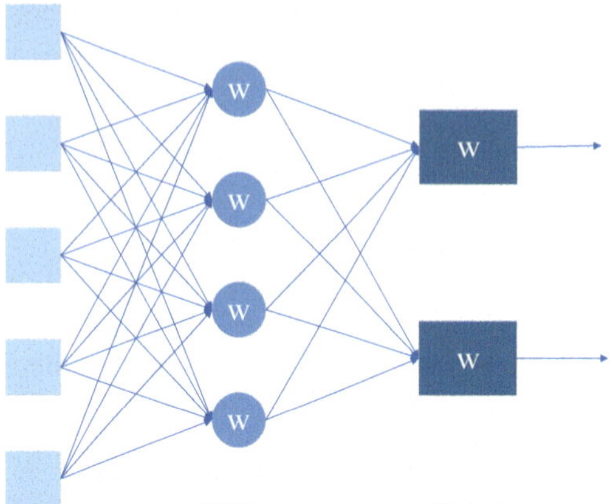

Figure 2.9 Artificial neural networks typical hierarchical structures.

"SOH" regression prediction task, the output layer typically consists of a single neuron, directly outputting the SOH value.

2.3.1.3 Typical neural network architectures

Depending on the connection patterns and functional emphasis, ANNs have given rise to various typical architectures, each suitable for different types of battery data and task requirements:

Feedforward neural network (FNN): Signals propagate unidirectionally from the input layer to the output layer without backward connections, representing the most fundamental NN architecture. Among these, the Multilayer perceptron (MLP) is a representative example, composed of an "input layer–fully connected hidden layer–output layer." Neurons in the hidden layer are fully connected to neurons in both preceding and succeeding layers, making it suitable for processing "tabular" battery data lacking obvious temporal or spatial correlations, particularly for predicting battery cycle life.

RNN: Designed for sequential data, RNN introduces "feedback connections" in the hidden layer, where the output of hidden-layer neurons feeds back to themselves or serves as input for the next time step, endowing the network with the capability to "remember" temporal dependencies. However, traditional RNNs suffer from the "long-term dependency forgetting" problem. Subsequent improvements, such as LSTM and GRU, employ "gating mechanisms" (e.g., forget gate, input gate, output gate) to selectively retain critical information over long sequences, making them more suitable for analyzing long-term sequential data from batteries.

CNN: CNN excels at extracting "spatial local features" from data through the combination of "convolutional layers + pooling layers." In the field of batteries, when analyzing microscopic images of battery electrodes, CNN can identify local features such as "particle morphology" and "crack distribution" by sliding convolution kernels across the image, thereby providing a basis for research on battery aging mechanisms and fault diagnosis.

2.3.1.4 Network connectivity and parameter learning

The core advantage of ANNs lies in their ability to learn data patterns through "weight adjustment." During network training, the loss function measuring the discrepancy between predicted and actual values, such as mean squared error or cross-entropy is employed. The backpropagation algorithm computes gradients in reverse, from the output layer to the input layer, and weights and biases of neurons across layers are updated via gradient descent, progressively bringing the network's predictions closer to the true values.

2.3.2 Convolutional neural networks

CNNs are a type of deep neural network model inspired by the structure and functional mechanisms of the biological visual cortex. Compared to traditional fully connected neural networks, CNNs employ specialized convolutional and pooling operations that enable effective extraction of spatial features from data. This has led to revolutionary advances in computer vision tasks such as image recognition, object detection, and image segmentation, while also demonstrating outstanding performance in other domains including natural language processing and speech recognition.

A typical CNN architecture generally consists of an input layer, convolutional layers, activation function layers, pooling layers, fully connected layers, and an output layer. Each layer performs a specific function, collectively accomplishing tasks such as feature extraction and classification.

(1) Input layer: The input layer is primarily responsible for receiving raw data. In image recognition tasks, the input typically consists of three-dimensional image data with dimensions height H, width W, channel count C. The channel count is 1 for grayscale images and 3 for RGB color images.

(2) Convolutional layer: The convolutional layer is the core component of a CNN, primarily responsible for extracting local features from input data through convolution operations. The specific process of convolution involves sliding the convolution kernel across the input feature map at a specified stride. Within each sliding window, the elements of the convolution kernel are multiplied with the corresponding elements of the input feature map, summed together, and then a bias term is added to produce the output feature map of the convolutional layer. The number of output feature maps from the convolutional layer equals the number of convolution kernels, with each kernel generating one output feature map that represents a distinct extracted feature.

In convolution operations, the concept of padding is also involved. When the convolution kernel slides along the edges of the input feature map, edge pixels participate in fewer convolutions, potentially causing a reduction in the output feature map's dimensions. To maintain the output feature map's size identical to that of the input feature map, or to adjust the output size according to specific requirements, padding can be applied to the edges of the input feature map. Common padding methods include zero padding.

(3) Activation function layer: The activation function layer typically follows the convolutional layer, serving to introduce nonlinearity into the network. Since convolution operations are inherently linear, complex data distributions cannot be effectively modeled using linear operations alone. Activation functions transform linear outputs into nonlinear outputs, enabling CNNs to learn intricate nonlinear relationships within the data. Commonly used activation functions include the sigmoid function, the tanh function, and the ReLU function. Among these, the ReLU function is widely adopted in CNNs due to its computational simplicity and resistance to the vanishing gradient problem; its mathematical expression is provided in Eq. (2.21).

$$f(x) = \max(0, x) \tag{2.21}$$

(4) Pooling layer: The pooling layer is typically positioned after the convolutional layer and the activation function layer. Its primary purpose is to down sample the feature maps extracted by the convolutional layer, thereby reducing the dimensionality of the feature maps and the number of parameters, further decreasing computational complexity. Additionally, it enhances the model's translational invariance, meaning that even when the input data undergoes minor translations, the pooled features remain largely unchanged. Common pooling operations include max pooling, average pooling, and sum pooling. Among these, max pooling is the most widely used; it selects the maximum value within a sliding window as the output after pooling.

(5) Fully connected layer: The fully connected layer is typically located at the end of a CNN network. Its function is to map the high-dimensional features extracted by preceding convolutional and pooling layers into a one-dimensional vector. By establishing connections with neurons in the subsequent layer, it enables comprehensive utilization of features and facilitates classification decisions. In the fully connected layer, each neuron is connected to all neurons in the previous layer, resulting in a relatively large number of parameters.

(6) Output layer: The output layer is configured according to specific task requirements. In classification tasks, the output layer typically employs the softmax activation function to produce a probability distribution over each category; in regression tasks, the output layer generally does not use an activation function and directly outputs the predicted values.

The training process of a CNN is similar to that of traditional neural networks, primarily employing the backpropagation algorithm and gradient descent to minimize the loss function. The specific steps are as follows: first, initialize all network parameters, including the weights and biases of convolutional kernels and fully connected layers; then, feed the training data into the network and compute the predicted output via forward propagation; next, calculate the value of the loss function based on the predicted output and the true labels; subsequently, use the backpropagation algorithm to compute gradients of the loss function with respect to each parameter; finally, update the network parameters using gradient descent to continuously reduce the loss function value. This procedure is repeated until convergence of the loss function or until a predefined number of training iterations is reached.

2.3.3 Recurrent neural networks and LSTM

RNNs are a class of neural network models specifically designed for processing sequential data. Unlike CNNs, which primarily handle spatial data, RNNs employ a recurrent architecture that leverages historical information to process current inputs, thereby enabling modeling of data with temporal or sequential order, such as time series and text sequences. The core characteristic of RNNs lies in their recurrent structure: at each time step during sequence processing, the network combines the current input with the previous hidden state to compute the current hidden state and output. This recurrent mechanism allows RNNs to retain historical information within the sequence and utilize it in making decisions at the current time step.

The recurrent unit in an RNN is not only connected to the input and output, but also features an internal loop. This loop enables information to be passed from one step in the network to the next. As illustrated in Fig. 2.10, a typical RNN architecture consists of an input layer, an output layer, and recurrent neural network units.

The architecture of RNNs can be classified into various types, commonly categorized into four structures based on the lengths of input and output sequences: one-to-one, one-to-many, many-to-one, and many-to-many, as shown in Fig. 2.11.

Although RNNs are theoretically capable of handling sequences of arbitrary length, traditional RNNs face the problems of vanishing or exploding gradients during actual training, making it difficult for them to learn long-term dependencies in long sequences. Vanishing and exploding gradients arise because, during backpropagation, gradients must be propagated through the recurrent structure of hidden layers; when the sequence length is long, these gradients tend to decay or grow exponentially over time steps.

Specifically, when computing the gradient of the loss function with respect to parameters at earlier time steps, the gradients from multiple time steps must be multiplied together. If the absolute value of each time

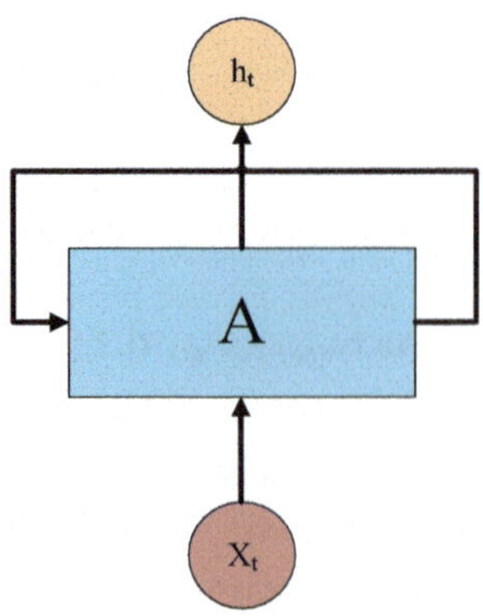

Figure 2.10 Typical RNN architecture.

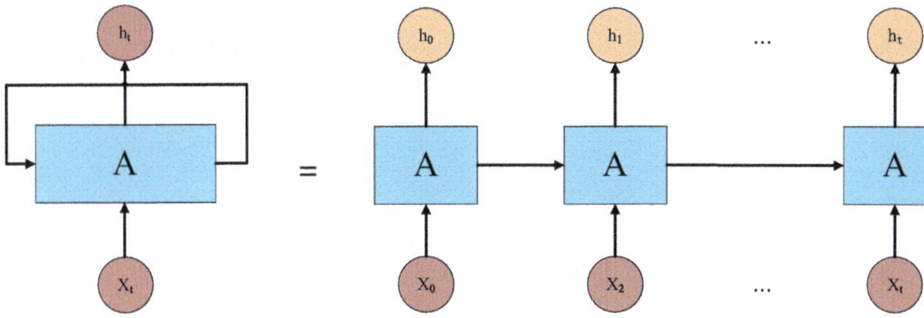

Figure 2.11 Expanded RNN architecture diagram.

step's gradient is less than 1, repeated multiplication across multiple steps will cause the gradient to gradually approach zero, resulting in the vanishing gradient problem. In this case, the network cannot effectively update parameters at earlier time steps, thereby failing to retain information from earlier stages of long sequences. Conversely, if the absolute value of each time step's gradient exceeds 1, repeated multiplication will cause the gradient to grow rapidly, leading to the exploding gradient problem, which results in excessively large parameter updates and instability during training.

To address the problems of gradient vanishing and gradient exploding in traditional RNNs, Hochreiter and Schmidhuber proposed the LSTM in 1997. LSTM effectively captures long-range dependencies in sequential data by introducing three gating units: input gate, forget gate, and output gate, and a cell state that regulates the inflow, outflow, and forgetting of information. The detailed architecture is illustrated in Fig. 2.12.

Forget gate: The function of the forget gate is to determine which information from the previous cell state should be discarded. It employs a sigmoid activation function, taking as input the hidden state from the previous time step, h_{t-1} and the current input x_t, producing an output value between 0 and 1. The computation is defined by Eq. (2.22), where W_f denotes the weight matrix of the forget gate and b_f represents its bias term, while σ stands for the sigmoid activation function.

$$f_t = \sigma\left(W_f\left[h_{t-1}, x_t\right] + b_f\right) \tag{2.22}$$

Input gate: The input gate primarily determines which parts of the input information at the current time step should be stored in the cell state. It consists of two components: first, a sigmoid activation function is applied to identify the portions of information that require updating, yielding i_t; second, a tanh activation function generates the candidate cell state $\widetilde{C}_t$, whose values range between -1 and 1. The computation is defined by the

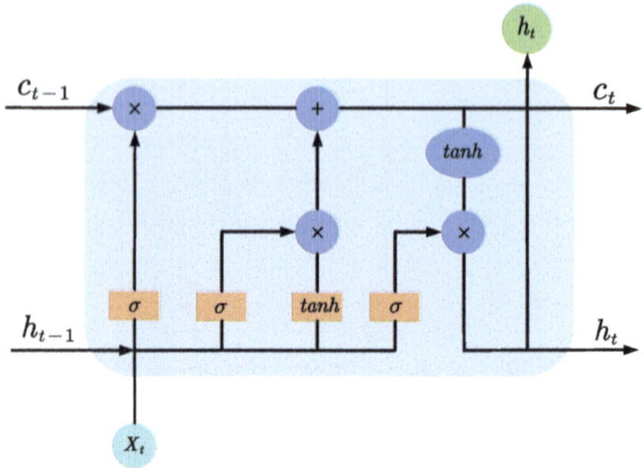

Figure 2.12 LSTM network architecture.

following Eqs. (2.23) and (2.24), where W_u and W_c denote the weight matrices for the input gate and candidate cell state, respectively, and b_u and b_c represent their respective bias terms.

$$i_t = \sigma(W_u \cdot [h_{t-1}, x_t] + b_u) \tag{2.23}$$

$$\widetilde{C}_t = \tanh(W_c \cdot [h_{t-1}, x_t] + b_c) \tag{2.24}$$

Cell state update: Based on the outputs of the forget gate and input gate, the cell state is updated. The previous cell state C_{t-1} is multiplied by the output of the forget gate f_t to discard irrelevant information; subsequently, the output of the input gate i_t is multiplied by the candidate cell state $\widetilde{C}$, yielding information that needs to be updated, which is then added to the cell state, resulting in the current cell state C_t.

$$C_t = f_t{}^* C_{t-1} + i_t{}^* \widetilde{C} \tag{2.25}$$

Output gate: The output gate determines which parts of the current cell state should be output to the hidden state h_t. First, a sigmoid activation function is applied to identify the portions of information to be output, yielding o_t.

$$o_t = \sigma(W_o[h_{t-1}, x_t] + b_o) \tag{2.26}$$

2.4 Other AI techniques

2.4.1 Particle swarm optimization

Particle swarm optimization (PSO) is an optimization algorithm originally proposed by Kennedy and Eberhart in 1995. The fundamental idea is to search for optimal solutions by mimicking the collective behaviors of insects, herds, flocks, and schools of fish. In this algorithm, each individual within the population continuously adjusts its search strategy based on its own experience as well as the experiences of other individuals, aiming to identify superior solutions.

Compared with other optimization algorithms, the PSO algorithm possesses several advantages. First, by continuously learning from the experiences of other members within the group, the PSO algorithm can locate the optimal solution more rapidly during the search process and exhibits strong global search capability. Second, the implementation of the PSO algorithm is relatively simple and does not require any assumptions or derivations regarding the objective function, thus enabling its broad application to various optimization problems. Furthermore, the PSO algorithm can adapt to different search spaces and types of objective functions, demonstrating good scalability and generality. Currently, it has successfully addressed numerous optimization problems in engineering fields such as power systems, signal processing, and battery model parameter identification.

The optimization process of the PSO algorithm can be described as follows. In PSO, a group of particles flies through a D-dimensional search space in pursuit of the optimal solution. The position of each particle can be represented as $X_i = (x_{i1}, x_{i2}, \ldots, x_{iD})$, and its velocity as $V_i = (v_{i1}, v_{i2}, \ldots, v_{iD})$. The fitness value of each particle is used to evaluate whether it is located at an optimal position. To achieve the optimal objective, all particles in the particle swarm optimization algorithm search within the search space based on their current positions and their relative positions with respect to the best-performing particle. Simultaneously, each particle continuously adjusts its position and velocity in the space based on its own best-known position and the globally best-known position among all particles, aiming to move toward a better solution. Then, during each iteration, the best position of each particle is denoted as $P\text{best}_i = [P\text{best}_{i1}, P\text{best}_{i2}, \ldots, P\text{best}_{iD}]$, and the best position among all particles is denoted as $G\text{best} = [G\text{best}_1, G\text{best}_2, \ldots, G\text{best}_D]$. After obtaining these two optimal values, the velocity and position can be updated respectively using Eqs. (2.27) and (2.28):

$$v_{id}(t+1) = v_{id}(t) + c_1 r_1 (P\text{best}_{id}(t) - x_{id}(t)) + c_2 r_2 (G\text{best}_{id}(t) - x_{id}(t)) \tag{2.27}$$

$$x_{id}(t+1) = x_{id}(t) + v_{id}(t+1) \tag{2.28}$$

In Eqs. (2.27) and (2.28), c_1, c_2 are the learning factor, typically set to 2; r_1, r_2 are a random number between 0 and 1; $x_{id}(t)$ and $v_{id}(t)$ represent the best position and velocity, respectively, of the i-th particle in the D-dimensional space at the t-th iteration; $\text{Pbest}_{id}(t)$ and $\text{Gbest}_{id}(t)$ denote the global best position in the D-dimensional space at the t-th iteration. The algorithm flowchart for solving optimization problems using PSO is illustrated in Fig. 2.13.

2.4.2 Introduction to genetic algorithms

Genetic algorithm (GA) is an optimization algorithm based on natural selection and genetic mechanisms, widely applied to solve complex optimization problems. NNs, as powerful machine learning models, perform exceptionally well in handling nonlinear problems. However, the training process of NNs typically relies on optimization methods such as gradient descent, which are prone to getting trapped in local optima. GAs, by simulating

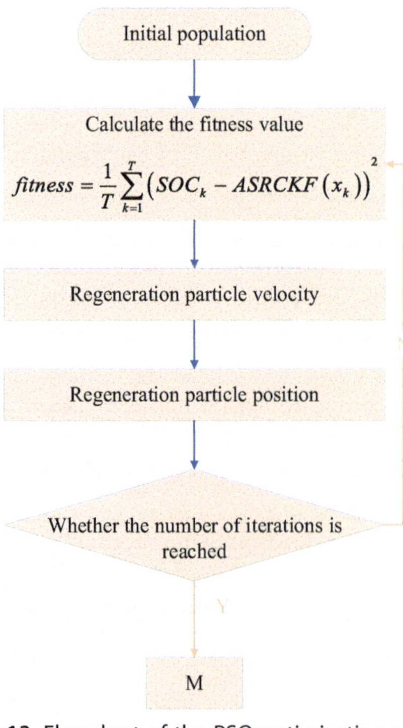

Figure 2.13 Flowchart of the PSO optimization procedure.

biological evolution processes, can search for optimal solutions across the global space; thus, they have been introduced into the optimization of NNs. The GA primarily consists of the following steps:

1) Initialize the population: Randomly generate a set of initial solutions (individuals), with each individual representing a potential solution.
2) Fitness evaluation: Evaluate the fitness of each individual according to the objective function of the problem.
3) Selection: Choose individuals with high fitness to undergo reproduction.
4) Crossover: New individuals are generated through the crossover operation, simulating sexual reproduction in biological organisms.
5) Mutation: Random mutation is applied to individuals to enhance the diversity of the population.
6) Iteration: Repeat the above steps until a termination condition is met.

The primary advantage of GAs lies in their strong robustness and adaptability to genetic operations, population dynamics, and candidate solutions, enabling them to effectively address nonconvex and nonlinear problems. Regarding their limitations, similar to other learning-based techniques such as ANNs, they exhibit high computational costs and possess a black-box structure, rendering the resulting control actions difficult to interpret. Nevertheless, as a powerful optimization tool, GAs demonstrate significant advantages in the structural optimization, weight optimization, and hyperparameter tuning of neural networks. By integrating GAs with neural networks, model performance can be effectively enhanced and complex real-world problems can be addressed. In the future, with advancements in computational power and algorithmic improvements, the application prospects of GAs in NNs are expected to expand even further.

2.5 Trends in AI applications

2.5.1 Multi-state coupling modeling

Currently, artificial intelligence and other algorithms have been widely applied in battery state estimation, including SOC, state of energy (SOE), state of power (SOP), SOH, as well as battery charging/discharging control and balancing control. In the field of lithium-ion battery state-of-charge estimation, the application of artificial intelligence algorithms is particularly comprehensive. Compared with battery modeling and filtering methods, data-driven battery SOC prediction algorithms can accurately estimate the SOC by merely utilizing parameters such as battery voltage, temperature, and current; the parameters of these data-driven models require self-adjustment. Some researchers have achieved accurate SOC estimation by optimizing existing backpropagation neural networks and radial basis function neural networks. In recent years, some scholars have employed novel Deep neural network (DNN) approaches for SOC estimation. By applying driving cycle loads under various environmental temperatures to lithium-ion batteries, DNNs are capable of adapting to diverse temperature conditions.

Since 1986, BP neural networks have been widely applied in numerous fields and have also been employed for predicting the state of charge, state of energy, state of power, and state of health of batteries. In BP neural networks, since the numbers of input and output neurons are predetermined, the number of hidden-layer neurons should be determined according to the following empirical formula, as shown in Eq. (2.29).

$$\begin{cases} m = \sqrt{n+l} + \alpha \\ m = \log_2 n \\ m = \sqrt{nl} \end{cases} \tag{2.29}$$

In Eq. (2.29), m represents the number of hidden layer nodes, n represents the number of input layer nodes, and l represents the number of output layer nodes. The value of m is determined to be 11, selected from the range of 1–10. The BP neural network model for estimating the state of charge, state of energy, state of power, and state of health of lithium-ion batteries is illustrated in Fig. 2.14.

After determining the basic structure of the BP network, the next step is to select the training function for the model. In MATLAB, commonly used functions include traingd (gradient descent), which exhibits slow convergence; traingda (gradient descent with adaptive learning rate), which adjusts the learning rate dynamically; and traingdm (gradient descent with momentum), an adaptive learning algorithm that converges more rapidly.

State of charge, state of energy, state of power, and state of health prediction based on random forest: Random forest (RF) is a decision tree algorithm that employs random sampling; it broadly belongs to the bagging ensemble approach. Its primary characteristic lies in constructing diverse trees through random sampling. For classification problems, the final result is determined by majority voting among multiple trees, selecting the most frequently occurring outcome. For regression problems, the final prediction is obtained by averaging the outputs from each individual tree. The

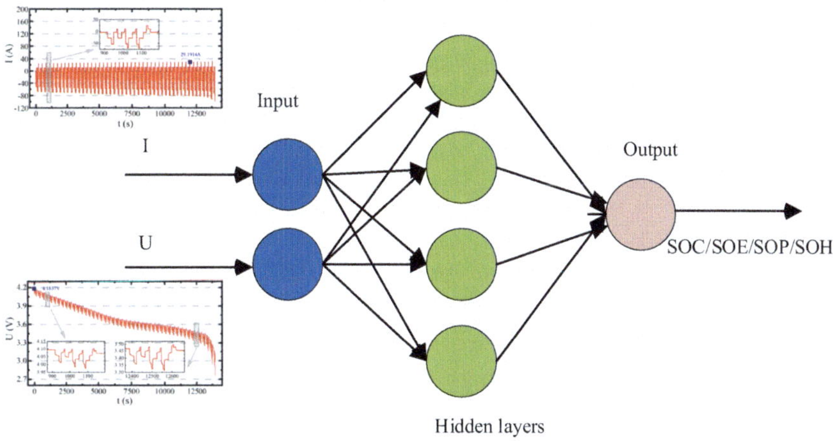

Figure 2.14 Backpropagation neural network model.

prediction of battery state of charge, state of energy, state of power, and state of health falls under regression tasks.

Random forest is well-integrated in Python through the sklearn machine learning library. In this study, the RandomForestRegressor from the sklearn.ensemble module is employed to train the battery dataset. However, since this algorithm has been shown to overfit on datasets with high noise levels, the specific parameter configurations were determined through multiple experimental iterations, thereby ensuring, to a certain extent, the accuracy of the final prediction results.

SOC prediction based on XGBoost: As a type of boosting tree, XGBoost possesses advantages such as parallel computation, high accuracy, and ease of implementation, and has become one of the most widely used decision tree models. In contrast to Gradient boosting decision trees (GBDT), which perform only first-order expansion of the loss function, XGBoost leverages second-order Taylor expansion, thereby achieving a closer approximation to actual loss error. Furthermore, column sampling in the objective function and the introduction of regularization terms in XGBoost help prevent model overfitting. The parameters in XGBoost are primarily categorized into three types: general parameters, booster parameters, and learning objective parameters.

A multistate prediction method for lithium-ion batteries based on a CNN-LSTM model. The CNN and LSTM neural network are employed to learn the nonlinear relationships between multistate parameters and lithium-ion battery discharge data, as well as the long-term dependencies present in sequences of multi-state parameters.

A CNN is a deep neural network featuring a convolutional architecture that can extract higher-level and more abstract data features from raw data, and is widely applied in fields such as image classification and face recognition. Among these, one-dimensional CNNs are primarily used for time-series data. Fig. 2.15 illustrates the architecture of a one-dimensional CNN.

Recurrent neural networks (RNNs) take the output from the previous time step as input for the current time step and maintain a hidden state within the hidden layer that encapsulates historical features of the data, thereby offering significant advantages in time-series forecasting. Fig. 2.16 illustrates the network architecture of the RNN model.

Although RNNs employ a highly deep feedforward network architecture, both theoretical analysis and empirical evidence indicate that their predictive performance deteriorates significantly when dealing with long sequences. As a variant of RNNs, the LSTM neural network addresses the issues of vanishing and exploding gradients commonly encountered during RNN training. Fig. 2.17 illustrates the internal structure of an LSTM unit. By incorporating a gating mechanism, LSTM networks gain the capability to retain information from past states, making them well-suited for handling long-term sequential problems, such as estimating battery state of charge, state of energy, state of power, and state of health.

Beyond their application in state-of-charge, state-of-energy, state-of-power, state-of-health, lifetime prediction, and charge/discharge and balancing control of lithium-ion batteries, these artificial intelligence

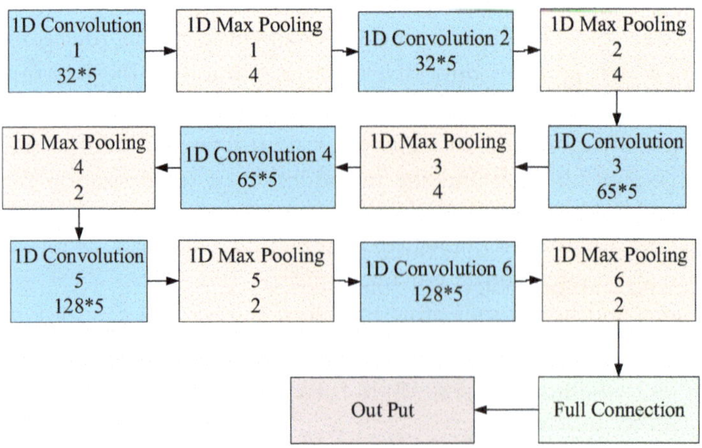

Figure 2.15 Structure of 1D-CNN.

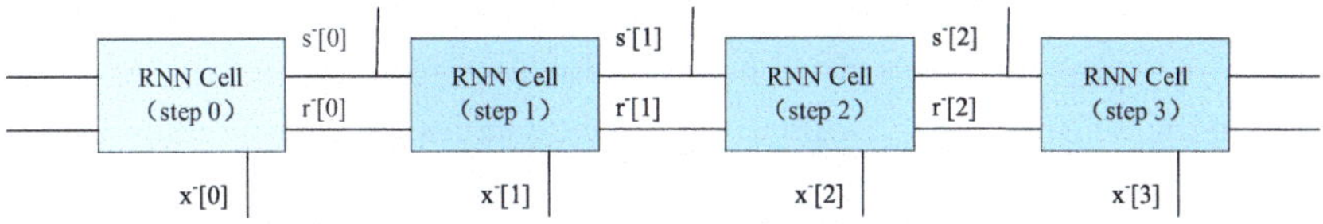

Figure 2.16 Structure of the RNN unit.

algorithms can also be employed in parameter identification for equivalent models of power lithium-ion batteries.

2.5.2 Physics-informed AI

Physics-informed AI, as an emerging modeling paradigm, is fundamentally based on a simple "model-data" fusion approach that integrates data-driven methods with physical models in a serial or parallel manner. These fusion strategies have demonstrated superior performance compared to single-method approaches in specific scenarios, thereby validating the preliminary value of combining mechanistic understanding with data. However, such approaches remain essentially superficial fusions and are subject to fundamental limitations. Their fusion logic often relies on heuristic or empirical design, requiring laborious manual tuning of parameters and exhibiting poor adaptability to varying operating conditions [5]. More critically, these methods fail to achieve deep intrinsic coupling between mechanism and

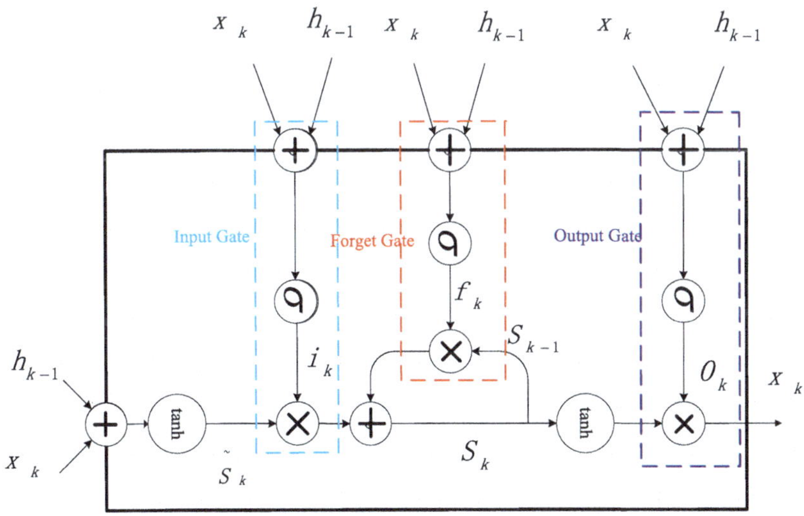

Figure 2.17 Structure of the LSTM unit.

data: physical information is merely treated as input or compensation rather than being rigorously embedded into the learning core in the form of strict mathematical constraints. Consequently, they cannot fundamentally guarantee the physical plausibility and generalization capability of estimation results. Therefore, Physics-informed neural networks (PINNs), as a novel deep fusion methodology integrating mechanism and data, have gradually emerged. The core innovation of PINNs lies in incorporating theoretically verified or experimentally confirmed physical laws into the AI model design process through mathematical formalization: this can be achieved either by regularizing the model's loss function with physical constraints, thereby guiding the model to learn statistical patterns in data while strictly adhering to physical laws; or by reconstructing the model architecture using physical priors to fundamentally constrain the solution space.

2.5.2.1 Basic principles and framework of the PINN method

The core idea of PINN is to embed the governing equations that describe a physical system into the learning process of a neural network, thereby enabling the network to simultaneously satisfy both data fitting and physical law constraints during training. This approach is grounded in the universal approximation theorem of neural networks, and leverages Automatic differentiation (AD) to efficiently compute partial derivatives of network outputs with respect to inputs. These derivatives are then used to construct residual terms of the physical equations, which are incorporated into the loss function [6]. In terms of optimization strategy, PINN adopts a multi-objective loss function formulation, composed of weighted data-fitting and physics-constraint loss components. The total loss function is explicitly expressed as shown in Eq. (2.30).

$$L_{\text{total}} = \lambda_{\text{data}} L_{\text{data}} + \lambda_{\text{physics}} L_{\text{physics}} \tag{2.30}$$

In Eq. (2.30), λ_{data} and λ_{physics} denote the weighted hyperparameters for the data loss function and physical loss function, respectively, with the optimization of these hyperparameters achieving an optimal balance between data-driven learning and physical constraints. This "data-driven plus physical constraint" mechanism enables PINNs to perform reasonable extrapolation in data-sparse regions based on physical laws, while enabling fine-fitting of nonlinear dynamics in data-rich regions. As a result, high-precision modeling and state estimation are achieved while maintaining physical consistency. However, the effectiveness of this approach critically depends on the accuracy and applicability of the embedded physical knowledge.

2.5.2.2 Construction of the loss function

The design of the loss function is a crucial component in realizing the core concept of the PINN framework, integrating data-driven approaches with physical mechanism constraints through multiobjective optimization.

2.5.2.2.1 Construction of the data loss function

The data loss term serves as a bridge connecting model predictions with experimental observations, fundamentally aiming to minimize the discrepancy between network predictions and experimental measurements, thereby providing a reliable fitting baseline for incorporating physical constraints. In SOC estimation, this loss term primarily focuses on fitting the terminal voltage, typically employing mean squared error as the criterion to ensure basic fitting accuracy with respect to observed data. However, in practical applications, the design of the data loss must also account for the influence of measurement noise. Furthermore, addressing the issue of nonuniform data distribution under dynamic operating conditions, an adaptive sampling strategy that emphasizes optimization of critical operational intervals can significantly enhance estimation accuracy.

2.5.2.2.2 Construction of the physical loss function

The physical loss term represents the essence of PINNs in distinguishing them from purely data-driven approaches. By embedding the governing equations of the battery model as soft constraints into the loss function, it guides the network solution to adhere to physical laws. The specific form of the physical loss varies significantly depending on the adopted battery model. For Equivalent circuit models (ECMs), the physical loss primarily arises from residuals of the state equations, which are concise and computationally efficient. In contrast, physical losses based on electrochemical models are more complex, for instance, in the Single particle model (SPM), they involve residuals of solid-phase diffusion partial differential equations along with corresponding boundary condition residuals. While these formulations offer more accurate descriptions of internal mechanisms, they substantially increase training complexity. This difference reflects an inherent trade-off between accuracy and efficiency, necessitating careful model selection according to specific application requirements. Despite differences in model complexity, both approaches require a multiobjective fusion strategy to achieve coordinated optimization with data-driven losses.

2.5.2.3 Challenges and prospects of the PINN method

In PINNs for battery state estimation, balancing the weights of individual components in the multiobjective loss function poses a primary challenge. Fixed-weight strategies struggle to adapt to the dynamic operating conditions of batteries. Moreover, the training process often suffers from gradient instability, particularly when solving stiff differential equations in electrochemical models, leading to difficulties in convergence or excessively long training times. Second, the performance of PINNs critically depends on the accuracy of the embedded physical models; errors inherent in simplified electrochemical models are introduced as systematic biases into the learning process. Battery systems involve multiple dynamical processes spanning time scales from seconds to hours, presenting significant challenges for standard PINN architectures in capturing such complex multitime-scale dynamics. Battery management systems typically require millisecond-level response times, yet the relatively complex network architecture and high computational cost of forward propagation in PINNs hinder their real-time deployment on resource-constrained embedded hardware. Furthermore, high-quality battery operational data are often scarce and unevenly distributed, especially under extreme operating conditions where data acquisition is particularly challenging.

To address the aforementioned challenges, future research can advance collaboratively across three dimensions: algorithmic innovation, application expansion, and engineering deployment.

1) At the algorithmic level, innovations include the development of adaptive loss weighting adjustment strategies and a multitime-scale PINN architecture to address dynamic operating conditions and internal coupling processes. Compact network designs are proposed, such as leveraging attention mechanisms to focus on critical state variables and exploring transfer learning frameworks, thereby enhancing model efficiency and generalization capability.
2) At the application level, explore embedding key physical parameters as trainable variables within the PINN framework, developing multiphysics coupled models to enhance robustness across all climatic conditions. Integrate uncertainty quantification methods into the PINN framework to provide confidence intervals for BMS decision-making,

and investigate the potential of zero-shot learning for inference under novel operating conditions.

3) At the deployment level, to address the three major bottlenecks faced by battery management systems: computational demand, real-time requirements, and embedded system compatibility, model compression and efficiency enhancement can be achieved through techniques such as network pruning, parameter quantization, and knowledge distillation. Reliability validation should be conducted using hardware-in-the-loop simulation platforms. In the long term, exploration of an edge-cloud collaborative computing framework is recommended, where complex training is performed in the cloud and lightweight models are deployed on-board vehicles, thereby enabling intelligent battery state management with high reliability and adaptability.

In summary, PINNs, as an emerging paradigm of "deep integration of mechanism and data," offer a promising solution to overcome the bottlenecks of traditional AI methods in battery state estimation. By embedding physical laws into model learning in the form of mathematical constraints, PINNs delicately balance accuracy in data fitting with physical consistency. Although challenges remain in loss function balancing, training efficiency, and engineering deployment, continuous innovation at the algorithmic, application, and deployment levels positions PINNs to lead the advancement of battery management technologies toward greater precision, reliability, and intelligence. Ultimately, this will lay a solid theoretical foundation for achieving intelligent, highly reliable, and adaptive battery state control.

References

[1] R. Yi, G. Dimitrios, L. Xing, et al., Mechanics-informed, model-free symbolic regression framework for solving fracture problems, J. Mech. Phys. Solid. 194 (2025) 105916.

[2] H. Zongchao, L. Kewen, X. Zhifeng, et al., STP-Model: a semi-supervised framework with self-supervised learning capabilities for downhole fault diagnosis in sucker rod pumping systems, Eng. Appl. Artif. Intell. 135 (2024) 108802.

[3] G. Leonie, C. Jonathan, C. Raffaella, A new ordinal mixed-data sampling model with an application to corporate credit rating levels, Eur. J. Oper. Res. 314 (3) (2024) 1111–1126.

[4] P. Nicolás García, M.J.M. Cuevas, G. Aida de Haro, Evolutionary simultaneous under and oversampling of instances for dealing with class-imbalance datasets in multilabel problems, Appl. Soft Comput. 159 (2024) 111618.

[5] K. Lawal Zaharaddeen, H. Yassin, C.L.D. Teck, et al., Physics-informed neural network (PINN) evolution and beyond: a systematic literature review and bibliometric analysis, Big Data Cogn. Comput. 6 (4) (2022) 140.

[6] H.A.A. Ali, L.H.J. Raijmakers, H. Tempel, et al., A hybrid electrochemical multi-particle model for Li-ion batteries, J. Electrochem. Soc. 171 (11) (2024) 110523.

CHAPTER THREE

Modeling methods for lithium-ion batteries

3.1 Equivalent circuit models

3.1.1 Model structures

The structural design of equivalent circuit models (ECMs) for lithium ion batteries (LIBs) is grounded in the battery's internal core electrochemical processes. By topologically combining circuit elements, these models abstractly simulate complex physicochemical phenomena. During charging, lithium ions in the positive electrode are driven by an electric field to detach from the lattice, enter the electrolyte, and migrate toward the negative electrode, while electrons flow through the external circuit to complete a loop. The three core phenomena governing the battery's external electrical characteristics: ohmic losses, polarization effects, and energy storage, are thus central targets in model development. Ohmic losses arise from instantaneous energy dissipation due to current flow through internal conductive media, and their "instantaneous response" behavior forms the basis for incorporating series resistance in the model. Polarization effects result in voltage lagging behind equilibrium potential and include charge transfer polarization at the electrode electrolyte interface and concentration polarization caused by lithium ion concentration gradients within the electrolyte. The "dynamic lag" characteristics of these two phenomena necessitate the inclusion of RC parallel networks in the model to simulate their decay behavior. Energy storage corresponds to lithium ions intercalating into the negative electrode lattice, converting electrical energy into chemical energy, a fundamental function of batteries represented in ECMs by an ideal open circuit voltage (OCV) source strongly dependent on state of charge (SOC), serving as a critical reference for SOC estimation. During discharge, these processes reverse, with only changes in energy and electron flow direction. Depending on the degree of simplification applied to

Modelling, Estimation and AI Applications for Lithium-Ion Battery Management Systems
ISBN: 978-0-443-45347-2
https://doi.org/10.1016/B978-0-443-45347-2.00003-0

electrochemical processes, ECMs are classified into three types: basic models focusing on core loss and storage, improved models incorporating polarization effects, and high precision models accounting for parameter nonlinearity.

3.1.1.1 Internal resistance model

The Rint model was proposed by the Iowa Laboratory in the United States. It is the simplest and most fundamental type among ECMs for LIBs as shown in Fig. 3.1. Its core consists of an ideal OCV source connected in series with an equivalent series internal resistance, resulting in a straightforward and intuitive topological structure that enables rapid preliminary simulation of the battery's macroscopic electrical characteristics. The configuration of the Rint model is illustrated in the figure, where Uoc denotes the ideal voltage source corresponding to the OCV, u_t represents the terminal voltage, and R_0 stands for the battery's ohmic internal resistance. Due to its structural simplicity, the Rint model serves as the foundation for another commonly referred to as the "ideal model," known as the equivalent electrical circuit model (EECM). However, it should be noted that the Rint model is suitable only for ideal batteries with minimal dynamic parameters; it fails to account for phenomena such as load transfer polarization and diffusion polarization. Under high rate charge and discharge conditions, significant discrepancies may arise between simulation results obtained using the Rint model and actual measurements. Furthermore, this model cannot accurately capture the dynamic characteristics of batteries under varying operating conditions. Given the complex electrochemical reactions occurring within LIBs, both ohmic resistance and OCV are subject to continuous variation. In contrast, the Rint model considers only static

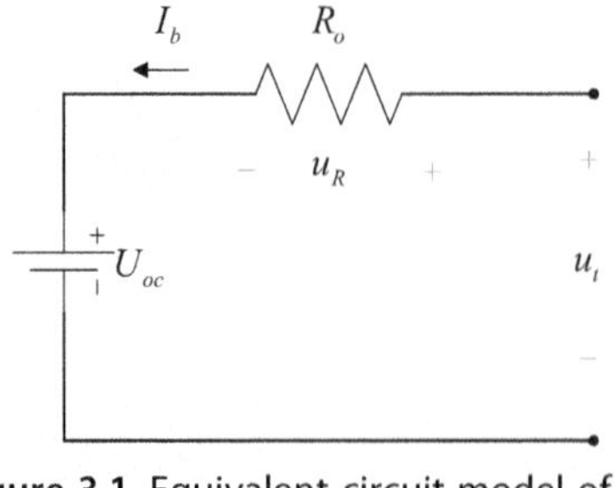

Figure 3.1 Equivalent circuit model of Rint.

battery characteristics and neglects polarization effects induced by environmental factors: particularly pronounced during battery aging or significant temperature variations, thereby limiting its accuracy and rendering it unsuitable for practical applications.

3.1.1.2 Thevenin model

The Thevenin model is derived from the Rint model, and during this derivation, its applicable range has been extended not only to encompass internal heat dissipation within the battery but also to address the issue of polarization effects inside the battery. To simulate the inherent nonlinear behavior during charging and discharging processes, an RC network is introduced. In this RC network, R_1 represents the total polarization resistance, consisting of two components: ohmic polarization resistance and concentration polarization resistance, as shown in Fig. 3.2. The ohmic polarization arises from the combined contributions of electrode material resistance, electrolyte resistance, separator resistance, and contact resistances among various components. Concentration polarization results from mass transport limitations in both the electrolyte and solid phase electrode materials, as well as slow electrochemical reaction kinetics. C_1 denotes the equivalent capacitance of the electrode, commonly referred to as the polarization capacitance. The corresponding equivalent circuit diagram of the Thevenin model is illustrated in the figure. The Thevenin model is often referred to as a first order RC model, capable of characterizing both ohmic and concentration polarization behaviors of LIBs, effectively capturing their dynamic characteristics during charge and discharge cycles. Although this model features a relatively simple structure and high computational efficiency, it should be noted that its accuracy in estimating SOC tends to degrade with increasing battery aging.

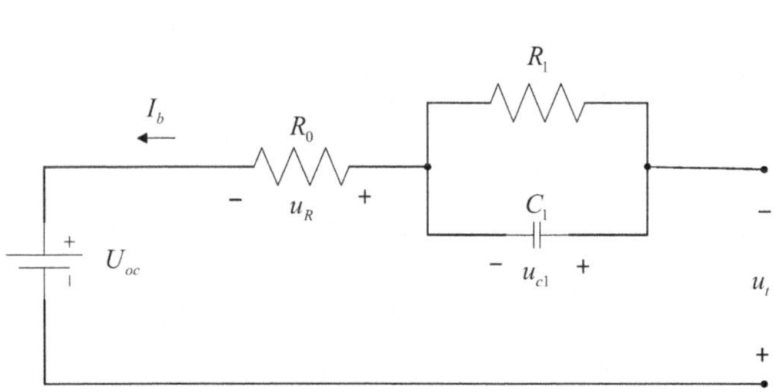

Figure 3.2 Thevenin model.

3.1.1.3 RC model

The RC model is the most widely used foundational architecture among ECMs for LIBs. It effectively simulates the battery's polarization effects and dynamic responses through a combination of resistors and capacitors, achieving a favorable balance between accuracy and computational efficiency, thus becoming the mainstream approach for battery state estimation and performance simulation. From a structural perspective, the RC model is based on the core framework of "ideal voltage source + ohmic internal resistance + RC network," as shown in Fig. 3.3: the ideal voltage source represents the OCV, which is strongly correlated with the SOC; the ohmic resistance (R_0) accounts for fundamental resistive losses in electrodes and electrolyte, while the RC network constitutes the model's core, where resistance represents polarization resistance and capacitance mimics the lag characteristics of polarization processes. The series RC pairs capture dynamic responses at specific time scales.

3.1.1.4 nRC model

The nRC ECM represents a significant advancement in modeling the dynamic characteristics of LIBs. Its core lies in introducing "n series RC networks" within a fundamental model framework, aiming to more accurately capture the multitime scale dynamic responses during battery charging and discharging processes. This approach serves as a critical technical solution for transitioning from simplified models to high-precision models. The specific structure is illustrated in the figure. Here, R1\C17\R2\C2 ... \Rn\Cn denote multiple RC networks, each reflecting voltage responses at different time scales, as shown in Fig. 3.4. Additional RC networks are incorporated to describe charge transfer processes, double layer capacitance, and diffusion

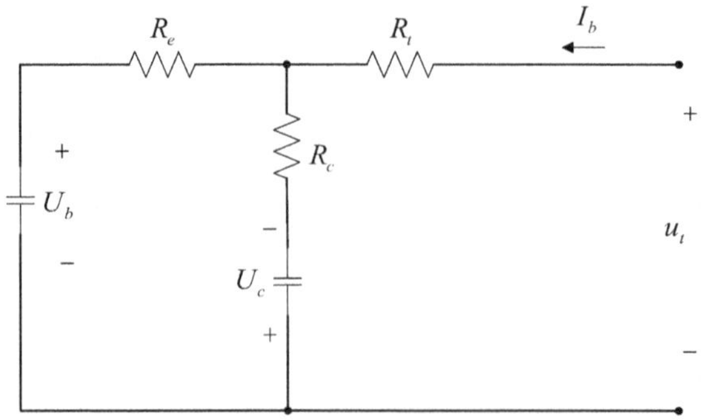

Figure 3.3 RC equivalent circuit model.

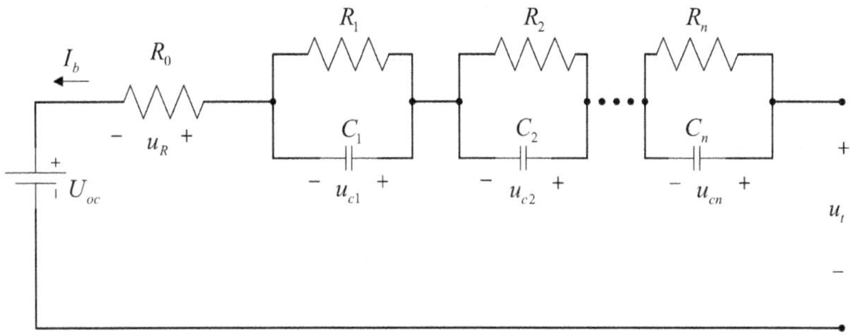

Figure 3.4 nRC equivalent circuit model.

phenomena within the electrolyte and electrode materials. Although theoretically increasing the number of RC pairs (i.e., adding more RC networks) can enhance the accuracy of the EECM, each additional RC network leads to a linear increase in both parameter identification dimensionality and computational complexity. This not only intensifies hardware computational demands but may also result in parameter redundancy and "overfitting", where the model becomes excessively tailored to specific test data, thereby reducing accuracy under real world operational variations. Consequently, increasing the number of RC networks does not always yield continuous improvements in EECM accuracy.

3.1.1.5 PNGV model

The partnership for a new generation of vehicles (PNGV) model, based on the Thevenin model, incorporates an additional capacitor in series with the original Thevenin structure, as shown in Fig. 3.5. This series capacitor is introduced to characterize the OCV variation induced by load current,

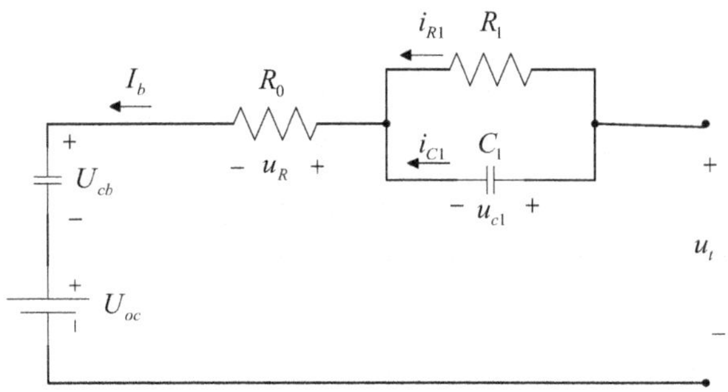

Figure 3.5 PNGV equivalent circuit model.

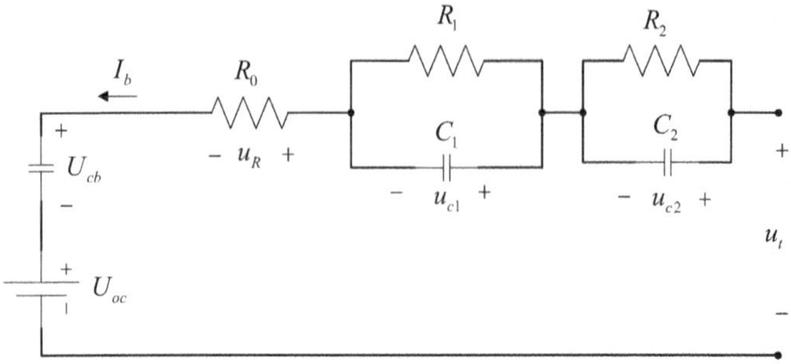

Figure 3.6 GNL equivalent circuit model.

thereby simulating the OCV changes in LIBs resulting from cumulative current effects. However, due to its limitation in modeling only the OCV variation caused by cumulative current effects, the PNGV model is applicable solely to single HPPC (hybrid pulse power characterization) tests or constant amplitude positive and negative pulse conditions, rendering it unsuitable for other pulse current scenarios, representing a significant drawback of this model. In the PNGV model, the capacitance value of C_b reflects both the battery's capacity characteristics and its DC response behavior, thus overcoming a key limitation of the conventional Thevenin model. The structural configuration of the PNGV model is illustrated in the figure. This model effectively captures the correlation between battery capacity and OCV, enabling accurate representation of DC response characteristics. To enhance curve fitting accuracy, an additional set of RC networks is introduced to account for polarization effects within the battery, thereby improving fidelity to the battery's internal structure and dynamic behavior. Generally, compared with the Rint and standard Thevenin models, the PNGV model demonstrates superior accuracy.

3.1.1.6 General nonlinear model

The GNL model, an important ECM, is an enhanced architecture developed from the PNGV model, as shown in Fig. 3.6. Its core innovation lies in increasing the number of RC networks to better capture battery polarization effects while preserving accurate characterization of the nonlinearity in OCV. This makes the GNL model particularly valuable in simulating dynamic behaviors and estimating states of LIBs. The GNL model inherits the PNGV model's advantage of using a large capacitance (C_b) to represent the cumulative voltage variation due to current, while introducing additional RC networks to refine the description of polarization processes. This design enables the model to simultaneously capture static capacity characteristics and accurately simulate multitime scale dynamic responses such as charge transfer and ion diffusion. The GNL model demonstrates significantly higher fitting accuracy for voltage curves compared to traditional models such as Rint and Thevenin, especially under high rate pulse charge/discharge conditions, where its capability for dynamic tracking of polarization voltage is particularly prominent. Moreover, the flexibility of the GNL model allows its application across various battery types and systems. However, its use is constrained by computational complexity: each additional RC network increases both parameter identification dimensionality and real-time computational load. Therefore, in resource constrained environments, a trade off between accuracy and efficiency must be carefully managed. For batteries exhibiting high hysteresis in terminal voltage, a single RCEECM combined with hysteresis modeling is typically adopted, whereas energy storage systems generally prefer two RCEECM configurations. The structural layout of the GNL model is illustrated in the figure.

3.1.2 Parameter identification

The accuracy of an LIB model in reflecting the battery's operational characteristics largely depends on the model parameters; therefore, model parameter identification is of great importance. Battery model parameter identification methods are primarily categorized into offline and online identification techniques. Offline parameter identification methods determine model parameters based on the battery's operational characteristics, whereas online parameter identification methods estimate parameters in real time using measured voltage and current data. The most widely used online parameter identification method is the recursive least squares (RLS) algorithm.

3.1.2.1 Offline parameter identification

After establishing the LIBs model, it is necessary to identify the corresponding parameters. Battery parameters are identified using an offline method. Before parameter identification, a pulse discharge experiment must be conducted on the battery. Fig. 3.7 shows a locally magnified curve of pulse discharge at a battery SOC of 0.8. As can be seen from the figure, the curve from pulse discharge to battery rest and until voltage reaches equilibrium can be divided into four stages.

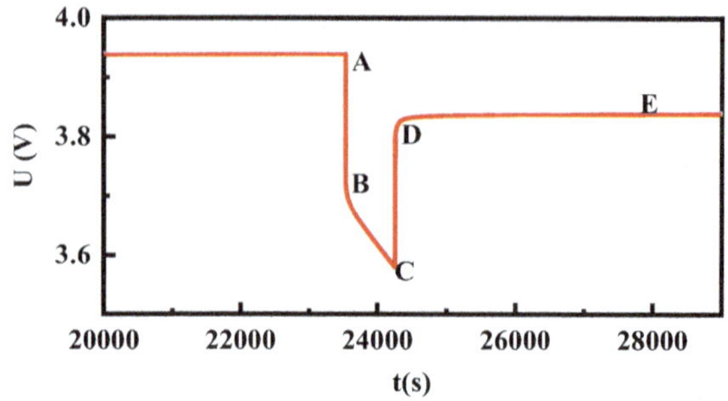

Figure 3.7 Enlarged curve of pulse discharge.

(1) Stage A–B: Since a load is applied from a quiescent state, the polarization capacitor voltage cannot change abruptly. The sudden voltage change observed at this stage is due to the ohmic internal resistance. Therefore, the ohmic internal resistance at this SOC can also be calculated based on the voltage difference during this phase. The calculation equation is as shown in (3.1):

$$R_0 = \frac{U_{AB}}{I_L} \tag{3.1}$$

(2) B–C phase: As the LIBs continue to discharge, the internal polarization effect gradually increases and reaches its peak at point C. During the discharge from B to C, the RC circuit can be regarded as a zero state response. Taking point B as t = 0 and point C as the end of charging, the voltage output equation across the LIBs can be expressed as shown in Eq. (3.2):

$$\begin{aligned} U(t) = U_{\mathrm{OCV}}(t) - I_L(t)R_0 - I_L(t)R_{p1}\left(1 - e^{-T/\tau_1}\right) \\ - I_L(t)R_{p2}\left(1 - e^{-T/\tau_2}\right) \end{aligned} \tag{3.2}$$

(3) C–D stage: Similar to the AB stage, this phase is generally considered to be caused by ohmic resistance.

(4) D–E stage: The internal electrochemical reactions of the LIBs cease, polarization gradually diminishes, and the voltage across the LIBs slowly recovers to UOCV. During this phase, there is no current input, which can be regarded as a zero input response. The output equation for the voltage at the terminals of the LIBs can be expressed as shown in Eq. (3.3):

$$U(t) = U_{OCV}(t) - U_{p1}(0)e^{-T/\tau_1} - U_{p2}(0)e^{-T/\tau_2} \tag{3.3}$$

where τ is the time constant, which can be obtained by fitting the corresponding curve using MATLAB.

3.1.2.2 Online parameter identification

Parameter identification of LIBs models typically relies on the intrinsic characteristics between voltage and current. In practical applications, offline parameter identification is usually achieved through extensive experimental analysis. However, parameters of the ECM vary significantly under different conditions, resulting in poor universality. Moreover, this method suffers from a longtime cycle and weak dynamic response, making it unsuitable for real world applications of power LIBs.

The accurate acquisition of LIBs model parameters depends on the selection of the model and the effectiveness of the tracking algorithm. The least squares (LS) method is widely applied in parameter identification for LIBs. As the most commonly used method for LIBs parameter identification, it primarily employs the principle of minimizing the sum of squared errors to achieve optimal function matching, thereby solving for the parameters.

Based on the voltage and current relationship of the 3ECM, the ECM can be expressed as a discrete-time system.

$$A(z^{-1})y(k) = B(z^{-1})u(k) + e(k) \tag{3.4}$$

In this Eq. (3.4), $A(z^{-1}) = 1 + a_1 z^{-1} + a_2 z^{-2} + \ldots\ldots + a_n z^{-n}$, $B(z^{-1}) = 1 + b_1 z^{-1} + b_2 z^{-2} + \ldots\ldots + b_n z^{-n}$, $u(k)$ represents the system input, $y(k)$ represents the system output, and $e(k)$ represents the system noise.

Subsequently, rewriting the aforementioned equation in difference equation form yields:

$$y(k) = -\sum_{i=1}^{n} a_i y(k-i) + \sum_{i=1}^{b} b_i u(k-i) + e(k) \tag{3.5}$$

Expressing Eq. (3.5) in the general form of LS: $y(k) = \Phi(k)\theta(k) + e(k)$, where $\Phi(k) = [-y(k-1), \ldots, -y(k-n), u(k-1), \ldots, u(k-n)], \theta(k) =$

$[a_1, ..., a_n, b_1, ..., b_n]$, $\theta(k)$ is the vector to be estimated, $a_1, ..., a_n, b_1, ..., b_n$ represents the parameters within the vector to be estimated, and $[\cdot]$ denotes a matrix.

According to the LS principle, the functional can be expressed as shown in Eq. (3.6).

$$J = \sum_{k=1}^{N} \left| e(k)^2 \right| \tag{3.6}$$

Subsequently, $\widehat{\theta} = \left(\Phi^T \Phi\right)^{-1} \Phi^T y$ the parameters to be estimated can be calculated.

The parameter estimates in the LS algorithm are obtained by continuously correcting the previous estimates, requiring minimal memory and computational resources. However, during the later stages of algorithm operation, the accumulation of prior computational data may lead to a "data saturation" issue, causing the newly obtained results to fail in accurately and effectively reflecting the battery's current operational state.

3.1.3 Applications and improvements

The ECM has become the most widely adopted mathematical framework in battery management systems (BMS) within the current industrial sector, owing to its excellent balance between computational efficiency and modeling accuracy. In practical engineering applications, circuit networks such as the second order Thevenin model and the PNGV model are extensively integrated into embedded microprocessors to perform real-time iterative estimation of SOC and SOP based on terminal voltage constraints. However, as electric vehicles demand increasingly precise range estimation and battery aging characteristics become progressively more complex throughout their full life cycle, traditional fixed parameter ECMs have gradually revealed their limitations. To overcome these shortcomings, both academic research and industrial development have pursued in depth exploration from two primary perspectives: adaptive handling of time varying parameters and physical enhancement of model structure.

First, due to the highly nonlinear and time varying characteristics of battery parameters, parameter dynamization and online identification techniques have become the core approach for improving the accuracy of ECMs. Traditional ECMs typically assume that ohmic resistance and polarization capacitance are constant, which clearly overlooks the physical reality that electrochemical reaction rates within batteries vary with external conditions. To more accurately describe battery behavior, modern modeling strategies generally establish multidimensional parameter mapping surfaces that incorporate SOC, temperature, and state of health (SOH). In practical operation, baseline parameters are obtained via look up tables, while realtime voltage and current response data are used in conjunction with recursive least squares (RLS) or dual extended Kalman filter (DEKF) algorithms to perform microsecond level online parameter correction. This hybrid strategy of "offline calibration + online correction" effectively addresses impedance drift caused by battery aging and significantly enhances model robustness throughout the battery's entire lifecycle.

3.2 Electrochemical models

3.2.1 Theoretical foundations

The construction of electrochemical models is based on an accurate description of the multiphysics coupling processes within batteries, primarily encompassing four key physical phenomena: reaction kinetics at the electrode surfaces, lithium ion diffusion within solid phase particles, ion transport in the liquid electrolyte, and charge conservation.

3.2.1.1 Electrode kinetics

At the solid liquid interface where active particles of the positive and negative electrodes contact the electrolyte, the insertion and extraction of lithium ions constitute the fundamental source of energy conversion in batteries. The rate of this process is not constant but is driven by "overpotential" at the interface. Overpotential can be understood as the difference between the actual potential and the equilibrium potential, which enables overcoming the activation energy barrier required for the reaction. Theoretically, there exists a highly nonlinear relationship between reaction rate and overpotential: as current increases, greater overpotential must be consumed to sustain reaction rates, leading to a deviation of voltage from its equilibrium value. Furthermore, reaction rates are directly constrained by the concentration of reactants, lithium ion concentration at the interface making this concentration dependence a key factor limiting battery performance under low SOC or high rate conditions.

3.2.1.2 Solid-phase diffusion

The transport of lithium ions within active material particles is generally regarded as the primary "rate limited step" under high rate charge and discharge conditions. In physical modeling, active materials are typically simplified into numerous tiny spherical particles. During battery charging and discharging, lithium ions must diffuse from the particle surface toward the core. This process is driven by concentration gradients and follows Fick's law of diffusion. Under fast charging conditions, a substantial concentration gradient develops between the particle surface and its center: the surface may already be saturated, while the interior remains largely depleted. This microscale phenomenon of "diffusion lag" explains why high current charging significantly reduces the battery's available capacity.

3.2.1.3 Liquid-phase transport

Within the pores of porous electrodes and the separator, lithium ion transport in the electrolyte becomes significantly more complex. It is governed simultaneously by two forces: "diffusion motion" driven by concentration gradients, which seeks to eliminate concentration differences, and "migration motion" driven by electric fields, which induces directional movement of charged ions. Due to the porous nature of both electrodes and separator, ions cannot move in straight lines but must traverse tortuous pore pathways, a geometric effect that substantially reduces the effective ion transport rate. Under prolonged high rate discharge conditions, the electrolyte concentration near the positive electrode may drop drastically, leading to a sharp increase in internal resistance, this critical characteristic can be captured by electrochemical models but remains beyond the descriptive capability of simple circuit models.

3.2.1.4 Charge conservation and electric potential distribution

As a closed system, a battery must strictly satisfy the principle of charge conservation internally at all times. Current within the battery is transported through two parallel pathways: one consists of electron flow, which travels via solid particle networks and current collectors, constrained by the material's electronic conductivity; the other comprises ion flow, which moves through electrolyte pores, limited by ionic conductivity. At any cross section inside the battery, the sum of these two current components must always equal the load current in the external circuit. By calculating the distribution ratio of these two current components at different positions, the physical model can precisely reconstruct the potential distribution along the electrode thickness direction, thereby accurately predicting the battery's terminal voltage output.

3.2.2 Porous electrode theory

The electrode of a typical LIB is a porous composite material composed of active substances, conductive additives, and binders, with the internal pores filled by a liquid electrolyte. This porous structure forms an interconnected network of solid phase frameworks and liquid phase channels: the solid phase facilitates electron transport, while the liquid phase enables ion migration, with electrochemical reactions occurring at the extensive solid/liquid interface. The theory of porous electrodes constitutes a set of governing equations describing the coupled interactions between electrochemical reactions and mass transport processes within this complex architecture, serving as the foundation for constructing high fidelity electrochemical models of LIBs.

The classical Newman model is a pseudo two-dimensional model (P2D), as illustrated in Fig. 3.8. The electrolyte solution consists of lithium ions (Li^{+}), anions (X^{-}), and solvent, while the electrodes are composed of porous intercalation materials, binders, and conductive additives. The positive and negative electrodes are separated by a porous separator. One

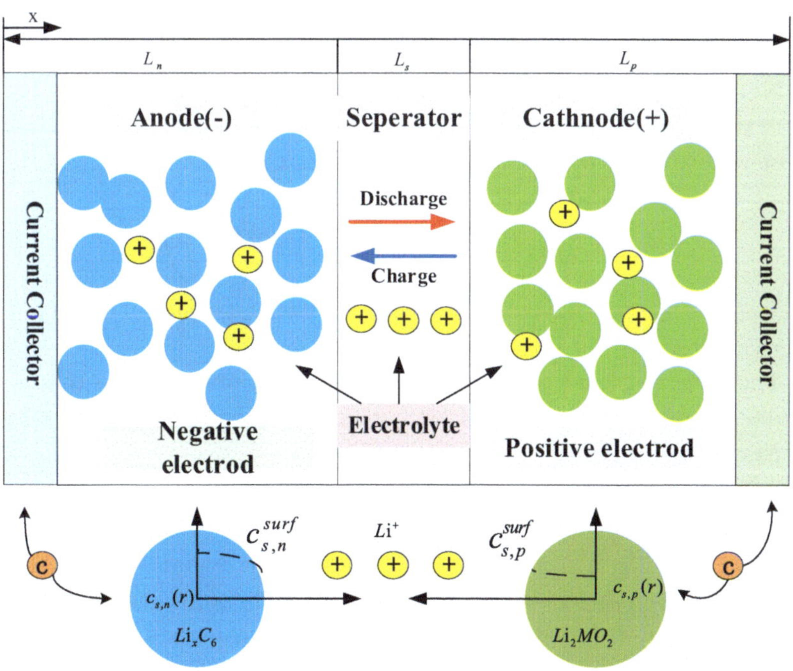

Figure 3.8 Schematic diagram of the P2D model structure.

dimension corresponds to the radial direction of the active particles, r, primarily involving physical processes such as lithium insertion/extraction reactions at the particle surface and lithium diffusion within the particle bulk. The other dimension corresponds to the electrode thickness direction, x, mainly encompassing physical processes including mass transport in the liquid-phase electrolyte and electronic conduction in the solid phase [1].

Newman's theory simplifies active particles into spherical shapes, considering only the diffusion of lithium atoms within the sphere and employing Fick's law to describe ion transport inside the active particles:

$$\frac{\partial c_s}{\partial t} = D_s \left[\frac{\partial^2 c_s}{\partial r^2} + \frac{2}{r} \frac{\partial c_s}{\partial r} \right] \tag{3.7}$$

In Eq. (3.7), c_s denotes the solid-phase concentration of lithium ions, D_s represents the solid phase diffusion coefficient, t is time, and r refers to the radial direction of the active particle radius.

Newman's theory describes electron transport within the solid phase conductive network using Ohm's law:

$$\mathrm{i}_s = -\sigma_s \nabla \Phi_s \tag{3.8}$$

In Eq. (3.8), i_s is the solid phase current density, σ_s is the solid phase electrical conductivity, and Φ_s is the solid phase electric potential.

Newman's theory employs the Butler Volmer equation, based on transition state theory and linear free energy relationships, to describe charge transfer reactions at the surface of active materials:

$$j_n = j_n^0 \left[\exp\left(\alpha \frac{F\eta}{RT} \right) - \exp\left(-(1-\alpha) \frac{F\eta}{RT} \right) \right] \tag{3.9}$$

In Eq. (3.9), j_n is the current density, j_n^0 is the exchange current density, η is the overpotential, and α is the transfer coefficient.

The focus of Newman's theory lies in the description of mass transport in concentrated solutions. To this end, it employs the multicomponent Stefan Maxwell equations:

$$c_i \nabla \mu_i = RT \sum_{j \neq i} \frac{c_i c_j}{c_T \mathrm{D}_{ij}} \left(v_j - v_i \right) \tag{3.10}$$

In Eq. (3.10), c_i is the concentration of particle i, $\nabla \mu_i$ is the electrochemical potential gradient of particle i, v_i is the velocity of particle i, and D_{ij} is the collective diffusion factor of the electrolyte. The difference between this equation and Fick's diffusion equation lies in that this formulation accounts for coupled multicomponent transport, whereas Fick's diffusion equation describes diffusion of a single component only. Starting from this equation, the equivalent diffusion factor of the electrolyte, can be derived as follows: $D = \mathrm{D}\frac{c_T}{c_0}\left(1 + \frac{\partial \ln \gamma \pm}{\partial \ln m}\right)$ (where c_T is the total concentration of solute and solvent, c_0 is the solvent concentration, $\gamma_{\pm}$ is the total activity coefficient of positive and negative ions, m is the molality, and $t_+ = 1 - \mathrm{t}_- = \frac{z_+ \mathrm{D}_{0+}}{z_+ \mathrm{D}_{0+} - z_- \mathrm{D}_{0-}}$ is the transference number of positive ions (where $z_{+(-)}$ is the charge carried by either positive or negative ions, and $\mathrm{D}_{0+(-)}$ is the independent diffusion factor of positive or negative ions). Neglecting convection or assuming local charge neutrality, and applying mass conservation, the following can be derived:

$$\frac{\partial c_e}{\partial t} = \nabla \cdot \left(D\left(1 - \frac{\partial \ln c_0}{\partial \ln c}\right)\nabla c_e\right) - \frac{\mathrm{i}_e \cdot \nabla t_0^+}{F} \tag{3.11}$$

In Eq. (3.11), i_e is the liquid phase current density, t_0^+ is the transference number of positive ions. According to charge conservation, the liquid-phase current density i_e can be expressed as follows, where κ_e is the liquid phase conductivity:

$$\mathrm{i}_e = -\kappa_e \nabla \Phi_e + \frac{2\kappa_e RT}{F}\left(1 + \frac{\partial \ln f_{\pm}}{\partial \ln c}\right)\left(1 - t_+^0\right)\nabla \ln c_e \tag{3.12}$$

Eqs. (3.11) and (3.12) constitute the microscopic equation set describing the transport of liquid phase substances. By averaging the microscopic equations over a representative volume element (RVE) that encompasses all phases, a coarse grained macroscopic model (volume averaging method) can be obtained. The RVE must be sufficiently "large" to capture microscopic scale information, yet sufficiently "small" to allow the use of continuum equations at the macroscopic level. After volume averaging, the multiphase porous electrode liquid phase mass transfer equations take the following form:

$$\varepsilon \frac{\partial c_e}{\partial t} = \nabla \cdot \left(D_{eff}\left(1 - \frac{\partial \ln c_0}{\partial \ln c}\right)\nabla c_e\right) - \frac{\mathrm{i}_e \cdot \nabla t_+^0}{F} \tag{3.13}$$

In Eq. (3.13), D_{eff} is the effective diffusion coefficient. Newman's theory generally employs the Bruggeman relationship to equivalently account for the influence of porous structures on the diffusion coefficient, D_{eff} expressing it as $D_{eff} = D\varepsilon^{\gamma}$, where ε denotes the porosity and γ represents the Bruggeman factor related to the porous structure.

Based on the (3.13) volume averaging relationship, the current density j_n in the equation is related to the liquid phase current density i_e as follows:

$$aj_n = \nabla \cdot i_e \tag{3.14}$$

In Eq. (3.14), a represents the interfacial area per unit volume (with units of m^2/m^3). Accordingly, based on Eq. (3.14), the charge transfer equation for the liquid phase can be written as follows:

$$aj_n = \nabla \cdot \left(-\kappa_e \nabla \Phi_e + \frac{2\kappa_e RT}{F} \left(1 + \frac{\partial \ln f_\pm}{\partial \ln c} \right) \left(1 - t_+^0 \right) \nabla \ln c_e \right) \tag{3.15}$$

Eqs. (3.13) and (3.15) constitute the macroscopic liquid-phase mass transfer equations accounting for the influence of porous structures. The effects of porous structures on solid phase charge transport and diffusion can also be represented using effective transport coefficients in the form of $X_{eff} = X\varepsilon^{\gamma}$ (with X as the transport parameters). Together, these equation sets form the fundamental equations of Newman's porous electrode theory.

Fig. 3.8 illustrates the schematic structure of the P2D model, which depicts the electrochemical reaction process of LIBs by describing the migration of lithium ions between the anode, separator, and cathode during charging and discharging, as well as the concentration distribution within the electrode particles. Newman's contribution to porous electrode theory lies in accurately describing the majority of physicochemical processes within LIBs, as well as the influence of the porous electrode structure on mass transport.

3.2.3 Numerical solution methods

As previously mentioned, the electrochemical model of LIBs constitutes a complex system composed of coupled nonlinear partial differential equations and algebraic equations. In most cases, analytical solutions to this system cannot be obtained; therefore, developing efficient and stable numerical solution methods is a key step toward enabling the simulation and application of electrochemical models. This section systematically presents the core numerical techniques for solving this model.

Numerical solution models typically follow a standard procedure, the fundamental concept of which is to discretize continuous spatiotemporal problems into discrete algebraic problems. The basic procedure is as follows:

1) Model discretization: Discretizing the governing equations in both spatial and temporal domains, transforming partial differential equations into algebraic equations.
2) Initial and boundary condition specification: The initial state of the system (e.g., a uniform lithium ion concentration distribution) and physical boundary conditions (e.g., input current at the current collector, flux continuity at the separator) are prescribed.
3) Solution of nonlinear equation systems: At each time step, solve the large nonlinear algebraic equation system formed after discretization.
4) Time stepping: Employ a time integration algorithm to progressively advance the solution until the simulation concludes.

3.3 AI-based modeling approaches

3.3.1 Limitations of traditional models

Conventional battery models, whether ECMs or electrochemical models, exhibit inherent and structural limitations when confronted with four core challenges demanded by modern BMSs: high precision nonlinear dynamic characterization, full life cycle parameter time varying tracking, realtime computation on embedded platforms, and cross condition and cross batch generalization adaptability. ECMs suffer from being "oversimplified," whereas electrochemical models are plagued by being "overly complex." These limitations cannot be fully overcome through mere algorithmic improvements or parameter optimizations; rather, they reveal the ceiling faced by purely physics based and structurally fixed modeling paradigms at the current stage.

Therefore, the field of battery modeling urgently requires the introduction of new paradigms and methodologies. Data driven artificial intelligence approaches, with their powerful nonlinear mapping capabilities, inherent ability to automatically learn patterns from data, and flexibility unconstrained by fixed physical structures, offer a promising new avenue for addressing the aforementioned challenges. However, purely "black box" AI models also suffer from issues such as physical inconsistency, poor interpretability, and high data dependency. This naturally leads to the development of hybrid modeling strategies that integrate physical models with data driven AI models, aiming to leverage the strengths of both while mitigating their respective limitations.

3.3.2 Machine learning for battery modeling

3.3.2.1 Support vector machine

Support vector machine (SVM) is a type of generalized linear classifier that performs binary classification through supervised learning. Its core principle lies in determining the maximum margin hyperplane in the feature space by solving for the optimal separating boundary based on training samples [2]. This approach relies not only on a small number of support vectors located near the decision boundary to define the position and orientation of the hyperplane, but also maximizes the geometric margin to enhance the generalization capability of the classifier. Mathematically, this process can be formulated as a convex optimization problem, typically solved using Lagrange multipliers and duality theory. Furthermore, by incorporating kernel functions, SVM can be extended to handle nonlinear classification, enabling effective separation of samples from different classes even in high-dimensional feature spaces.

Assume we have a training sample set:

$$\{(x_i, y_i)\}_{i=1}^{N}, x_i \in \mathbb{R}^n, y_i \in \{-1, +1\} \tag{3.16}$$

We need to find a hyperplane:

$$w^T x + b = 0 \tag{3.17}$$

where w is the normal vector, b is the bias term, and the classification function is:

$$f(x) = \text{sign}(w^T x + b) \tag{3.18}$$

To ensure that the sample points are correctly classified, $y_i(w^T x_i + b) \geq 1$, the hyperplane must maintain a margin of separation $\frac{2}{\|w\|}$ from both classes of samples, and the objective is to maximize this margin; therefore, the optimization problem reduces to finding the minimum $\frac{2}{\|w\|}$.

Kernel functions $K(x_i, x_j)$ are introduced to address nonlinearly separable problems by mapping data into a higher dimensional space, making them linearly separable. The dual formulation is:

$$\begin{aligned} &\max_{\alpha} \sum_{i=1}^{N} \alpha_i - \frac{1}{2} \sum_{i,j=1}^{N} \alpha_i \alpha_j y_i y_j K(x_i, x_j) \\ &\text{s.t.} \sum_{i=1}^{N} \alpha_i y_i = 0, 0 \leq \alpha_i \leq C \end{aligned} \tag{3.19}$$

In Eq. (3.19), α_i is the Lagrange multiplier, and C is the penalty coefficient, controlling the trade off between maximization and classification error. Eq. (3.19) represents the maximization of the objective function $\sum_{i=1}^{N} \alpha_i - \frac{1}{2} \sum_{i,j=1}^{N} \alpha_i \alpha_j y_i y_j K(x_i, x_j)$ under the given constraints $\sum_{i=1}^{N} \alpha_i y_i = 0, 0 \leq \alpha_i \leq C$.

In battery modeling, SVM and support vector regression (SVR) are primarily employed for critical tasks such as SOC, SOH, remaining useful life (RUL) estimation, and fault diagnosis. Their core advantage lies in effectively handling small sample datasets, high noise environments, and nonlinear features by mapping input features into a high dimensional space via kernel functions, thereby enabling accurate regression or classification predictions. In practical applications, raw signals such as voltage, current, temperature, and cycle count are first subjected to feature engineering, including sliding window statistical features, accumulated capacity, internal resistance estimation, and residual features; these features are then normalized. Subsequently, an appropriate kernel function and hyperparameters are selected for model training, with model generalization capability evaluated through time series crossed validation or rolling window validation. Finally, the model is deployed with consideration of engineering constraints, such as inference speed, number of support vectors, realtime update capability, and robustness to anomalous inputs, optimized accordingly. Moreover, SVM/SVR is frequently integrated with physics based models through residual learning or weighted fusion strategies to enhance prediction accuracy and reliability. This integration enables high precision, robust, and interpretable SOC/SOH/RUL prediction and fault diagnosis within BMS.

3.3.2.2 Random forest

Random forest (RF) is an ensemble learning method whose core principle lies in enhancing model accuracy and stability by aggregating the predictions of multiple decision trees. As a typical representative of the bagging (bootstrap aggregating) approach, RF introduces randomness at both the sample and feature levels to train a set of weak classifiers, ultimately producing the final prediction through voting or averaging. This method effectively reduces the risk of overfitting inherent in individual decision trees while maintaining strong generalization performance on large datasets and high dimensional features. In classification tasks, the prediction outcome is determined by majority voting; in regression tasks, it is obtained as the average of predictions from all decision trees.

Assume the training dataset is:

$$\mathscr{D}=\{(x_i,y_i)\}_{i=1}^{N}, x_i\in\mathbb{R}^n, y_i\in\mathbb{R}, y_i\in\{-1,+1\} \tag{3.20}$$

RF generates B bootstrap sample sets $\mathscr{D}_b$ through the bagging technique:

$$\mathscr{D}_b \sim \text{Bootstrap}(\mathscr{D}), b=1,2,\ldots,B \tag{3.21}$$

A prediction function for the regression problem is obtained by training a decision tree $h_b(x)$ on each subset of the sample:

$$\widehat{f}(x)=\frac{1}{B}\sum_{b=1}^{B}h_b(x) \tag{3.22}$$

The prediction function for classification problems is:

$$\widehat{y}(x)=\text{argmax}_k\sum_{b=1}^{B}|(h_b(x)=k) \tag{3.23}$$

In Eq. (3.23), $|(\cdot)$ is the indicator function.

During the splitting of each tree node, RFs do not consider all features for selection; instead, they randomly select m ≪ n features as candidates and choose the optimal splitting feature from among them. This approach enhances the diversity among individual trees while reducing the correlation among models.

The expected prediction error of a random forest can be approximately decomposed as:

$$E=\rho\cdot\sigma^2+(1-\rho)\frac{\sigma^2}{B} \tag{3.24}$$

In Eq. (3.24), ρ represents the correlation among base learners, σ^2 represents the variance of individual base learners, and B represents the number of trees. Therefore, reducing the correlation among trees and increasing the number of trees can both effectively reduce prediction error.

In battery modeling, the RF algorithm is primarily employed in critical stages such as SOC estimation, SOH prediction, RUL inference, and fault diagnosis. Its advantages lie in its ability to handle complex nonlinear relationships and high dimensional features, while demonstrating strong robustness against noise and overfitting. Compared to single models, RF enhances prediction stability and generalization performance by aggregating results from multiple decision trees, and it can output feature importance scores that facilitate the identification of key factors, such as voltage, current, temperature, and internal resistance, that most significantly influence battery performance. In engineering practice, by extracting statistical features, cyclic features, and health indicators from raw operational data, RF enables accurate and reliable battery state estimation with relatively low computational cost, thereby providing strong support for the safety, stability, and interpretability of BMS.

3.3.2.3 K-nearest neighbors algorithm

The K-nearest neighbors (KNN) algorithm is an instance based supervised learning method whose fundamental principle is as follows: for a given sample to be classified (or regressed), the algorithm identifies the K most similar neighboring samples in the training set using a distance metric, and then makes predictions based on the classes or values of these neighbors. This approach does not rely on an explicit training process; instead, it stores the training samples directly as the model. Consequently, KNN is characterized by its simplicity and ease of interpretation. In classification tasks, KNN typically employs majority voting, whereas in regression problems, it commonly uses a weighted average of neighbor values.

Assume the training set is:

$$\left\{\left(x_i, y_i\right)\right\}_{i=1}^{N}, x_i \in \mathbb{R}^n, y_i \in \{-1, +1\} \text{ or } \mathbb{R} \tag{3.25}$$

In Eq. (3.25), x_i represents the sample feature vector; y_i represents the sample label.

A commonly used similarity measure is the Euclidean distance:

$$d(x, x_i) = \sqrt{\sum_{j=1}^{n} \left(x_j - x_{i,j}\right)^2} \tag{3.26}$$

Moreover, Manhattan distance, Minkowski distance, or cosine similarity may also be employed, depending on the nature of the features.

For classification problems, the predicted class of sample x is the one that appears most frequently among its KNN:

$$\widehat{y} = \mathrm{argmax}_{c \in \mathscr{C}} \sum_{i \in N_K(x)} \mathbb{I}\left(y_i = c\right) \tag{3.27}$$

In Eq. (3.27), c denotes the class set; $N_K(x)$ denotes the KNN of sample x; and $\mathbb{I}(\cdot)$ is an indicator function.

For regression problems, the prediction value of KNN is the average of its neighbors:

$$\hat{y} = \frac{1}{K} \sum_{i \in N_K(x)} y_i \tag{3.28}$$

If distance weighting is taken into account, the formula becomes:

$$\hat{y} = \frac{\sum_{i \in N_K(x)} \frac{1}{d(x, x_i)} y_i}{\sum_{i \in N_K(x)} \frac{1}{d(x, x_i)}} \tag{3.29}$$

In battery modeling, KNN is primarily employed for tasks such as SOC estimation, SOH prediction, RUL inference, and fault diagnosis. The approach involves extracting statistical features, capacity degradation characteristics, and internal resistance information from battery operational data, including voltage, current, temperature, and cycle count, then computing similarity between the current sample and historical samples using KNN. The nearest neighbors are identified, and predictions are derived via majority voting or weighted averaging. For SOC estimation, KNN enables rapid matching of battery data under similar operating conditions and delivers accurate state of charge predictions. In SOH and RUL modeling, KNN infers health status and lifespan by comparing degradation trajectories. In fault diagnosis, KNN facilitates rapid identification of operating conditions most similar to known fault patterns. Advantages include simplicity of implementation and strong interpretability; however, drawbacks include high computational cost and sensitivity to high dimensional features, necessitating dimensionality reduction techniques, indexing structures, or data compression to enhance efficiency.

3.3.2.4 Gaussian process regression

Gaussian process regression (GPR) is a probabilistic, nonparametric Bayesian regression method. Its fundamental idea is to assume that the target function $f(x)$ is drawn from a Gaussian process (GP), meaning that the joint distribution of function values at any finite set of input points follows a multivariate Gaussian distribution. By specifying a mean function and a covariance function, GPR infers the distribution at unobserved points given observed data, thereby providing not only the predicted mean but also an estimate of prediction uncertainty.

A Gaussian process is defined by a mean function $m(x)$ and a covariance function $k(x, x')$:

$$f(x) \sim \mathscr{GP}\left(m(x), k\left(x, x'\right)\right) \tag{3.30}$$

In Eq. (3.30), $m(x) = E[f(x)]$ denotes the mean function; $k(x, x') = \mathbb{E}[(f(x) - m(x))(f(x') - m(x'))]$ denotes the covariance function.

The joint distribution of function values for training samples $X = \{x_i\}_{i=1}^N$ and test points x_* is:

$$\begin{bmatrix} f(X) \\ f(x_*) \end{bmatrix} \sim \mathcal{N}\left(0, \begin{bmatrix} K(X, X) & K(X, x_*) \\ K(x*, X) & K(x_*, x_*) \end{bmatrix}\right) \tag{3.31}$$

Given training data (X, y), the predictive distribution for a test point $x*$ is Gaussian with mean and variance given by:

$$\begin{aligned} \mu(x_*) &= K(x_*, X)\left[K(X, X) + \sigma_n^2 I\right]^{-1} y \\ \sigma^2(x_*) &= K(x_*, x_*) - K(x_*, X)\left[K(X, X) + \sigma_n^2 I\right]^{-1} K(X, x_*) \end{aligned} \tag{3.32}$$

In Eq. (3.32), $\mu(x_*)$ is the predicted mean, $\sigma^2(x_*)$ is the predicted variance, K is the covariance, and σ_n^2 is the variance of the observation noise.

In battery modeling, Gaussian process regression (GPR) is primarily employed for SOC, SOH, and RUL modeling. Its advantages include the absence of explicit assumptions regarding battery degradation mechanisms, enabling the use of kernel functions to capture complex nonlinear relationships among battery characteristics. Under limited experimental data, GPR delivers reliable predictions. It not only provides estimates of SOC and SOH but also outputs confidence intervals, thereby offering risk 预警 for BMSs. Furthermore, by employing different kernel functions or their combinations, GPR can be adapted to meet diverse modeling requirements under varying operating conditions.

In practical applications, GPR is typically combined with input features such as voltage, current, temperature, and cycle count to train kernel function parameters, demonstrating high prediction accuracy in SOC estimation and providing robust uncertainty information for SOH and RUL prediction tasks, thus holding significant value in BMSs.

3.3.3 Deep learning–based modeling techniques

3.3.3.1 Convolutional neural network

Convolutional neural networks (CNNs) are a class of feedforward neural networks characterized by mechanisms such as local receptive fields, weight sharing, and pooling, primarily designed for processing data with grid like structures (e.g., images, time-series signals). The core idea lies in extracting local features through convolution operations, followed by nonlinear transformations and hierarchical feature integration, enabling the representation of increasingly abstract features from low level to high level. CNNs have found extensive applications in image recognition, speech processing, and multidimensional data analysis in battery modeling.

In the convolutional layer, the convolution operation between the input feature map X and the convolution kernel W produces the output feature map:

$$z_{i,j}^{(k)} = \sum_{m=1}^{M} \sum_{n=1}^{N} w_{m,n}^{(k)} \cdot x_{i+m,j+n} + b^{(k)} \tag{3.33}$$

In Eq. (3.33), $w_{m,n}^{(k)}$ denotes the weight of the k th convolution kernel; $b^{(k)}$ represents the bias term; and $z_{i,j}^{(k)}$ refers to the convolution output.

Convolutional outputs are typically mapped through nonlinear activation functions, such as ReLU:

$$a_{i,j}^{(k)} = \max\left(0, z_{i,j}^{(k)}\right) \tag{3.34}$$

Pooling layers are used to reduce feature dimensions, decrease the number of parameters, and enhance model robustness. Common pooling operations include the following:

$$p_{i,j}^{(k)} = \max_{(m,n)\in\Omega} a_{i+m,j+n}^{(k)} \tag{3.35}$$

In Eq. (3.35), Ω denotes the pooling window.

At the end of the network, the feature maps are flattened and fed into a fully connected layer:

$$y = f(Wx + b) \tag{3.36}$$

In Eq. (3.36), the function $f(\cdot)$ can be either softmax or the identity function.

For classification problems, the cross entropy loss is commonly used:

$$L = \frac{1}{N}\sum_{i=1}^{N}\left(y_i - \widehat{y}_i\right)^2 \tag{3.37}$$

In battery modeling, CNNs can process multidimensional time series data (such as voltage, current, and temperature sequences) and automatically extract latent features. Their primary applications include the following:

SOC estimation: Construct battery operating data into two dimensional or sequential features as input to a CNN, which automatically learns the nonlinear relationship between SOC and the signals.

SOH prediction: Employing CNN to extract degradation curve features and model the evolution of battery health status;

RUL prediction: CNN can extract key features related to life degradation through historical cyclic data and multidimensional operating condition signals;

Fault diagnosis: CNNs exhibit excellent performance in pattern recognition and can be employed to detect anomalous signals and classify different types of battery faults.

The advantage of CNN lies in its ability to automatically extract features, suitability for large scale data modeling, and strong fitting and generalization capabilities; its drawbacks include high demands for data volume and computational resources, and a tendency toward overfitting in small sample scenarios. Consequently, in BMS, CNN is often combined with traditional methods or physical models to enhance accuracy and reliability.

3.3.3.2 Recurrent neural networks

Recurrent neural networks (RNNs) are a class of neural networks specifically designed to handle sequential data. Their core characteristic lies in the introduction of "recurrent connections" within the network architecture, whereby the hidden state at the current time step depends not only on the current input but also on the hidden state from the previous time step. This enables RNNs to capture temporal dependencies in sequence modeling, making them well suited for applications such as time series forecasting, natural language processing, and battery state modeling.

The computation process of an RNN is iteratively unrolled at each time step t, with the input being, the hidden state being, and the output being.

$$ht = f\left(W_{xh}x_t + W_{hh}h_{t-1} + b_h\right) \tag{3.38}$$

In Eq. (3.38), W_{xh} is the weight matrix for input to the hidden layer; W_{hh} is the recurrent weight matrix from the hidden layer to itself; b_h is the bias term; tanh or ReLU is commonly used as the activation function.

At each time step, the output of the RNN is computed from the hidden state:

$$y_t = g(W_{hy}h_t + b_y) \tag{3.39}$$

In Eq. (3.39), the activation function $g(\cdot)$ is typically either Softmax or the identity function.

Assuming the sequence length is T, t the target output is $\{y_t\}_{t=1}^{T}$, the loss function is the average or weighted sum of errors across all time steps. The commonly used crossed entropy loss is given by $L = -\sum_{t=1}^{T}\sum_{c=1}^{C} y_{t,c}\log\left(\widehat{y}_{t,c}\right)$. For regression tasks, the Mean squared error (MSE) is typically employed as $L = \frac{1}{T}\sum_{t=1}^{T}\left(y_t - \widehat{y}_t\right)^2$.

Standard RNNs are prone to the vanishing or exploding gradient problems when handling long sequences. To address this issue, improved architectures such as LSTM and GRU have been proposed, which effectively capture long-range dependencies through gated mechanisms.

In battery modeling, RNNs are widely employed due to their proficiency in handling time series data. RNNs can leverage historical sequences of signals such as voltage, current, and temperature to capture the dynamic characteristics of battery state evolution over time, thereby enabling more accurate SOC estimation. Additionally, RNNs can learn long term evolution patterns such as capacity degradation and internal resistance changes, facilitating SOH prediction and RUL inference. In fault diagnosis, RNNs are capable of identifying anomalous patterns in time series, aiding in the early detection of potential failures. Overall, RNNs possess inherent advantages in BMSs by effectively modeling the temporal dependencies inherent in battery operation, thus improving the accuracy of state estimation and lifetime prediction. However, their high training complexity often necessitates integration with enhanced architectures such as LSTM or GRU networks to improve performance and stability.

3.3.3.3 Long short-term memory network

The LSTM network is an enhanced RNN architecture specifically designed to address the issues of vanishing and exploding gradients commonly encountered in traditional RNNs when processing long sequences.

Compared with standard RNNs, LSTM incorporates gated mechanisms within its network structure, employing a set of learnable gating units to dynamically regulate information flow, determining which information should be forgotten, updated, or output. Through this mechanism, LSTM can effectively store and utilize critical information over extended temporal spans, thereby capturing complex long term dependencies. Its fundamental architecture consists of a forget gate, an input gate, an output gate, and a cell state; these components work in concert to enable LSTM to demonstrate superior stability and expressive capacity in sequence modeling and time series forecasting tasks.

The forget gate controls how much information from the previous memory cell state should be retained.

$$f_t = \sigma\left(W_f \cdot [h_{t-1}, x_t] + b_f\right) \tag{3.40}$$

In Eq. (3.40), h_{t-1} represents the hidden state at the previous time step; x_t represents the input at the current time step; W_f and b_f denote the weight matrix and bias term of the forget gate, respectively; $\sigma(\cdot)$ denotes the sigmoid function, which maps the output to the interval [0,1]. When the value f_t is close to 1, it indicates that almost all information from the previous step is retained; when it is close to 0, it indicates that almost all information from the previous step is forgotten.

The input gate controls how much information from the current time step is written into the cell state, encompassing both the input gate activation value i_t and the candidate state C_t:

$$i_t = \sigma(W_i \cdot [h_{t-1}, x_t] + b_i) \tag{3.41}$$

The cell state C_t is updated jointly by the forget gate and the input gate:

$$C_t = f_t \odot C_{t-1} + i_t \odot \widetilde{C}_t \tag{3.42}$$

In Eq. (3.42), $\odot$ denotes element wise multiplication.

The output gate determines the value of the hidden state h_t as

$$h_t = \sigma(W_o \cdot [h_{t-1}, x_t] + b_o) \odot \tanh(C_t) \tag{3.43}$$

In battery modeling, LSTM networks are widely employed for SOC estimation, SOH prediction, RUL inference, and fault diagnosis. Due to the pronounced temporal dependencies in battery operational data, such as voltage, current, and temperature, LSTM effectively captures both short term and long term dynamic characteristics through its gated architecture. In SOC estimation, LSTM leverages historical charging and discharging data to achieve stable and accurate predictions. For SOH and RUL modeling, it learns long term degradation patterns associated with capacity fade and increasing internal resistance. In fault diagnosis, LSTM identifies complex temporal patterns, enabling early detection of anomalies. Compared to traditional RNNs, LSTM demonstrates superior performance in modeling long sequences, thereby holding significant application value in BMS.

3.3.4 From physical to hybrid modeling

The core task of a BMS is to accurately estimate key internal states reflecting battery performance and health through realtime monitoring and analysis of measurable external variables, thereby managing the battery's charge and discharge power accordingly. As illustrated in Fig. 3.9, these input and output variables collectively constitute the fundamental elements of battery modeling and state estimation problems.

However, due to the complex electrochemical processes within batteries, two dominant paradigms have long prevailed in the field of LIB modeling: physics based mechanistic models and data driven models based on data. Each paradigm has its own distinctive features, yet also exhibits inherent limitations.

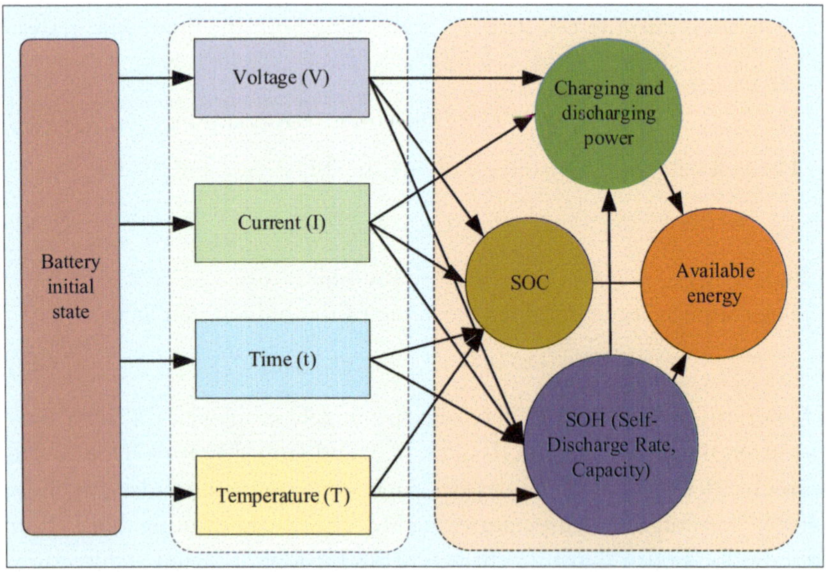

Figure 3.9 Battery modeling and state estimation diagram.

Hybrid modeling emerged precisely under such circumstances. It aims to integrate the mechanistic interpretability of physical models with the environmental adaptability of data driven models, thereby constructing "gray box" models that adhere to fundamental physical laws while also learning unmodeled dynamics and time varying characteristics from data. This modeling approach is not a mere compromise but rather achieves synergistic advantages through organic integration, seeking an optimal balance among model accuracy, robustness, and computational efficiency.

3.3.4.1 Physics-guided data-driven models

Physics informed data driven models are primarily based on data-driven approaches, enhanced by incorporating physical knowledge as constraints or priors to improve scientific interpretability and data efficiency. Physics-informed neural networks (PINNs) serve as a paradigmatic example of such methods. In PINNs, the loss function design includes not only conventional data fitting error terms but also a "physical residual" term derived from governing equations. This formulation compels the neural network during training to simultaneously satisfy both data fidelity and physical law constraints, thereby maintaining physical plausibility even in regions with sparse data.

Another important approach is physical constraint embedding, in which known physical relationships are incorporated as hard constraints directly into the model structure or training process. For instance, in a state estimation model, the monotonic relationship between OCV and SOC can be imposed as a constraint to ensure that model predictions consistently adhere to this fundamental physical property. Such methods not only enhance the model's extrapolation capability but also significantly reduce the demand for training data.

3.3.4.2 Physics-informed models enhanced by data-driven approaches

Data driven enhanced physical models use physical models as their structural backbone, leveraging data driven techniques to address their limitations. Within this framework, the physical model is responsible for capturing the primary dynamic characteristics of the system, while the data driven component serves to compensate for unmodeled dynamics and time varying parameters.

AI assisted parameter identification is one of the key implementation approaches. By training machine learning models with battery operating states as inputs, critical parameters in physical models can be predicted online in real time, thereby endowing originally static physical models with adaptive capabilities. For instance, neural networks can be constructed to identify parameters such as resistance and capacitance in ECMs in real time, which vary with temperature and aging states.

Residual learning represents another effective strategy. This approach constructs a cascaded architecture consisting of a "physics based model + AI residual corrector." The physics model provides an initial prediction that aligns with mechanistic trends, while the AI model learns the residual between the physics model's predictions and the actual observations. This division of labor preserves the interpretability of the physics model while enhancing overall accuracy through data-driven refinement.

3.3.4.3 Highly integrated gray-box model

In deeply integrated gray box models, physical knowledge and data driven techniques are organically combined, making it difficult to strictly distinguish between them [3]. State space hybrid models serve as typical representatives of such approaches. Within this framework, the state equation is defined by a simplified physical model that describes the evolution of the system's internal states, while the observation equation is represented by a neural network, establishing the relationship between states and observations. This structure preserves the physical interpretability of state evolution while fully leveraging the powerful nonlinear representation capabilities of neural networks [4].

Neural network differential equations represent another approach to deep fusion. They treat the output of a neural network as the right hand side function of a differential equation, leveraging the neural network to learn and represent complex dynamical systems. This method is particularly suitable for systems in which certain physical laws are partially known, but some dynamics are difficult to describe precisely, offering a novel perspective for constructing high accuracy battery dynamic models.

References

[1] V. Kumar, S.A. Kumar, Development of an electrochemical-thermal model and methodology for optimizing number of tabs in cylindrical lithium-ion batteries, Appl. Therm. Eng. 279 (PC) (2025) 127576.

[2] S. Pijush, Application of least square support vector machine (LSSVM) for determination of evaporation losses in reservoirs, Engineering 3 (4) (2011) 431–434.

[3] B. Jennifer, G. Rainer, W.G. Bessler, A grey-box model with neural ordinary differential equations for the slow voltage dynamics of lithium-ion batteries: application to single-cell experiments, J. Power Sources 614 (2024) 234918.

[4] G. Zelin, L. Yiyan, Z. Zhenghao, et al., A virtual-component-embedded equivalent circuit model for lithium-ion battery state estimation, Int. J. Electr. Power Energy Syst. 170 (2025) 110896.

CHAPTER FOUR

State of charge (SOC) estimation

4.1 Definition and significance

State of Charge, abbreviated as SOC, is a key technical parameter that appears simple at first glance but is rich in implications. It precisely defines the "condition" of a battery at a specific moment, serving as an information bridge connecting users, devices, and the battery management system. From the most rigorous academic and engineering perspectives, the SOC of a lithium-ion battery is defined as the ratio between the remaining charge in the battery under given conditions and its rated total capacity under the same conditions. Its mathematical expression is given by (4.1):

$$SOC(t) = \frac{Qremaining(t)}{Qrated} \times 100\% \tag{4.1}$$

In Eq. (4.1), SOC(t) represents the state of charge at time t, a quantity that dynamically varies over time. Qremaining(t) represents the remaining dischargeable capacity of the battery at time t. Qrated represents the rated capacity of the battery, that is, the total electrical charge that can be released when discharging from a fully charged state to the cutoff voltage under standard specified conditions.

This definition clearly indicates that SOC is a relative quantity, expressed as a percentage, typically ranging from 0% to 100%. A value of 0% signifies that the battery has discharged to its preset minimum safe limit (far from being fully "depleted," as over-discharging can cause permanent damage), whereas 100% indicates that the battery has been charged to its designed maximum capacity.

State of charge is a key parameter characterizing the remaining available energy of a battery, and its accurate estimation holds foundational significance for Battery management systems (BMS). It not only provides the fundamental guarantee for safe and long-term battery operation, but also serves as the cornerstone for enhancing user experience and system efficiency, as well as enabling advanced intelligent management functionalities.

Modelling, Estimation and AI Applications for Lithium-Ion Battery Management Systems
ISBN: 978-0-443-45347-2
https://doi.org/10.1016/B978-0-443-45347-2.00004-2

Accurate SOC estimation is a fundamental prerequisite for ensuring battery systems operate within safe limits, preventing misuse, and extending their service life. The accuracy of SOC directly determines whether overcharging or overdischarging will occur: overcharging may trigger violent side reactions, leading to thermal runaway, and even combustion or explosion; overdischarging can cause irreversible structural damage to electrode materials, accelerating battery degradation. A reliable and conservative SOC estimation strategy serves as the foundation for designing all battery protection logic and represents the first line of defense against premature battery aging.

At the application level, high-accuracy SOC estimation directly translates into superior user experience and enhanced system economic efficiency. For electric vehicle users, it ensures a reliable remaining driving range displayed on the instrument panel, effectively alleviating "range anxiety." For energy storage power station operators, it enables precise energy scheduling and maximizes battery asset utilization. The accuracy of SOC estimation directly determines the usable capacity range of batteries; an inaccurate SOC estimate forces the BMS to reserve excessive buffer margins, thereby reducing actual driving range or diminishing the effective throughput capacity of the energy storage system.

SOC is the most critical state parameter in BMS, serving as the fundamental "input" for all advanced algorithms and optimization control strategies. Whether implementing active cell balancing, establishing optimal fast-charging curves, or recursively estimating State of health (SOH), accurate and real-time SOC values are indispensable. Consequently, overcoming the complex nonlinear and time-varying relationship between battery voltage and SOC, and achieving robust SOC estimation across the entire lifecycle and under all operating conditions, remains a core technical challenge and primary research focus for both academia and industry.

4.2 Traditional SOC estimation methods

In battery management systems, accurately estimating the SOC of a battery is crucial. SOC reflects the amount of remaining charge in the battery and plays a pivotal role in rational battery utilization, lifespan extension, and ensuring the proper operation of equipment. Traditional SOC estimation methods each possess distinct characteristics; several common approaches will be detailed below.

4.2.1 Open circuit voltage method

The Open-circuit voltage (OCV) method estimates the SOC based on the established relationship between the battery's open-circuit voltage and its SOC. When a battery is in an open-circuit state, its voltage is determined solely by internal electrochemical reactions and electrode conditions, independent of external loads. For most battery types, such as lithium-ion batteries, the open-circuit voltage exhibits a predictable variation with changes in SOC, as shown in Fig. 4.1. For instance, the OCV of a lithium-ion battery typically decreases gradually as SOC declines, with the rate of voltage drop varying across different SOC intervals. By establishing a reference curve that maps OCV to SOC, usually derived through experimental measurements, one can determine the corresponding SOC value during actual operation by measuring the battery's open-circuit voltage and referring to this curve.

The open-circuit voltage method estimates the SOC based on the fixed relationship between the open-circuit voltage and SOC of lithium-ion batteries under static conditions. This method is applicable when the battery has been left idle for a prolonged period, allowing the voltage to stabilize. The corresponding formula is:

$$V_{OCV} = f(\mathrm{SOC}) \tag{4.2}$$

In Eq. (4.2), f is a mapping function specific to battery chemistry, obtained through experimental calibration.

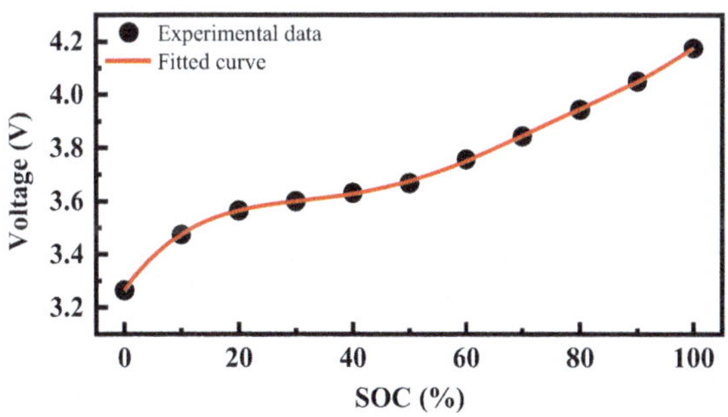

Figure 4.1 OCV–SOC curve.

The advantage of the open-circuit voltage method lies in its simplicity and ease of implementation. It requires only the measurement of the battery's open-circuit voltage, without the need for complex equipment or computational procedures. When the battery is at rest and free from self-discharge effects, the relationship between open-circuit voltage and SOC is relatively accurate, enabling reliable SOC estimation. However, this method also presents notable drawbacks. First, it necessitates that the battery be in an open-circuit state at rest, whereas in practical applications batteries are frequently subjected to charging and discharging cycles, making it difficult to ensure sufficient resting time and thereby compromising estimation accuracy. Second, battery aging and temperature variations can cause shifts in the open-circuit voltage and SOC relationship, leading to errors in estimation when based on fixed calibration curves.

In certain application scenarios where real-time dynamic performance of batteries is not critical, such as small portable electronic devices during standby mode or battery energy storage systems subject to periodic maintenance checks, the OCV method can be employed for a rough estimation of SOC, providing a general understanding of battery status. For example, the voltage of a lithium-ion battery typically ranges from 3.0 to 4.2 V, corresponding to an SOC from 0% to 100%. Although this method is simple, it is significantly affected by temperature and battery aging.

4.2.2 Coulomb counting

The ampere-hour integration method estimates the SOC by measuring the current during the battery's charging and discharging processes and integrating the current. The fundamental principle involves starting from the initial SOC value, adding the charge (in ampere-hours) that has flowed into the battery during charging, and subtracting the charge that has flowed out during discharging, thereby determining the current SOC value. This method estimates SOC by integrating current to calculate changes in charge, and is suitable for continuous charging and discharging processes. Core formula:

$$\mathrm{SOC}(t) = \mathrm{SOC}(t_0) + \frac{1}{C_n}\int_{t_0}^{t} \eta I(\tau)\,d\tau \tag{4.3}$$

In Eq. (4.3), $\mathrm{SOC}(t_0)$ is the initial state of charge, C_n is the rated capacity, I is the current (negative for charging, positive for discharging), and η is the coulombic efficiency (charge-discharge efficiency). This method requires a high-precision current sensor and accurate calibration of the initial SOC.

As one of the most widely employed methods for SOC estimation in battery management systems, the ampere-hour integration method derives both its value and limitations from its distinctive operational principle. Its core advantage lies in the simplicity and efficiency of its methodology: by directly integrating current over time to quantify net charge variation, it offers clear physical interpretation, straightforward mathematical modeling, and ease of implementation with stable operation in embedded systems. This "accounting-style" charge tracking mechanism enables seamless adaptation to the real-world dynamic operating conditions of batteries, whether during acceleration and braking in electric vehicles or frequent charge/discharge cycles in electronic devices, providing continuous, real-time SOC trajectories with rapid response, thereby delivering critical insights into power consumption for users.

However, this simplicity has directly led to its inherent, structural defects. The method is essentially an "open-loop" system, a fundamental characteristic that determines its three core challenges:

Initial value dependency syndrome: the entire computational framework is entirely built upon the foundation of the initial SOC value. If the initial value is biased, all subsequent calculation results will systematically deviate from the true values.

Irreversible accumulation of errors: the inherent minor zero-point drift, sampling noise, and imprecise Coulomb efficiency settings of high-precision current sensors are continuously amplified during integration. These errors accumulate without reduction, akin to a clock that constantly runs fast, causing the SOC estimate to gradually diverge significantly from reality over time.

High cost and lack of self-healing: To mitigate error accumulation, high-precision, low-drift current sensors must be employed, directly increasing the system's hardware cost. Moreover, the method itself lacks any intrinsic mechanism for self-detection and correction.

For this reason, the coulomb counting method is by no means employed in isolation in practical applications. Its so-called "fatal shortcomings" precisely define its role as a "core algorithm" rather than the sole algorithm. To ensure long-term estimation reliability, it must be integrated with measurement techniques capable of providing absolute reference standards, forming a closed-loop correction system. The most classic and effective approach is periodic integration with the open-circuit voltage method. When the system detects that the battery has been in a prolonged idle state, it measures the open-circuit voltage and consults its fixed correlation with SOC, thereby obtaining a highly accurate "true" SOC value. This value is then used to forcibly reset or smoothly correct the coulomb counting result, effectively "zeroing out" accumulated errors in one step. This master-slave integration strategy, leveraging complementary strengths, represents the industry-standard solution for achieving high reliability and high-precision battery state management.

4.2.3 Internal resistance method

The internal resistance of a battery refers to the resistance within the battery that impedes the flow of current, and it is influenced by multiple factors such as the SOC, battery aging, and temperature. Generally, as the SOC decreases, the internal resistance gradually increases; furthermore, battery aging also leads to an increase in internal resistance. Therefore, by measuring the battery's internal resistance and establishing a relationship model between internal resistance and SOC, it becomes possible to estimate the SOC based on measured internal resistance values. Several methods exist for measuring battery internal resistance, including the AC impedance method and DC internal resistance method. The AC impedance method involves applying a small-amplitude AC voltage signal to the battery and measuring the resulting AC voltage and current responses at the terminals to calculate the internal resistance. In contrast, the DC internal resistance method determines internal resistance by measuring instantaneous voltage and current variations during charging or discharging processes.

The internal resistance method is an indirect estimation approach based on the correlation between battery internal parameters and SOC. Its core principle lies in the systematic and measurable variation of battery internal resistance as the SOC changes. This method estimates the SOC by leveraging the correlation between internal resistance and SOC, traditionally applied in specific scenarios such as lead-acid battery estimation. Its fundamental mathematical model can be expressed as:

$$\mathrm{SOC} = g(R_{\mathrm{int}}) \tag{4.4}$$

In Eq. (4.4), R_{int} represents the battery's overall internal resistance under specific conditions, and function g characterizes the mapping relationship between internal resistance and SOC. This function is typically calibrated using extensive experimental data and exists in the form of lookup tables or fitted curves. Under ideal circumstances, by precisely measuring the battery's internal resistance, the corresponding SOC value can be uniquely determined through this relationship.

To enhance estimation robustness, the internal resistance method is often combined with the OCV method, forming a hybrid estimation strategy. Its enhanced model can be expressed as:

$$SOC = h(U_{OC}, R_{int}) \tag{4.5}$$

In this Eq. (4.5), the open-circuit voltage is denoted U_{oc}. The function h becomes a bivariate function, simultaneously utilizing both the open-circuit voltage and internal resistance to jointly constrain the state of charge, thereby partially compensating for estimation bias arising from reliance on a single parameter.

As revealed by the analysis of Fig. 4.2, the internal resistance method for estimating SOC exhibits fundamental limitations. The image clearly demonstrates a nonlinear relationship between battery internal resistance and SOC, with pronounced non-monotonic behavior observed at room temperature: a single resistance value corresponds to multiple SOC points, thereby preventing the establishment of a unique mapping relationship.

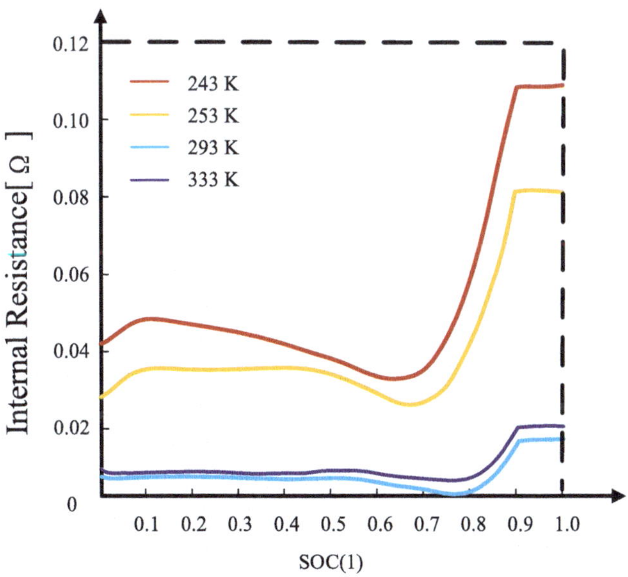

Figure 4.2 Relationship curves between battery internal resistance and SOC at different temperatures.

More critically, temperature emerges as a key interfering factor: at −10°C (263 K), the internal resistance increases significantly, and the curve morphology undergoes a complete transformation, exhibiting substantial deviation from the characteristics observed at room temperature. This high sensitivity to both temperature variations and aging renders fixed internal resistance, SOC models inherently prone to large estimation errors under varying temperature conditions. Empirical evidence indicates that while the internal resistance method can reflect changes in battery state, its performance is influenced by multiple coupled factors, making it unsuitable as a standalone approach for SOC estimation in lithium-ion batteries.

4.3 SOC estimation based on CKF and improved algorithms

Accurate SOC estimation is the fundamental basis for battery management systems to achieve functions such as energy management, state-of-health assessment, and cell balancing. As discussed in Section 4.2, conventional coulomb counting methods suffer from initial error accumulation, while OCV-based approaches are unable to provide online real-time estimation. Model-based methods, particularly Kalman filter family algorithms and their derivatives, have become the prevailing research direction for SOC estimation due to their ability to effectively suppress noise interference through recursive algorithmic principles and achieve optimal estimation. Among these, the Extended Kalman filter (EKF) was among the first applied; however, its reliance on first-order Taylor expansion for linearizing nonlinear systems introduces significant truncation errors when dealing with highly nonlinear battery systems, and it requires complex Jacobian matrix computations, resulting in limited stability and accuracy that still need improvement.

To overcome the inherent limitations of the EKF, the CKF, a nonlinear filtering method with superior theoretical properties, has been introduced into the field of battery SOC estimation. Building upon this foundation, various improved algorithms tailored for different application scenarios have emerged, significantly enhancing estimation accuracy, convergence speed, and robustness. This section systematically elaborates on the fundamental principles of CKF, with particular emphasis on reviewing and analyzing several representative improvement strategies, including adaptive noise adjustment, stability enhancement via singular value decomposition, and integration with other intelligent algorithms. Furthermore, by comparing relevant literature results, the specific performance improvements achieved by these algorithms in enhancing SOC estimation are demonstrated.

The core idea of the spherical-radial cubature Kalman filtering algorithm originates from the "spherical-radial" cubature rule, which accurately computes the mean and covariance of a nonlinear function passed through a Gaussian distribution using a set of cubature points with equal weights. Unlike the EKF, which relies on local linearization, the CKF employs numerical integration techniques and theoretically achieves approximation accuracy of third order or higher for nonlinear systems. This provides a natural advantage in characterizing the complex electrochemical behaviors of batteries under dynamic operating conditions.

The execution process of the CKF algorithm primarily consists of two key steps: time update (prediction) and measurement update (correction). During the time update stage, the algorithm predicts the current state vector and error covariance based on the previous time step's state estimate using the state equation. In the measurement update stage, the algorithm uses the difference between actual measured terminal voltage values and model-predicted values, referred to as the innovation sequence along with the Kalman gain to perform a weighted correction of the predicted state, thereby obtaining an optimal state estimate. The CKF employs a set of carefully designed cubature points (twice the number of state dimensions) to capture the probability distribution of state variables. After these points are propagated through a nonlinear system, their statistical properties are used to compute prediction means and covariances with high accuracy. This deterministic sampling strategy avoids Jacobian matrix computations required by EKF and eliminates the substantial computational burden associated with stochastic sampling in particle filtering, achieving a favorable balance between estimation accuracy and computational efficiency. This makes CKF particularly suitable for real-time applications in vehicle-mounted BMS. The flowchart for battery state estimation based on cubature Kalman filtering is illustrated in Fig. 4.3 below.

Although the standard CKF algorithm theoretically outperforms the EKF, its practical performance highly depends on predefined process and measurement noise covariance matrices. These noise statistical characteristics are not constant in real-world, complex battery operating environments. If the noise covariance matrices are fixed, a mismatch between actual and assumed noise statistics may lead to reduced filtering accuracy or even filter divergence. To address this issue, researchers have proposed

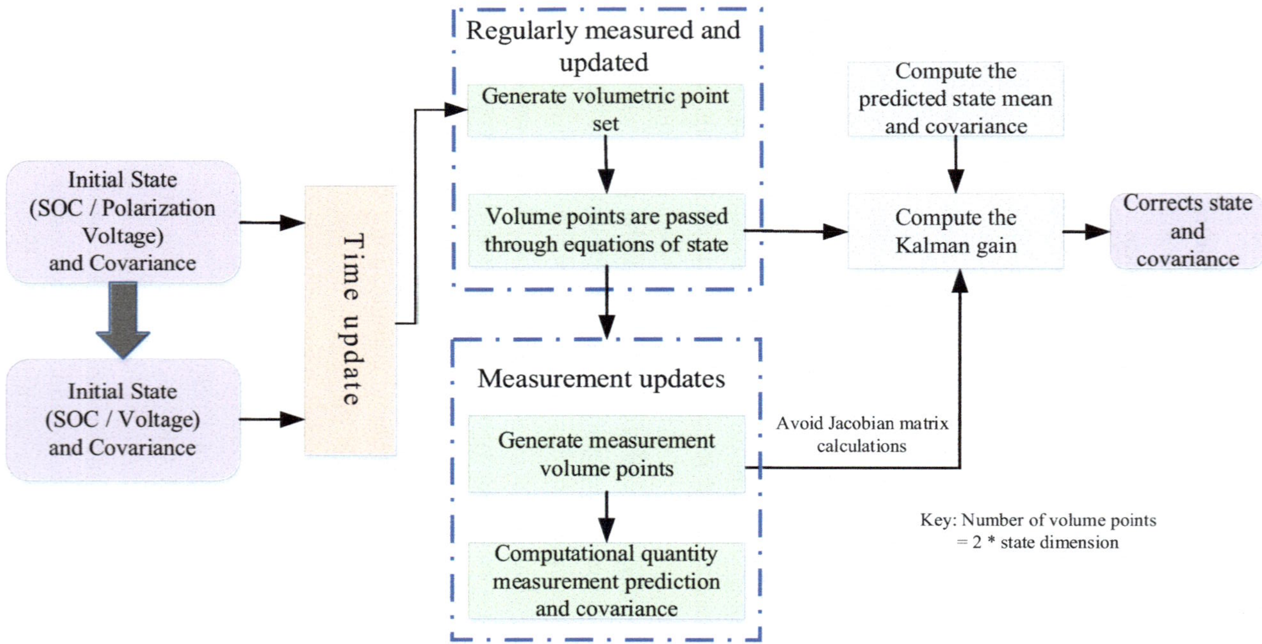

Figure 4.3 Flowchart of battery state estimation based on volume Kalman filtering.

various adaptive CKF algorithms that enhance algorithm robustness by updating noise statistical properties in real time. One study [1] targeting the time-varying nature of parameters in aged lithium-ion batteries, proposed an adaptive CKF algorithm that introduces a fading factor to gradually diminish the influence of historical data, thereby emphasizing recent observations. This enables rapid tracking of changes in battery model parameters and significantly improves the accuracy of SOC estimation under varying health states of aged batteries. Another study [2] presents a more comprehensive approach by integrating Variable forgetting factor recursive least squares (VFFRLS) with noise-adaptive CKF: VFFRLS online identifies battery model parameters, while the adaptive module dynamically adjusts the CKF's noise covariance matrix. This dual-adaptive mechanism ensures excellent estimation stability and accuracy under dynamic loading conditions and varying temperature environments. The flowchart of the VFFRLS-noise adaptive CKF algorithm is illustrated in Fig. 4.4 below.

To further enhance the numerical stability of the CKF algorithm, particularly under conditions involving non-positive definite or ill-conditioned covariance matrices, Singular value decomposition (SVD) has been incorporated into the CKF framework, resulting in the SVD-CKF algorithm. Standard CKF may lose symmetry and positive definiteness of the predicted error covariance matrix during recursive computation due to accumulated rounding errors, potentially leading to filter divergence. The SVD-CKF replaces the conventional Cholesky decomposition with SVD to compute the square root of the covariance matrix, leveraging SVD's greater robustness to ill-conditioned matrices and thereby ensuring numerical stability. Building upon this foundation, fuzzy logic control was further introduced [3], leading to a Fuzzy adaptive SVD-CKF (FASVD-CKF) algorithm. The innovation of this algorithm lies in its use of a fuzzy inference system to adaptively adjust the measurement noise covariance matrix based on the innovation sequence and its covariance. When abnormal fluctuations occur in the innovation sequence, the fuzzy logic system intelligently modulates the noise level, effectively suppressing interference from anomalous measurements, while SVD ensures robust numerical computation throughout the filtering process. Experimental results demonstrate that under complex operating conditions such as Dynamic stress testing (DST) and Beijing bus dynamic stress testing (BBDST), the FASVD-CKF algorithm achieves a convergence speed improvement of approximately 25%–40% compared to standard CKF, while maintaining high estimation accuracy, highlighting the significant potential of integrating intelligent control theory with

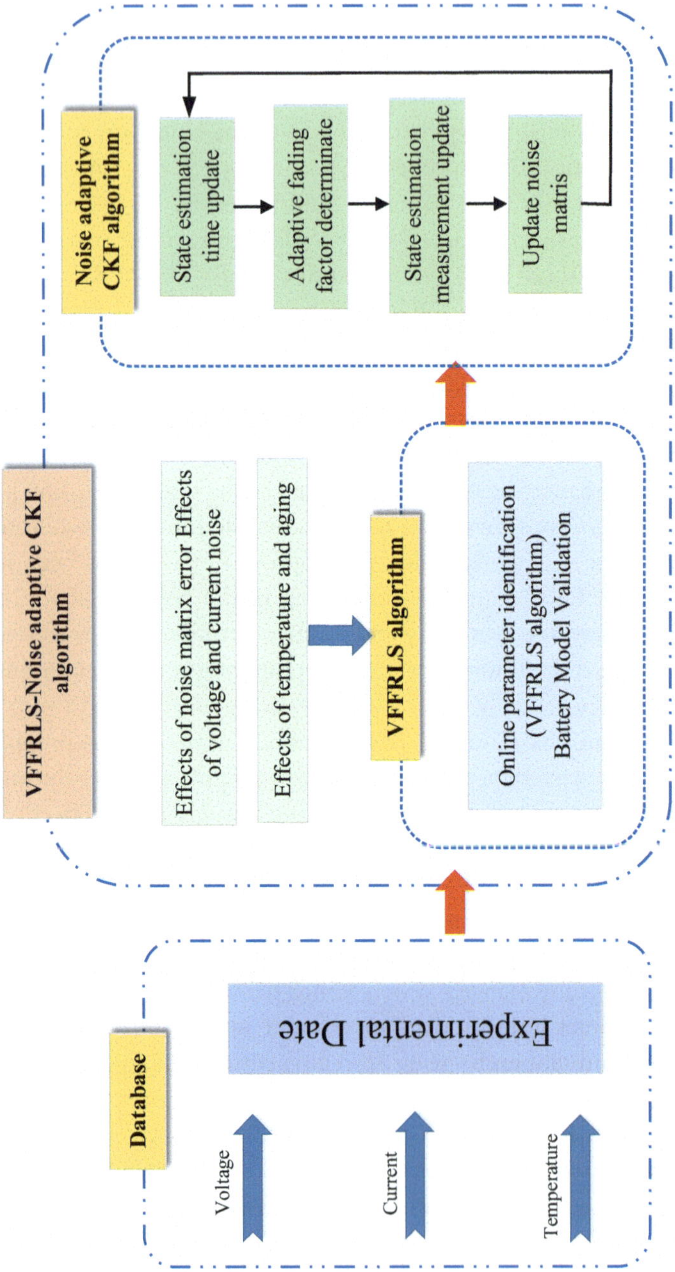

Figure 4.4 Flowchart of the VFFRLS-noise adaptive CKF algorithm.

advanced filtering algorithms. This integration enables the algorithm to maintain high estimation accuracy and reliability even in the presence of sudden sensor noise or outlier data points. The battery state estimation procedure based on fuzzy modeling and singular value decomposition within a Kalman filtering framework is illustrated in Fig. 4.5 below.

Another important direction for improvement involves integrating the CKF with other estimation or intelligent algorithms to leverage their respective strengths and compensate for their weaknesses, thereby forming a superior hybrid estimation strategy. Literature [4] has explored a joint estimation framework combining CKF with Particle filtering (PF) for simultaneous estimation of battery SOC and model parameters. In this framework, PF exploits its strong capability in handling non-Gaussian and nonlinear problems to perform online identification of battery model parameters—such as ohmic resistance, polarization resistance, and capacitance, fully accounting for the dynamic behavior of the battery. Subsequently, the CKF utilizes the accurately updated model parameters provided by PF in real time to rapidly and efficiently estimate system states, including SOC. This division of labor effectively capitalizes on the computational efficiency of CKF and the strong adaptability of PF in dealing with complex time-varying systems, achieving precise co-estimation of states and parameters.

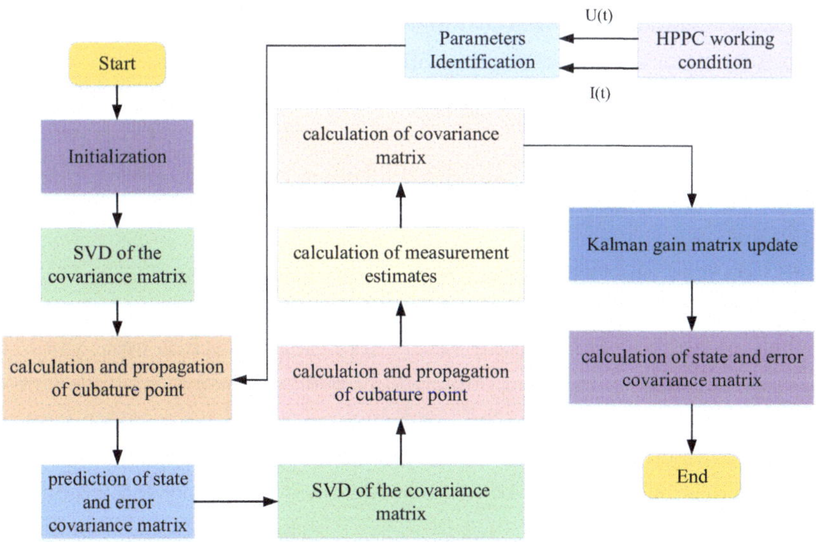

Figure 4.5 Flowchart of Kalman filter-based battery state estimation using fuzzy modeling and singular value decomposition.

In contrast, another study [5] proposes a novel approach by integrating an Extreme learning machine (ELM)—a single-hidden-layer feedforward neural network with exceptionally fast learning speed, into the CKF framework, resulting in the ELM-CKF algorithm. In this method, ELM is employed to establish a nonlinear black-box model between battery voltage and SOC, replacing traditional equivalent circuit models and thereby avoiding the complex process of model parameter identification. The CKF then performs SOC estimation based on the model constructed by ELM. This hybrid data-driven and model-driven approach is particularly suitable for complex battery systems that are difficult to model precisely. The combination of ELM's rapid nonlinear fitting capability with CKF's excellent filtering performance ensures high accuracy while significantly simplifying the estimation procedure.

In addition to the aforementioned strategies integrating filters and neural networks, research has also explored novel criteria based on information theory to further enhance the performance of the CKF under non-ideal noise conditions. Although existing studies [6] primarily focus on the EKF, their core idea: the Maximum correntropy criterion (MCC), provides a significant direction for addressing non-Gaussian noise. As a local similarity measure, MCC exhibits stronger robustness against outliers and non-Gaussian noise, particularly impulsive noise. By incorporating the MCC criterion into the CKF framework, constructing a cost function based on correntropy, and solving it accordingly, one can replace the conventional minimum mean square error criterion in traditional Kalman filter frameworks. This enables stable and accurate state estimation even when systems are severely contaminated by non-Gaussian noise. This approach offers a valuable improvement path for future applications of CKF in more stringent real-world engineering environments where noise characteristics are unknown.

In summary, the SOC estimation methods based on the CKF and their improved algorithms have effectively addressed the challenges encountered by the standard CKF in practical battery applications, such as time, varying noise, model inaccuracies, and numerical instability, through multiple technical approaches, including the introduction of adaptive mechanisms, enhanced numerical stability, and integration with other algorithms. These improved algorithms have, to varying degrees, enhanced the accuracy, convergence speed, and robustness of SOC estimation, enabling better adaptation to complex real-world operating conditions and battery degradation throughout their entire lifecycle. Nevertheless, it should be acknowledged that algorithmic complexity tends to increase alongside performance improvements. Therefore, in future research, optimizing computational efficiency while maintaining estimation performance, so as to facilitate implementation within embedded BMS with limited computational resources, remains a significant direction worthy of further investigation. Moreover, deeper integration of these enhanced strategies with data-driven approaches such as deep learning discussed in Section 4.4, may represent a key pathway toward achieving next-generation SOC estimation technologies characterized by higher intelligence and stronger adaptability.

4.4 SOC estimation based on CNN and improved algorithms

4.4.1 The compatibility of CNN's core theory with SOC estimation

The core theoretical advantage of Convolutional neural networks (CNNs) stems from three key mechanisms: local receptive fields, parameter sharing, and hierarchical feature extraction, which are highly compatible with the temporal data characteristics inherent in SOC estimation for batteries. SOC, as a state variable representing the remaining capacity of a battery, exhibits significant "locality" in its temporal dependencies with input features such as voltage, current, and temperature, specifically, the SOC value at any given moment primarily depends on the recent temporal changes in these features rather than random variations across the entire time sequence. The local receptive field mechanism of CNNs enables precise capture of such short-range temporal correlations through sliding convolution operations, thereby effectively mitigating interference from irrelevant time steps and noise in SOC estimation.

From the perspective of data structure adaptation, SOC estimation takes one-dimensional time-series data, such as voltage and current signals continuously sampled over time as input. Conventional spatial CNNs require structural modifications to accommodate such data: two-dimensional spatial convolution kernels are reduced to One-dimensional (1D) kernels, eliminating spatial dimension redundancy and enabling convolution operations to focus on temporal feature extraction. The essence of 1D convolution lies in performing local weighted summation over time-series data, where the weight matrices are automatically learned through training to capture temporal dependencies. This eliminates the need for manually designed feature extraction rules, thereby overcoming the limitations of traditional SOC estimation methods that rely heavily on manual feature engineering.

Parameter sharing is the key theoretical foundation enabling CNNs to handle large-scale data in SOC estimation. In conventional fully connected networks, each input feature corresponds to independent weights, causing the number of parameters to grow exponentially with feature dimensionality. In contrast, CNNs achieve significant reduction in parameter count by sharing weights across temporal dimensions through sliding convolutional kernels. This not only reduces model training complexity but also enhances generalization across diverse operating conditions, identical convolutional kernels can learn common temporal features under varying conditions, thereby mitigating model overfitting induced by operational variability.

The hierarchical feature extraction mechanism aligns with the requirement of "feature abstraction" in SOC estimation: shallow convolutional layers can extract low-dimensional temporal features such as voltage fluctuations and current rate of change; deep convolutional layers, through nonlinear combinations of shallow features, generate high-dimensional abstract features that more closely approximate the complex mapping relationship between SOC and input variables. For instance, deep convolution can abstract the low-dimensional pattern of "slow voltage rise + stable current" into a high-dimensional pattern representing the "constant-current charging phase," a mode whose correspondence with SOC is more explicit, thereby enhancing estimation accuracy.

4.4.2 The estimated theoretical implementation process of SOC by CNN

The input data for SOC estimation must undergo preprocessing to align with the theoretical characteristics of CNNs, with the core logic centered on "data consistency," "temporal integrity," and "feature validity." First, the theoretical basis for outlier removal lies in the "continuity assumption" of SOC time-series data, under normal operating conditions, signals such as battery voltage and current vary smoothly over time. Outliers disrupt this continuity, leading CNNs' local convolutions to capture erroneous temporal correlations. Therefore, outliers deviating from the data distribution must be identified and removed using statistical criteria to ensure the authenticity of temporal features.

The theoretical necessity of data normalization stems from the gradient descent optimization mechanism in CNNs. Differences in the scales of input features lead to significant disparities in their numerical ranges. If fed directly into the model, features with larger values will dominate the update of convolutional kernel weights, causing critical features with smaller magnitudes, such as voltage change rates to be overlooked. By applying min-max normalization to map all features into the [0,1] interval, each feature is ensured equal influence during gradient updates, enabling the CNN to learn the mapping relationship between multiple features and SOC in a balanced manner and eliminating scale-related biases that could otherwise interfere with model training.

Temporal data reconstruction is the core step for adapting sequential data to the two-dimensional input format of CNNs, with the theoretical key lying in the rational partitioning of "time windows." The selection of time windows involves balancing two conflicting requirements: first, temporal information integrity windows that are too small result in fragmented local features and prevent the capture of complete short-term temporal correlations; second, computational windows that are too large increase the dimensionality of input data, leading to a dramatic rise in computational load during convolution operations. Theoretically, window length should match the "characteristic period" of SOC variation. For instance, during battery charging and discharging processes, each cycle exhibiting a significant voltage fluctuation can serve as a basis for window segmentation, ensuring that each window contains a complete set of short-term temporal features and thereby providing effective local inputs for CNN-based convolutional operations.

The design of the CNN architecture must follow a hierarchical logic of "feature extraction, feature compression, and feature mapping," with each layer's functional design aligned with the theoretical requirements of SOC estimation. The architectural diagram of the convolutional neural network is shown in Fig. 4.6.

The core function of the convolutional layer is to transform raw temporal features into high-dimensional representations. The theoretical design hinges critically on the "receptive field size" of the convolutional kernel, this receptive field must be aligned with the temporal scale associated with SOC. If the receptive field is too small, only features from a single time step can be captured, failing to reflect temporal trends; if it is too large, irrelevant time steps containing noise may be inadvertently included. Theoretically, the receptive field of the convolutional kernel should cover the "sensitive temporal interval" of SOC variation, the N preceding time steps most relevant to a given SOC value at a specific moment, ensuring that the

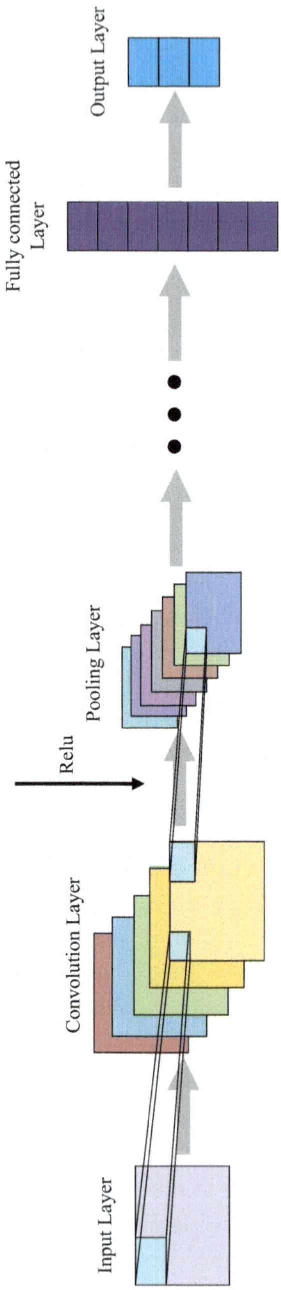

Figure 4.6 Schematic diagram of the convolutional neural network architecture.

convolution operation effectively extracts short-range temporal correlation features.

The theoretical value of pooling layers lies in "feature compression and enhanced generalization capability." Feature maps output by convolutional layers contain substantial redundant information; pooling layers reduce data dimensionality while preserving critical features through down sampling techniques such as max pooling and average pooling. From the theoretical perspective of SOC estimation, max pooling is more suitable for extracting "peak information" from temporal features, as such information exhibits a more pronounced correlation with SOC; in contrast, average pooling is better suited for smoothing features, thereby mitigating the impact of transient noise on feature representation.

The design of the fully connected layer and output layer aims to achieve "mapping from high-dimensional features to SOC values." The theoretical function of the fully connected layer is to transform the high-dimensional feature vector output from the pooling layer into a one-dimensional vector, and to fit the complex nonlinear relationship between features and SOC through nonlinear activation functions, since the mapping between SOC and input features is inherently nonlinear, nonlinear activation functions can effectively model this intricate relationship. The theoretical basis for employing a linear activation function in the output layer lies in the fact that SOC is a continuous value; a linear activation function directly outputs estimated values that match the actual range of SOC, thereby avoiding output range shifts caused by nonlinear activation functions.

Moreover, the introduction of regularization mechanisms is theoretically essential. CNNs are prone to overfitting during training due to an excessive number of parameters or limited training data, resulting in the model overly learning condition-specific features from the training set and failing to generalize to unseen operating conditions. Dropout enforces the model to learn more robust features by randomly deactivating a subset of neurons, thereby reducing reliance on local training data and enhancing the generalization capability of SOC estimation under complex operating conditions.

4.4.3 Theoretical limitations of basic CNNs for SOC estimation

Although the foundational CNN possesses the aforementioned theoretical advantages, it still faces three major theoretical limitations in SOC estimation under complex operating conditions: first, insufficient capture of long-term temporal correlations. The local convolutional mechanism of CNNs can only focus on short-time window temporal features, failing to capture temporal correlations exceeding the window length. Since battery SOC is a cumulative state variable, long-term temporal correlations are critical for estimation accuracy, particularly under dynamic operating conditions; the absence of temporal information beyond the window leads to significant estimation bias.

Second, equalization of key feature weights. The baseline CNN assigns equal weights to all input features and time steps, failing to account for the "information value differences" among various features and time steps in SOC estimation. For instance, the rate of voltage change carries significantly higher information value than the rate of current change within the SOC-sensitive range (e.g., SOC <20%), yet the baseline CNN cannot adaptively enhance such critical features. As a result, irrelevant or low-value features interfere with SOC estimation, particularly under extreme operating conditions where the differences in feature information value become more pronounced, leading to a noticeable decline in estimation accuracy.

Third, gradient propagation decay in deep networks. As the number of layers in a CNN increases, gradients during backpropagation are prone to "vanishing gradients", shallow layer gradients tend toward zero after multiple layers of transmission, preventing shallow convolutional kernels from effectively learning low-dimensional features. Since low-dimensional features form the foundation of SOC estimation, insufficient shallow feature extraction directly compromises the quality of deep high-dimensional features, ultimately leading to difficulties in model convergence and an inability to improve estimation accuracy.

4.4.4 Theory of SOC estimation for CNN hybrid network based on improved algorithm

In response to the theoretical limitations of conventional CNNs, hybrid networks must be constructed through module integration and mechanism optimization. The core improvement strategies involve "enhancing temporal correlation capture," "achieving adaptive focusing on critical features," and "optimizing gradient propagation." Theoretical analysis is subsequently conducted on CNN-BiLSTM-Attention hybrid networks and CNN hybrid networks incorporating optimized algorithms.

4.4.4.1 Theoretical construction of the CNN-BiLSTM-attention hybrid network

The core theoretical value of the CNN-BiLSTM-Attention hybrid network lies in multi-mechanism synergy. By integrating CNN's local feature extraction, BiLSTM's bidirectional temporal modeling, and Attention's focus on salient features, it establishes a three-tiered feature processing framework, "local, global, and salient", enabling precise mitigation of the inherent theoretical limitations of conventional CNNs. From the fundamental requirements of SOC estimation, SOC dynamics rely simultaneously on "local short-term features" and "global long-term temporal patterns," with significant disparities in information value across different features and time steps. A single CNN cannot concurrently satisfy these three demands; however, multi-module fusion leverages complementary strengths among mechanisms to achieve more comprehensive feature modeling.

Specifically, the fusion of CNN and BiLSTM exhibits clear theoretical compatibility: the local high-dimensional features extracted by CNN can serve as input temporal sequences for BiLSTM, thereby avoiding noise interference that would otherwise occur if BiLSTM directly processed raw temporal data. Transient noise present in the original data may lead BiLSTM to overlearn irrelevant temporal correlations. In contrast, the hierarchical feature extraction performed by CNN effectively denoises and abstracts the original data, providing BiLSTM with high-quality temporal feature inputs and ensuring that BiLSTM can focus on learning long-term temporal dependencies.

The integration of BiLSTM with Attention addresses the issue of "information redundancy" in sequential modeling: while BiLSTM can capture bidirectional temporal dependencies, it assigns equal weights to temporal features across all time steps, failing to distinguish between critical and non-critical time steps. In contrast, the Attention mechanism can dynamically allocate weights to emphasize contributions from key time steps and salient feature channels, thereby achieving "amplification of useful information and suppression of irrelevant information." This process is entirely data-driven and automatically accomplished without manual intervention, aligning well with the dynamic nature of feature information value in SOC estimation. The architectural diagram of the CNN-BiLSTM-Attention neural network is illustrated in Fig. 4.7.

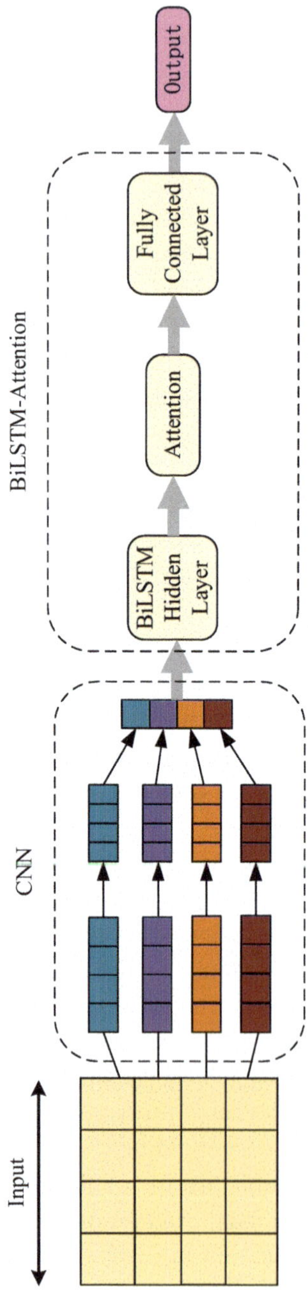

Figure 4.7 Structure diagram of the CNN-BiLSTM-Attention neural network.

4.4.4.2 Theoretical advantages and roles of each module

CNN assumes the role of "basic feature preprocessing" in the hybrid network, with its theoretical functions manifesting in two aspects: first, noise filtering, through the synergistic operation of convolution and pooling, transient noise in the original time-series data is eliminated. The local weighted summation performed by convolution smooths out noise, while pooling further preserves feature trends, preventing noise from propagating into subsequent time series modeling modules; second, feature abstraction, transforming low-dimensional raw features such as voltage and current into high-dimensional abstract features. These abstract features more closely reflect the intrinsic mapping relationship between SOC and input variables, thereby reducing the complexity of temporal modeling for BiLSTM. Consequently, BiLSTM is relieved from learning fundamental features directly from raw data and can instead focus exclusively on capturing long-term temporal dependencies.

Theoretically, the quality of features extracted by CNN directly determines the estimation accuracy of the hybrid network: if CNN fails to effectively extract key low-dimensional features, the long-term temporal dependencies learned by BiLSTM will be based on erroneous foundational features, leading to biased SOC estimation; conversely, high-quality CNN features enable BiLSTM to perform temporal modeling more efficiently and rapidly converge to optimal temporal correlation patterns.

The core theoretical advantage of BiLSTM lies in its bidirectional temporal information fusion. Its architecture consists of two parallel components: a forward LSTM and a backward LSTM. The forward LSTM learns temporal dependencies in chronological order, capturing the influence of "historical features on the current SOC"; the backward LSTM learns temporal dependencies in reverse chronological order, capturing the corrective effect of "future features on the current SOC." This bidirectional modeling mechanism addresses the "temporal information bias" inherent in conventional CNNs and unidirectional LSTMs, where conventional CNNs only capture short-range local temporal patterns, and unidirectional LSTMs only model forward long-term temporal dependencies. In contrast, BiLSTM simultaneously encompasses both short- and long-term, forward- and backward-directed temporal information, more accurately reflecting the sequential characteristics of SOC as a cumulative state variable.

From the theoretical perspective of LSTM's gating mechanism, BiLSTM achieves "selective memory and forgetting" of temporal information through the collaborative operation of input, forget, and output gates: for temporal features strongly correlated with SOC, the forget gate remains open, enabling long-term retention of such information; for irrelevant temporal features, the forget gate closes, thereby preventing interference. This selective mechanism enables BiLSTM to effectively preserve critical information and suppress redundant data during long-term temporal modeling, further enhancing the accuracy of capturing temporal dependencies.

The Attention mechanism plays a "key feature adaptive focusing" role in hybrid networks, with its theoretical foundation rooted in a weight allocation model. By computing "importance scores" for features, it assigns differentiated weights to distinct feature channels and temporal steps, thereby enhancing critical features. In response to the requirements of SOC estimation, the Attention mechanism must be developed from dual perspectives, both "channel dimension" and "temporal dimension"—necessitating the construction of a dual-attention mechanism spanning "temporal-channel" dimensions.

From the theoretical perspective of channel attention, the strength of association between different input features and SOC varies significantly: the rate of voltage change exhibits much stronger correlation with SOC in the sensitive SOC range compared to temperature, whereas temperature demonstrates a markedly increased correlation under extreme operating conditions. Channel attention aggregates global information across feature channels derived from BiLSTM outputs, extracts global statistical characteristics for each channel, and then employs a fully connected layer combined with nonlinear activation to learn dependencies among channels. Ultimately, it generates channel-wise weights, channels with strong associations receive weights approaching 1, while those with weak associations receive weights approaching 0, thereby achieving focused identification of key features along the channel dimension.

From the theoretical perspective of temporal attention, the contributions of temporal features at different time steps to the current SOC estimation vary: temporal features within SOC-sensitive intervals contribute significantly more than those within stable intervals. Temporal attention computes scores for the temporal sequence output by BiLSTM, then normalizes these scores into time-step weights via Softmax activation. This process assigns substantially higher weights to critical time steps compared to non-critical ones, ensuring that the model prioritizes high-contribution temporal information during SOC estimation while minimizing interference from less relevant time steps.

From a mathematical perspective, the computation of Attention weights essentially constitutes a "measure of feature similarity", the weight magnitude is determined by calculating the similarity between the current feature and the target feature of the SOC, where higher similarity results in larger weights. This process is entirely data-driven and adaptively captures variations in feature importance across different operating conditions, thereby eliminating the subjectivity and limitations inherent in manually predefined weights.

4.4.4.3 Theoretical framework of CNN hybrid networks with optimization algorithms

The core theoretical value of residual connections lies in addressing the vanishing gradient problem. Their design logic is based on the concept of "residual mapping": the output of a shallow network is directly passed to a deeper network via a "shortcut path," thereby shifting the learning objective of the deep network from learning a complete feature mapping to learning only the residual mapping. This transformation significantly reduces the difficulty of training deep networks while ensuring that gradients can effectively propagate backward to shallow layers.

From the mathematical theory of gradient propagation, in the gradient transmission process of a basic CNN, gradients must be propagated through a product of multiple convolutional layers and activation functions. When the network has many layers, this product tends to approach zero. In contrast, residual connections enable gradients from shallow layers to be directly transmitted to deep layers via shortcut paths, bypassing multiple convolutional operations. This ensures that gradients in shallow layers remain sufficiently large, allowing shallow convolutional kernels to effectively learn low-dimensional features (such as voltage fluctuations and current trends), thereby establishing a solid foundation for deep feature extraction.

In the Res–CNN–BiLSTM hybrid network, the theoretical role of residual connections manifests on two levels: first, optimizing feature extraction in the CNN, by introducing residual connections between convolutional layers of the CNN, shallow low-dimensional features are effectively learned, preventing insufficient feature extraction in shallow layers due to vanishing gradients; second, enhancing feature propagation from the CNN to the BiLSTM, residual connections directly transmit shallow features from the CNN to the input of the BiLSTM, supplementing the temporal feature input of the BiLSTM. This enables the BiLSTM not only to learn deep high-dimensional features from the CNN but also to leverage shallow low-dimensional features, thereby achieving more comprehensive modeling of temporal dependencies.

From the perspective of theoretical requirements for SOC estimation, the introduction of residual connections can further enhance the model's adaptability to extreme operating conditions. Under extreme conditions, battery voltage and current characteristics exhibit more complex variations, necessitating deeper network architectures to capture such intricate mappings. However, standard CNNs are limited in constructing deep networks due to the vanishing gradient problem. Residual connections overcome this limitation by enabling the construction of deeper CNN structures. Through multiple hierarchical levels of feature extraction, these deeper networks can learn complex feature patterns under extreme conditions, providing BiLSTM with more accurate temporal feature inputs and ultimately improving SOC estimation accuracy under extreme operating conditions.

4.4.4.4 CNN-transformer with adaptive learning rate optimization

The theoretical advantage of the CNN-Transformer hybrid network lies in its capability to capture long-range temporal dependencies, while adaptive learning rate optimization addresses the issue of "parameter update efficiency" during model training. By dynamically adjusting the learning rate, it ensures rapid convergence to optimal parameters and effectively suppresses overfitting. The architecture of the CNN-Transformer neural network is illustrated in Fig. 4.8.

From the theoretical perspective of the Transformer, the model leverages multi-head attention mechanisms to capture long-range temporal dependencies across time steps. Its attention computation is independent of temporal windows, enabling direct learning of feature dependencies across the entire temporal range. This advantage overcomes the modeling limitations of BiLSTM in ultra-long sequences, BiLSTM relies on sequential propagation of time steps for temporal modeling, leading to a sharp increase in computational complexity when handling ultra-long sequences. In contrast, the parallel attention computation in Transformers efficiently processes ultra-long sequences, providing more comprehensive long-term temporal information for SOC estimation.

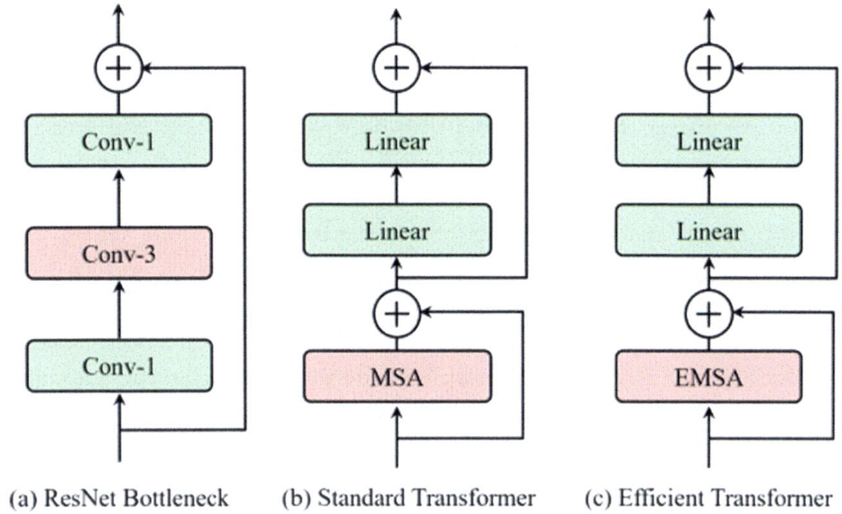

Figure 4.8 Schematic diagram of the CNN-BiLSTM-Attention neural network architecture.

In the integration of CNN and Transformer, the theoretical role of CNN is "feature preprocessing": through hierarchical feature extraction, CNN transforms raw sequential data into high-dimensional abstract features, thereby avoiding noise interference and excessively high computational complexity when the Transformer directly processes raw data. Raw data typically possess high dimensionality; direct input into a Transformer leads to exponential growth in attention computation cost. In contrast, CNN-based feature compression significantly reduces input dimensionality, ensuring that the Transformer can efficiently learn long-range temporal dependencies.

The theoretical foundation of Adaptive learning rate optimization (AdamW) lies in dynamic parameter updates. Traditional fixed learning rate algorithms (e.g., SGD) apply a uniform learning rate to all parameters, resulting in either excessively rapid or slow updates for sensitive parameters, excessively large learning rates cause parameter oscillations and hinder convergence, while excessively small rates lead to slow convergence. AdamW addresses this issue through two mechanisms: first, the momentum term (first-moment estimation), which accumulates historical parameter update directions to accelerate convergence; second, adaptive scaling (second-moment estimation), which dynamically adjusts the learning rate based on the magnitude of gradients, parameters with large gradients are assigned smaller learning rates to prevent oscillations, while those with small gradients receive larger learning rates to expedite updates. Furthermore, AdamW incorporates weight decay (L2 regularization), imposing penalization on parameters to prevent them from growing too large and thereby mitigating model overfitting. This mechanism is particularly crucial for large-scale networks such as CNN-Transformer architectures, ensuring that the model maintains strong generalization ability while effectively capturing long-range temporal dependencies during training.

From the perspective of theoretical demand estimation for SOC, the introduction of adaptive learning rate optimization enhances training stability under complex operating conditions: due to variations in the mapping relationship between SOC and input features across different operating conditions, models require dynamic parameter adjustment to adapt to these differences. The adaptive learning rate of AdamW can rapidly respond to such changes in mapping relationships, ensuring that parameter updates consistently occur at an optimal pace, thereby preventing difficulties in model convergence or overfitting caused by operational variability.

4.4.5 Theoretical comparison and applicable scenarios of improved hybrid networks

The theoretical advantages of different improved hybrid networks vary significantly, and their applicable scenarios require analysis based on the operating condition characteristics and temporal requirements of SOC estimation:

From the perspective of temporal correlation scale, CNN-BiLSTM-Attention is more suitable for medium-to-short-term SOC estimation (e. g., real-time SOC estimation during vehicle operation, where the temporal scale typically ranges from minutes to hours): the bidirectional temporal modeling capability of BiLSTM efficiently captures medium-to-short-term temporal correlations, while the attention mechanism's focus on key features enables rapid response to feature variations in SOC-sensitive regions, thereby ensuring high accuracy in real-time estimation. In contrast, CNN-Transformer is better suited for ultra-long-term SOC estimation (e.g., long-term SOC monitoring in energy storage systems, with temporal scales spanning several days or even weeks): the multi-head attention mechanism of Transformer efficiently handles ultra-long temporal sequences without relying on fixed time windows, enabling the capture of long-range dependencies across the entire temporal range. Additionally, adaptive learning rate optimization ensures stable convergence during prolonged training periods.

From the perspective of operational condition complexity, Res–CNN–BiLSTM is more suitable for extreme operating conditions: residual connections enable the construction of deeper CNN architectures, facilitating the extraction of complex feature patterns under extreme conditions through multiple hierarchical layers, while simultaneously alleviating the vanishing gradient problem and ensuring robust feature extraction capability in deep networks. In contrast, CNN-BiLSTM-Attention is better suited for dynamic operating conditions: bidirectional temporal modeling allows rapid adaptation to temporal feature fluctuations caused by operational changes, while the attention mechanism dynamically adjusts the weights of critical features in real time, thereby maintaining high accuracy in SOC estimation during operational transitions.

From the perspective of feature dependency, all improved hybrid networks outperform the baseline CNN. The fundamental reason lies in the fact that the baseline CNN can only learn local, short-range feature dependencies, whereas the improved networks, through multi-module fusion, are capable of simultaneously capturing multi-scale feature dependencies across "local/global," "forward/backward," and "short-range/long-range" dimensions. Furthermore, they can adaptively adjust the weights of critical features. This capability closely aligns with the multidimensional feature dependency characteristics of SOC as a cumulative state variable: SOC variation depends not only on local short-range features but also on global long-range temporal patterns, and the strength of dependency among different features and time steps dynamically varies with operating conditions. A single baseline CNN cannot meet such multidimensional and dynamic feature dependency requirements. In contrast, the improved hybrid networks achieve more comprehensive and precise feature modeling through optimized theoretical mechanisms.

This section theoretically analyzes the mechanisms and advantages of CNN and improved hybrid networks for SOC estimation: the base CNN achieves adaptive extraction of SOC temporal features through local receptive fields, parameter sharing, and hierarchical feature extraction. However, inherent theoretical limitations, including insufficient capture of long-term temporal dependencies, equal weighting of critical features, and gradient vanishing, restrict estimation accuracy under complex operating conditions. The improved hybrid networks effectively address these limitations: the CNN-BiLSTM-Attention architecture enables comprehensive multi-scale feature modeling through synergistic local feature extraction, bidirectional temporal modeling, and focused attention on key features; the Res–CNN–BiLSTM enhances deep feature extraction capability by optimizing gradient propagation via residual connections; the CNN-Transformer achieves long-range temporal dependency modeling and efficient training through multi-head attention mechanisms and adaptive learning rates.

From a theoretical perspective, the core logic of all improved algorithms revolves around achieving a better alignment with the state characteristics of SOC: as a cumulative state variable representing the remaining capacity of a battery, SOC exhibits three fundamental properties: temporal dependency, sensitivity, and complexity. No single algorithm can simultaneously satisfy the modeling requirements for all three characteristics. In contrast, hybrid networks based on multi-module fusion leverage complementary strengths among different mechanisms to achieve more accurate characterization of SOC features. This represents the fundamental theoretical advantage of improved networks over baseline CNNs.

Future theoretical research on CNN-based SOC estimation may be conducted along three directions. First, multimodal feature fusion theory, current networks primarily rely on electrical features such as voltage, current, and temperature; future work may incorporate physical features such as battery impedance and cycle count to construct a hybrid CNN network based on multimodal feature fusion, and investigate weighting allocation mechanisms among different modalities to further enhance the comprehensiveness of SOC estimation. Second, lightweight network theory, existing improved networks exhibit high theoretical complexity; thus, research is needed to simplify network architecture through methods such as pruning and quantization while preserving theoretical advantages, thereby adapting to the computational resource constraints of embedded BMS. The core challenge lies in balancing "theoretical accuracy" against "computational complexity." Third, cross-battery consistency theory, current modeling relies solely on data from individual batteries; future studies should explore transfer learning and related methodologies to achieve generalized SOC estimation across different battery types and aging levels, addressing model adaptation issues caused by variations in battery characteristics. This would provide a theoretical foundation for large-scale battery pack SOC estimation.

References

[1] T. Zheng, L. Zhang, Y. Hou, et al., State of charge estimation of aged lithium-ion batteries based on adaptive CKF, Energy Storage Sci Technol 9 (4) (2020) 1193.
[2] L. Wang, J. Han, L. Chang, et al., Lithium-ion battery state of charge estimation based on variable forgetting factor recursive least squares–noise adaptive cubature Kalman filter algorithm, Ind Eng Chem Res 61 (22) (2022) 7489–7503.
[3] X. Yang, S. Wang, X. Wenhua, et al., A fuzzy adaptive singular value decomposition cubature Kalman filter algorithm for lithium-ion battery state of charge estimation, Int J Circ Theor Appl 50 (2) (2022) 614–632.
[4] W. Xu, Y. Xiaofeng, Joint state of charge and parameter estimation of lithium-ion batteries using cubature Kalman filter and particle filter, J Power Electron 20 (1) (2020) 292–307.
[5] L. Dejun, S. Liu, D. Feng, Lithium battery state of charge estimation based on ELM-CKF algorithm, in: Proceedings of the 6th international conference on electronic information technology and computer engineering, 2022, pp. 596–599.
[6] C. Wu, W. Hu, M. Jinhao, et al., State of charge estimation of lithium-ion batteries based on maximum correntropy extended Kalman filter algorithm, Trans China Electrotech Soc 36 (24) (2021) 5165–5175.

CHAPTER FIVE

State of health (SOH) estimation

5.1 Definition and significance

In lithium ion Battery management systems (BMS), the State of health (SOH) is not defined as a single dimensional metric but rather as a multidimensional and comprehensive concept. Traditionally, SOH has often been assessed solely based on capacity retention, the ratio of the current usable capacity to the initial rated capacity. However, with increasing complexity in battery application scenarios and the diverse mechanisms of battery aging, this single capacity oriented definition of SOH can no longer fully reflect the actual health status of batteries. Consequently, the measurement framework for SOH has gradually evolved into a multidimensional indicator system encompassing capacity, power capability, internal resistance, safety performance, and thermal characteristics [1].

5.1.1 Capacity-related SOH

Capacity is one of the most direct indicators for assessing battery health. Capacity degradation primarily results from the loss of active lithium, degradation of electrode material structures, and by-products generated through side reactions. A commonly used formula for defining SOH in terms of capacity is

$$\mathrm{SOH}_C = \frac{C_n}{C_0} \times 100\% \tag{5.1}$$

In Eq. (5.1), C_n represents the discharge capacity measured during the nth cycle, and C_0 denotes the initial rated capacity of the battery. SOH based on capacity is commonly used in energy storage and electric vehicle applications; however, a limitation is that the battery's power performance and safety may already have deteriorated even before significant capacity degradation occurs.

Modelling, Estimation and AI Applications for Lithium-Ion Battery Management Systems
ISBN: 978-0-443-45347-2
https://doi.org/10.1016/B978-0-443-45347-2.00005-4

5.1.2 Power-related SOH

A core performance metric of batteries in application is their power output capability, i.e., the ability to deliver high current over short durations. Power degradation primarily results from increased internal resistance and enhanced electrode polarization. The general formula for measuring power SOH is typically expressed as

$$SOH_P = \left(1 - \frac{P_n}{P_0}\right) \times 100\% \tag{5.2}$$

In Eq. (5.2), P_n denotes the maximum output power under the current conditions, serving as the initial maximum power. Power SOH is more suitable for the requirements of electric vehicles during rapid acceleration conditions, reflecting the battery's capability to meet scenarios requiring high power output.

5.1.3 Impedance/resistance-based SOH

Increased internal resistance is another critical manifestation of battery degradation. As the number of charge/discharge cycles increases, the Solid electrolyte interphase (SEI) layer continuously thickens, and the conductive network may be disrupted, leading to rising ohmic and polarization internal resistances. Internal resistance-based SOH can be defined as

$$SOH_R = \frac{R_{EOL} - R_n}{R_{EOL} - R_0} \times 100\% \tag{5.3}$$

In Eq. (5.3), R_{EOL} denotes the internal resistance at the end of battery life, R_n represents the equivalent internal resistance at the n th cycle, and R_0 refers to the initial internal resistance. Since impedance increase often precedes capacity degradation, this metric exhibits high sensitivity during the early stages of battery life.

5.1.4 Safety-related SOH dimension

Battery safety constitutes an indispensable aspect of SOH assessment. Safety degradation is manifested in reduced tolerance to overcharging and overdischarging, diminished thermal stability, increased gas evolution, and elevated risk of internal short circuits. Although safety cannot be directly characterized by a single quantitative metric, it can be comprehensively evaluated through combined analysis of parameters such as the thermal runaway onset temperature, gas generation rate, or rate of pressure change. This dimension is particularly critical in electric vehicles and large scale energy storage applications.

5.1.5 Thermal-related SOH

The thermal management performance of batteries similarly affects their lifespan and reliability. During aging, increased thermal resistance and reduced heat dissipation efficiency may lead to thermal accumulation and localized overheating. Evaluation methods for SOH can be established from a thermal perspective using indicators such as thermal conductivity, rate of temperature rise, or uniformity of temperature distribution.

5.1.6 Integrated multidimensional SOH measurement system

In practical engineering applications, a single dimensional SOH indicator often fails to meet requirements. To achieve a more comprehensive description of battery health, it is typically necessary to fuse multiple indicators across various dimensions. For example, a composite SOH can be constructed using a weighted approach:

$$\mathrm{SOH}_{\mathrm{total}} = \omega_1 \mathrm{SOH}_C + \omega_2 \mathrm{SOH}_P + \omega_3 \mathrm{SOH}_R + \cdots \tag{5.4}$$

In Eq. (5.4), ω_i denotes the weight coefficients for each dimension need to be adjusted according to the specific application scenario.

SOH is a critical indicator for assessing the degree of performance degradation in lithium ion batteries. Its significance extends beyond merely reflecting changes in battery health, such as capacity fade and increased internal resistance, thereby ensuring operational safety. It also provides essential data for life prediction and Remaining useful life (RUL) estimation, supporting predictive maintenance and reliability management. Furthermore, SOH serves as a key input parameter for BMS, directly influencing charge/discharge control and balancing strategies. In engineering applications, accurate SOH assessment enables informed decision making regarding battery replacement, secondary utilization, and lifecycle economic analysis, while also laying the foundation for intelligent energy management and digital twin modeling of batteries. Consequently, SOH holds a central position in both academic research and industrial applications.

5.2 Empirical models from aging data

Empirical models based on aging data are among the most commonly used approaches for SOH and RUL prediction of lithium ion batteries. Unlike mechanistic models, which attempt rigorous derivation from electrochemical and material mechanisms, empirical models primarily rely on experimental data, employing statistical fitting of degradation characteristics, such as capacity, internal resistance, and power, under specific operating conditions to establish mathematical descriptions of degradation patterns. These models offer advantages of low computational complexity and ease of realtime implementation within battery management systems, thereby demonstrating broad practical value in engineering applications. Based on current mainstream empirical models, the following classifications exist.

5.2.1 Capacity degradation model

Capacity retention rate is the most direct indicator for assessing battery aging. Capacity fade models characterize the SOH by describing the variation pattern of a battery's available capacity over time or number of cycles. Common models include the following:

Linear model

$$Q(t) = Q_0 - k \tag{5.5}$$

In Eq. (5.5), Q(t) denotes the available capacity at the current time, Q_0 represents the initial capacity, t stands for the number of cycles or time elapsed, and k is the linear degradation rate constant. This model can reflect the approximately linear degradation trend of batteries during the early to middle stages of their life and is commonly used in engineering applications for battery life prediction.

Exponential model

$$Q(t) = Q_0 \cdot e^{-at} \tag{5.6}$$

In Eq. (5.6), Q(t) denotes the available capacity at the current time, t stands for the number of cycles or time elapsed, and a is the exponential decay coefficient. This model is more suitable for the accelerated degradation phase observed in batteries under high temperature conditions or during later stages of life.

Double exponential model

$$Q(t) = Q_0 - a\left(1 - e^{-bt}\right) - c\left(1 - e^{-dt}\right) \tag{5.7}$$

In Eq. (5.7),while parameters a and c correspond to the amplitudes of different degradation processes, and b and d represent the degradation rate constants. This model consists of a superposition of two exponential terms, enabling it to simultaneously capture two typical degradation mechanisms observed during long term cycling: one is a slow and gradual capacity fade caused by side reactions, and the other is a more rapid degradation resulting from loss of electrode active material or structural damage. Due to its ability to effectively distinguish and fit these two aging pathways, the double exponential formulation is widely applied in research on long term battery life and cycle life modeling.

Power law model

$$Q(t) = Q_0 - Kt^n, (n < 1) \tag{5.8}$$

In Eq. (5.8), K is the decay rate coefficient, and n is a power exponent less than 1. The power law model assumes that battery capacity loss over time or cycle number follows a decelerating decline pattern, rapid degradation in the early stages, gradually slowing down over time. This formulation effectively captures the nonlinear capacity fade behavior observed during the actual operation of electric vehicles, particularly suitable for long term life prediction. Consequently, this model is widely employed for SOH assessment and RUL prediction of electric vehicle batteries.

Semiempirical hybrid model

$$Q(t) = Q_0 - a\sqrt{n} - bn \tag{5.9}$$

In Eq. (5.9), the term $\sqrt{n}$ reflects the influence of diffusion controlled side reactions, and the linear term b_n corresponds to mechanical or chemical aging processes associated with cycling. This model exhibits good fitting performance in long term life tests and is widely applied in the analysis of accelerated life test data.

5.2.2 Internal resistance growth model

The internal resistance of the battery gradually increases with cycling, which directly leads to a decline in power performance. Common models for internal resistance growth include the following:

Linear model

$$R(t) = R_0 + k \cdot t \tag{5.10}$$

In Eq. (1.10), k is the resistance growth rate constant, and t denotes the number of cycles or usage time. This model assumes that under relatively stable operating conditions, side reaction rates are low, resulting in approximately uniform impedance increase over time. Therefore, the linear model is commonly used to characterize the slow degradation process of batteries during the early or midlife stages, particularly under standard operating conditions.

Exponential model

$$R(t) = R_0\left(1 + \alpha e^{\beta t}\right) \tag{5.11}$$

In Eq. (1.11), α and β are empirical parameters related to temperature, charge/discharge rate, and the intensity of side reactions. This model reflects the exponential rapid increase in internal resistance under harsh operating conditions, which results from accelerated side reactions, such as electrolyte decomposition, abnormal thickening of the SEI layer, and structural collapse of electrode materials. The model is particularly significant in studies concerning safety risks and precursors to battery thermal runaway.

Power law model

$$R(t) = R_0 + k \cdot t^n \tag{5.12}$$

In Eq. (1.12), k represents the growth coefficient and n the power exponent. Compared with the linear model, the power law model offers greater flexibility in fitting nonlinear growth characteristics: when $0 < n < 1$, impedance increases rapidly at first but gradually slows down, making it suitable for describing characteristics during the early stage of life; when $n > 1$, impedance growth accelerates, better reflecting behavior at the end of life. Therefore, this model is appropriate for comprehensive impedance modeling across the entire cycle life stage.

5.3 Machine learning–based state of health estimation

With the explosive growth of battery aging data, traditional empirical fitting models, constrained by their fixed mathematical structures, struggle to adapt to complex nonlinear aging trajectories. Consequently, data driven Machine learning (ML) approaches have emerged as a viable alternative. The core paradigm of these methods follows a "feature engineering + algorithmic mapping" framework: first, expert knowledge is leveraged to extract health indicators sensitive to aging from raw charge/discharge data; subsequently, regression algorithms are employed to establish a nonlinear mapping between these health indicators and SOH. By reducing reliance on detailed electrochemical mechanisms within the battery, this approach instead captures aging patterns through statistical learning theory, significantly enhancing the generalization capability and accuracy of SOH estimation.

At the feature engineering level, high quality input features are critical in determining the performance ceiling of a model [2]. Typically, key health indicators are extracted from Incremental capacity analysis (ICA) and Differential voltage analysis (DVA) curves, including peak heights, peak voltage positions, and capacity integrals over specific voltage intervals. These physically interpretable features are fed into classical algorithms such as Support vector regression (SVR). SVR employs kernel function techniques to map low dimensional nonlinear aging data into a high dimensional feature space, where an optimal hyperplane is identified for regression prediction. Due to its adherence to the principle of minimizing structural risk, SVR exhibits exceptional robustness when handling small sample, high dimensional battery aging data, effectively mitigating the overfitting problem.

In addition to SVR, Gaussian process regression (GPR) is another significant branch in the field of SOH estimation. Unlike deterministic models that provide only point predictions, GPR is grounded in Bayesian statistical theory and can output a probability distribution of predictions. This means that GPR not only delivers an estimated SOH value but also simultaneously quantifies the confidence interval of this estimate. Such capability is crucial for the safety of BMS, as it enables the BMS to identify when model outputs become unreliable. However, the limitations of these shallow machine learning approaches are also evident: they are highly dependent on the quality of manually engineered features, and their computational complexity often increases exponentially when handling large scale datasets, making it difficult to meet realtime processing requirements.

5.4 Advances in deep learning for state-of-health prediction

Due to the reliance of shallow machine learning on manual feature engineering, deep learning techniques, possessing powerful automatic feature extraction capabilities have gradually dominated the forefront of

SOH prediction research in recent years. By constructing deep neural network architectures, deep learning achieves an "end to end" mapping from raw voltage, current, and temperature data to SOH. This paradigm shift eliminates the subjective bias associated with manually designed features and enables the automatic discovery of subtle degradation patterns within raw data that are difficult for human experts to discern, particularly demonstrating a decisive advantage in handling battery lifetime data characterized by long term temporal dependencies.

Among various deep architectures, Recurrent neural networks (RNNs) and their variants: Long short-term memory networks (LSTMs) and Gated recurrent units (GRUs) are the preferred tools for handling battery aging time series data. Lithium ion battery aging is a cumulative process characterized by "path dependence," where the current SOH is not only influenced by the present cycling conditions but also implicitly constrained by prior abuse history. LSTMs effectively address the issues of vanishing and exploding gradients in long sequence training through a meticulously designed "gating mechanism" (forget gate, input gate, output gate), enabling the capture of long term aging dependencies spanning hundreds of cycles. Experimental results demonstrate that LSTMs achieve significantly higher accuracy than conventional time series models in tracking abrupt phenomena such as sudden capacity degradation.

In addition to temporal mining, Convolutional neural networks (CNNs) have been innovatively applied to SOH estimation. Although CNNs are traditionally used for image processing, in the context of battery systems, One dimensional convolution (1D-CNN) has been employed to scan across charging voltage curves, thereby extracting local features such as subtle slope variations in voltage plateaus. Recently, hybrid models incorporating attention mechanisms have emerged as a research hotspot. The attention mechanism enables the model to assign varying weights to different historical cycling segments, automatically focusing on critical time intervals that contribute most significantly to the current aging state, such as rapid charging under low temperature conditions, thereby substantially enhancing both model interpretability and prediction accuracy. Furthermore, addressing the challenges of lengthy battery aging test cycles and high data acquisition costs, transfer learning techniques are increasingly being integrated into deep networks. By transferring pretrained model weights from large scale public datasets to target batteries, this approach enables high-accuracy modeling even under limited data availability.

References

[1] T. Junjie, S. Wang, C. Wen, et al., A comprehensive review of state-of-charge and state-of-health estimation for lithium-ion battery energy storage systems, Ionics 30 (10) (2024) 5903–5927.

[2] C. Schulze Maxwell, F. Rodrigues Marco-Tulio, D. McBrayer Josefine, et al., Critical evaluation of potentiostatic holds as accelerated predictors of capacity fade during calendar aging, J. Electrochem. Soc. 169 (5) (2022).

CHAPTER SIX

State of power (SOP) estimation

6.1 Definition and significance

After entering this chapter, the research focus shifts from characterizing battery energy storage to characterizing power capability—specifically, the maximum charge and discharge power that a battery can achieve within a given time window under constraints related to voltage, temperature, and current. This capability is typically defined as the State of power (SOP), which characterizes the transient power boundaries of the battery system and serves as a critical basis for developing vehicle torque response, energy management, and safety protection strategies. In contrast to the State of charge (SOC), which emphasizes energy availability, SOP focuses on the upper limit of available power under multiple concurrent constraints and exhibits high sensitivity to factors such as temperature, aging, and polarization states, demonstrating significant time-variability and operating condition dependency. Therefore, establishing a consistent definition framework for SOP estimation—characterized by clearly defined constraints and capable of online implementation—is fundamental to Battery management systems (BMS) for tasks such as power allocation, power limitation control, and safety margin management.

Building upon this foundation, Section 6.1 presents the definition and engineering significance of SOC, standardizes the time window setting, input-output configuration, and evaluation metrics for SOC estimation, thereby establishing a unified benchmark for subsequent methodological comparisons. Section 6.2 focuses on an online estimation approach integrating models with filtering techniques, discussing a constraint-based computational strategy for SOC determination. Section 6.3 introduces sequential learning methods such as Long short-term memory (LSTM) networks, elaborating on data-driven mechanisms for SOC estimation. Finally, Section 6.4 summarizes multi-factor coupled SOC strategies and key

Modelling, Estimation and AI Applications for Lithium-Ion Battery Management Systems
ISBN: 978-0-443-45347-2
https://doi.org/10.1016/B978-0-443-45347-2.00006-6

engineering implementation considerations from perspectives of temperature effects, aging degradation, and multi-constraint coordination.

6.1.1 Definition of SOP, constraint specifications, and time windows

The SOP characterizes the transient power capability boundary of a battery under its current operating condition. To ensure computational feasibility and engineering comparability, this paper defines SOP as the maximum sustainable power that a battery can deliver within a given time window Δt while satisfying a set of operational constraints. Accordingly, SOP can be distinguished into discharge SOP and charge SOP. Although both exhibit formal symmetry—centered on the time window, constraint set, and maximum sustainable power—their constraint activation sequences and dominant boundaries may differ in engineering implementation. For instance, discharge is more likely to be limited by low-voltage and overcurrent constraints, whereas charging is more susceptible to over-voltage and low-temperature charging constraints.

The value of SOP is not an intrinsic constant of the battery but rather the boundary of the feasible region defined collectively by the constraint set. To facilitate alignment of SOP across different methods and operating conditions, the constraint set should preferably adopt a standardized, categorized representation, at least including the following four categories.

1) Voltage constraint. The terminal voltage must comply with minimum/maximum voltage limits (V_{min}/ V_{max}) to prevent loss of available capacity due to undervoltage or safety risks caused by overvoltage.
2) State of Charge Constraint. The SOC must remain within permissible limits (SOC_{min}/SOC_{max}) to ensure that power requests do not result in unacceptable violations of energy boundary constraints.
3) Current and power device constraints. Charging and discharging currents are jointly limited by the capabilities of the cell, battery pack, and power devices (I_{min}/I_{max}), and are also influenced by the rated values of components such as fuses, relays, and inverters.
4) Temperature and thermal safety constraints. Temperature must comply with an upper limit (T_{max}) and associated thermal safety constraints to prevent derating or failure risks caused by thermal accumulation.

The comparability of SOC-based SOP estimates primarily depends on the consistency of constraint definitions. If different studies or systems employ varying values for V_{min}/V_{max}, SOC windows, current limits, or thermal safety thresholds, the resulting SOP may differ significantly even under identical battery conditions and operating scenarios. Therefore, before any SOP results are presented, the corresponding constraint set and threshold settings must be explicitly defined.

The time window Δt is an indispensable element in the definition of SOP. SOP describes the capability to sustainably meet power demands over a future period, thus requiring a specific duration to be explicitly defined. For example, vehicle acceleration and regenerative braking typically correspond to transient events on the order of seconds, and power pulse tests commonly adopt typical windows such as 10 s or 30 s. On longer timescales, SOC drift and thermal accumulation significantly alter the feasible boundary, further restricting sustainable power. Generally, a shorter Δt results in an SOP that is closer to the instantaneous limits imposed by voltage and current constraints; conversely, a longer Δt makes thermal constraints and SOC boundaries more prominent in shaping the SOP. To avoid conceptual ambiguity, engineering formulations are recommended to explicitly denote the time window using the notation SOP@Δt, with corresponding constraint sets and thresholds provided simultaneously in sections or figures.

Under the unified definition and framework described above, the acquisition of SOP can generally be categorized into two approaches. The first is model-based SOP, which employs battery models to predict key variables—such as voltage, SOC, and temperature—over a future time horizon Δt. Subsequently, the maximum allowable current or maximum sustainable power is determined under a given set of constraints. This approach emphasizes physical consistency and interpretability, enabling explicit integration of constraints into the computational framework. The second approach is data-driven SOP, which leverages historical operational data to learn the mapping from measurable inputs—such as voltage, current, temperature, and their temporal features—to SOP outputs. This enables fast estimation even under complex nonlinearities and strong couplings. Although these two methods differ in their underlying mechanisms, they share a common objective under the unified framework presented in this chapter: to output a physically meaningful maximum sustainable charge or discharge power within a specified Δt and under defined constraint sets. This consistency provides a uniform basis for subsequent comparative analysis across accuracy, robustness, and real-time performance in later sections.

6.1.2 Engineering significance and application scenarios of SOP

During the operation of electric vehicles, driving conditions such as acceleration, overtaking, and climbing demand short-term high-power output from the battery; regenerative braking, on the other hand, requires short-term high-power absorption. These power demands exhibit pronounced transient characteristics and are often superimposed with extreme operating conditions such as cold-start, continuous acceleration, and prolonged downhill braking, causing the power boundaries to vary rapidly over time.

At the vehicle control level, SOP directly serves as the power upper bound constraint for the powertrain system, determining whether motor torque requests can be fulfilled and the extent to which derating must be applied. Typical decision variables include the maximum allocatable traction power upper bound, $P_{trac,max}$, and the maximum absorbable regenerative power upper bound, $P_{regen,max}$. At the energy management level, SOP also participates in power distribution and mode switching (e.g., traction/regeneration/coasting) rules, ensuring that upper-level control always operates within a feasible domain. For architectures incorporating multiple energy sources or hybrid systems, SOP further enables determination of the power-sharing ratio between the battery and other energy storage units.

If the SOP is overestimated, power requests may exceed the feasible operating domain, triggering protective actions such as undervoltage, overcurrent, or overtemperature protection. This manifests as sudden power drop, failure of regenerative braking, or repeated power derating, and in severe cases may lead to component overstress and safety hazards. The corresponding protective variables include the minimum terminal voltage constraint V_{min}, maximum current constraint I_{max}, and maximum temperature constraint T_{max}. Conversely, if the SOP is underestimated, excessive conservatism in power limitation occurs, resulting in degraded vehicle dynamic response, loss of regenerative energy recovery, and reduced driving range. The engineering consequence is a decline in availability rather than immediate system failure. Therefore, in electric vehicle applications, SOP establishes a consistent mapping between torque/power decision variables and voltage/current/temperature protection constraints through quantifiable power boundaries, enabling controlled derating and effective safety margin management.

On the power grid side or in industrial and commercial energy storage systems, batteries are required to frequently charge and discharge for tasks such as frequency regulation (Automatic generation control, AGC), peak shaving and valley filling, power smoothing, and backup support. These applications are characterized by power commands originating from dispatching systems, well-defined response requirements, and continuously varying temperature environments, aging levels, and cooling conditions during long-term operation.

SOP serves as a core parameter in energy storage scheduling, defining the set of allowable charge and discharge power levels at any given moment, thereby directly influencing the feasible region and constraint boundaries of scheduling optimization. Typical scheduling variables include the controllable frequency regulation power range P_{AGC}, the maximum discharge power $P_{dis,max}$ and maximum charge power $P_{chg,max}$ during peak shaving periods, as well as cluster-level power allocation weights based on a multi-cluster parallel battery architecture. Furthermore, SOP is coupled with thermal management and lifetime management strategies: the scheduling layer must dynamically adjust power commands according to SOP to avoid sustaining non-sustainable power tracking under cooling constraints or accelerated aging conditions.

If the SOP is overestimated, the scheduling system may continuously issue excessively high power commands, leading to voltage boundary violations, temperature rise, or component overloading—manifesting as frequent protection actions, failure to track AGC commands, or even triggering thermal runaway risks. Key protective variables remain focused on V_{min}/V_{max}, I_{max}, T_{max}, and thermal management capacity limits. Conversely, if the SOP is underestimated, the available dispatchable power range will be unnecessarily restricted, preventing energy storage systems from fully participating in frequency regulation and peak shaving. This results in degraded service capability and economic losses. Therefore, in energy storage applications, SOP transforms battery health and thermal constraints into dispatchable power variables while explicitly embedding safety boundaries within the power command generation and execution processes.

Portable devices, industrial power supplies, and aerospace power systems frequently encounter significant pulse loads or intermittent high-power demands, such as startup surges, motor start-stop cycles, and peak power consumption during communication transmissions. Compared to automotive and grid applications, these systems typically emphasize stringent volume/weight constraints and high reliability under operation, and may be subjected to broader ambient temperature ranges or more rigorous safety certification requirements.

At the system design and operational levels, the SOP determines whether transient loads can be supplied and the required level of redundancy. Typical decision variables include the allowable peak power value $P_{\text{peak, allow}}$, the setting of the pulse duration window Δt, and the power-sharing strategy between batteries and external power sources/supercapacitors in multi-power architectures. For industrial and aerospace systems, SOP is also used to define power limit curves and serves as boundary conditions for mission planning or load management algorithms, ensuring that critical loads remain prioritized under power-constrained conditions.

If the SOP is overestimated, it will directly lead to issues such as load power supply failure, undervoltage reset, device overcurrent, and excessive temperature rise; in severe cases, this may result in mission interruption or safety incidents. These risks are triggered and amplified by protective variables including voltage, current, and temperature. Conversely, if the SOP is underestimated, it will lead to excessive redundancy configuration or overly conservative operational strategies, increasing system weight and cost while reducing mission efficiency. In high-reliability scenarios, the SOP is not merely about defining a power upper limit; its conservatism and interpretability must also align with protective thresholds to support auditable power-limiting strategies and rational redundancy design.

6.1.3 Formal description of the SOP estimation problem, definition of inputs and outputs, and evaluation metrics

Treating SOP estimation as an engineering problem, the key issue lies not in the specific algorithmic form adopted, but in the ability to stably deliver power boundary outputs suitable for control and protection within clearly defined time windows and constraint conditions, thereby establishing a comparable and reproducible evaluation framework. To avoid non-comparable results arising from differences in definitions, constraints, and test setups across studies, this section provides a standardized description of SOP estimation from three perspectives: challenge origin, problem formulation, and evaluation metrics, thus establishing a unified benchmark for the methodological approaches presented in Sections 6.2 and 6.3.

SOP estimation initially confronts significant nonlinearity. In practical battery systems, the terminal voltage exhibits strong nonlinear relationships with current, polarization, and temperature, particularly under high-rate and low-temperature conditions, where ohmic voltage drop, concentration polarization, and interfacial kinetics jointly contribute to a nonlinear sensitivity of power boundaries to state variations. Second, the system demonstrates pronounced time-varying characteristics: fluctuations in ambient temperature, thermal state evolution, and aging-induced increases in internal resistance and capacity fade lead to temporal drift in model parameters and boundary conditions, rendering offline-calibrated static boundaries unsuitable for direct online application. Third, SOP is determined by the coupling of multiple state variables—SOC, SOH, temperature, and polarization states collectively influence terminal voltage response and thermal behavior, thereby jointly defining the sustainable power upper limit within a given time window. Finally, engineering implementation is constrained by observability: BMS typically measures only terminal voltage, current, and a limited number of temperature points, subject to sensor noise and drift; internal states remain unmeasurable directly, necessitating robust inference under information-limited conditions for SOP estimation.

To interface with BMS control and protection logic, SOP estimation should be formally defined in a manner that maps measurable inputs and inferred states to power boundary outputs. Generally, the estimator's inputs should include at least measured sequences of terminal voltage V, current I, and temperature T; where available, additional auxiliary information such as SOC, SOH, or historical window features may be incorporated. The output format is recommended to adopt one of two equivalent representations. The first consists of discharge SOP and charge SOP—defined as the maximum sustainable discharge power $P_{dis,max}$ and maximum sustainable charge power $P_{chg,max}$ under a given time window Δt and constraint set. The second representation expresses the result in terms of maximum allowable current: outputting the maximum permissible discharge current $I_{dis,max}$ and maximum permissible charge current $I_{chg,max}$, which can then be converted into power via voltage prediction or conservative voltage bounds. Compared to direct power output, current limits are more readily aligned with overcurrent protection mechanisms, power limit curves, and actuator constraints, thus offering superior interface compatibility in engineering implementation. Regardless of the chosen output format, it is essential to explicitly specify the corresponding time window Δt and constraint boundaries (e.g., V_{min}/V_{max}, SOC range, T_{max}), ensuring consistent physical interpretation of the SOP.

To enable comparable evaluation of different SOP estimation trajectories, an indicator system should be established simultaneously from four dimensions: accuracy, robustness, real-time performance, and generalizability.

1) Accuracy. This refers to the deviation between the estimation results and the reference baseline, which can be quantified using statistical metrics such as Mean absolute error (MAE), Root mean square error (RMSE), or Maximum Absolute Error. More directly applicable engineering metrics include power violation rate/voltage violation rate, defined as the proportion of instances where voltage undervoltage, overvoltage, or overcurrent constraints are triggered during power requests executed under the guidance of the estimated SOP.
2) Robustness. The stability of the estimator's output and its ability to avoid erroneous boundary identification under conditions such as noise disturbances, sensor bias drift, parameter mismatches, and insufficient temperature measurement points are evaluated. Robustness can be assessed through stress testing by introducing noise, bias, or altering parameter sets.
3) Real-time performance. Reflects the deploy ability of the estimator within the BMS operating cycle, with particular emphasis on the inference/update cycle and computational resource consumption. It typically requires updates to be completed within milliseconds to align with control and protection cycles.
4) Generalizability. The consistency of performance across different temperatures, individual cells, and operating conditions is evaluated. Typical operating conditions may include the Dynamic stress test (DST), the Beijing bus dynamic street test (BBDST), and pulse conditions with varying amplitudes and durations. The generalizability dimension emphasizes whether the estimation framework can maintain reliable outputs beyond the training/calibration conditions without exhibiting systematic bias.

The four categories of indicators above jointly constitute a unified evaluation framework for SOP estimation. Accuracy determines whether the boundary is trustworthy, robustness indicates reliability under abnormal or uncertain conditions, real-time performance determines whether the method is deployable in practice, and generalizability determines whether the method can be reused across different operating conditions. In the subsequent sections of this chapter, all methodological approaches shall be compared and analyzed under identical time window Δt and consistent constraint criteria, incorporating the aforementioned metrics to prevent conclusions based solely on a single error indicator.

6.2 SOP estimation based on UKF and improved algorithms

SOP is one of the most critical state parameters in BMS. Accurate and reliable estimation of SOP is decisive for ensuring battery safety, optimizing energy management strategies, and extending battery lifespan. However, real-time SOP estimation poses significant challenges. Batteries constitute highly nonlinear, time-varying electrochemical systems strongly influenced by operating conditions. Conventional SOP estimation methods, such as lookup table-based approaches or simple internal resistance models, often suffer from insufficient accuracy and struggle to handle dynamic operating conditions and battery aging effects.

Therefore, model-based filtering algorithms have become the mainstream approach in research on SOP estimation. Among these, the Kalman filter (KF) family has been widely favored due to its excellent state estimation capabilities. For nonlinear systems such as batteries, the Extended Kalman filter (EKF) was extensively employed. However, EKF suffers from inherent drawbacks, including significant first-order linearization errors and complex computation of Jacobian matrices. To address these issues, the Unscented Kalman filter (UKF) was developed. The UKF employs the Unscented transform (UT) to approximate the probability distribution of nonlinear functions with third-order Taylor expansion accuracy, without requiring Jacobian matrix calculations. As a result, it demonstrates higher estimation accuracy and stability than EKF in nonlinear systems.

This section will provide an in-depth exploration of SOP estimation methods for batteries based on the UKF and its improved algorithms. First, we will elaborate on the principles of the standard UKF algorithm and its application in joint SOC estimation. Subsequently, several improved UKF algorithms targeting the standard UKF will be analyzed in detail, including adaptive UKF, multi-time-scale UKF, and UKF integrated with other algorithms, with emphasis on their mechanisms for enhancing the accuracy and robustness of SOP estimation. Finally, a comparative performance analysis of these algorithms will be conducted through simulation or experimental case studies, followed by a discussion on future research directions.

6.2.1 Basic principles of the UKF algorithm

6.2.1.1 *Unscented transform*

The purpose of the UT is to construct sigma points that approximate a Gaussian distribution, thereby enabling the estimation of the transformed mean and covariance. Suppose χ is an n-dimensional vector following a certain distribution $\chi = N(x,P)$, and P is an $n\times n$ matrix. Applying the UT transformation yields $2n+1$ vectors (sigma points), whose mean satisfies:

$$\begin{cases} \sum_{i=0}^{2n} \chi_i = x(2n+1) \\ \sum_{i=0}^{2n} (\chi_i - x)(\chi_i - x)^T \end{cases} \tag{6.1}$$

The sigma points are analyzed using a symmetric strategy:

$$\begin{cases} \chi_i = x, i = 0 \\ \chi_i = x + \left[\sqrt{(n+\lambda)P}\right]_i, i = 1, 2, \sim n \\ \chi_i = x - \left[\sqrt{(n+\lambda)P}\right]_{i-n}, i = n+1, n+2, \sim 2n \end{cases} \tag{6.2}$$

Sigma-point weight analysis:

$$\begin{cases} \omega_m^0 = \dfrac{\lambda}{n+\lambda} \\ \omega_m^0 = \dfrac{\lambda}{(n+\lambda)+(1-\alpha^2+\beta)} \\ \omega_m^i = \omega_c^i = \dfrac{\lambda}{2(n+\lambda)}, i = 1 \sim 2n \end{cases} \tag{6.3}$$

In Eq. (6.3), $\lambda = \alpha^2(n+\kappa)-n$, α represents the degree of deviation of the sigma points from the state variable; λ denotes the scaling parameter for the sigma points; κ represents the auxiliary scaling parameter for the sigma points; and β indicates the distribution characteristics of the sigma points.

6.2.1.2 *Unscented Kalman filter algorithm*

The UKF (Unscented Kalman Filter) algorithm is a nonlinear system estimation method whose essence lies in an improved approach combining the UT with the KF algorithm. This method abandons the traditional idea of approximating nonlinear functions through linearization, instead employing the UT transformation to process data, thereby extending the analysis of system state variables. Subsequently, through complex system computations, it enables effective estimation of nonlinear system states. The following assumes that the spatial state equation of the system is represented as shown in the formula for analytical discussion.

$$\begin{cases} x_k = A_{k-1}x_{k-1} + B_{k-1}u_{k-1} + w_{k-1} \\ y_k = C_k x_k + D_k u_k + v_k \end{cases} \tag{6.4}$$

In Eq. (6.4), A_k denotes the system state transition matrix; B_k denotes the system input matrix; C_k denotes the system state observation matrix; D_k denotes the system input observation matrix; x_k represents the system state at time k; y_k represents the system measurement at time k; w_k represents the process noise of the system, with covariance matrix Q; v_k represents the measurement noise of the system, with covariance matrix R. For the nonlinear system described by the above Eq. (6.4), the iterative procedure of the UKF is as follows:

① Initialization

$$\begin{cases} \widehat{x}_0 = E(x_0) \\ P = E\left[\left(x - \widehat{x}_0\right)\left(x - \widehat{x}_0\right)^T\right] \end{cases} \tag{6.5}$$

② Prediction phase

UT transformation obtains sigma point reconstruction matrix representation:

$$\chi_{k-1} = \left[\widehat{x}_{k-1} \quad \widehat{x}_{k-1} + \sqrt{(n+l)P} \quad \widehat{x}_{k-1} - \sqrt{(n+l)P}\right] \tag{6.6}$$

Performing prediction analysis on the state Eq. (6.7) at time step k using the sigma points at time step $k-1$, as illustrated:

$$\begin{cases} \widetilde{\chi}_{i|k-1} = A_{k-1}\chi_{i|k-1} + B_{k-1}u_{k-1} \\ \widehat{x}_k = \sum_{i=0}^{2n} \omega_m^i \widetilde{\chi}_{i|k-1} \end{cases} \tag{6.7}$$

Acquisition of the state quantity covariance matrix:

$$P_{xx}^{-} = \sum_{i=0}^{2n} \omega_c^i \left[\widetilde{\chi}_{i|k-1} - \widehat{x}_k\right]\left[\widetilde{\chi}_{i|k-1} - \widehat{x}_k\right]^T + Q_k \tag{6.8}$$

③ Update phase

Substitute the sigma-point state representation quantities from Eq. (6.3) into the system observation equation to analyze the state of the observation quantity:

$$\begin{cases} \widetilde{y}_{i,k-1} = C_{k-1}\widehat{\chi}_{i|k-1} + D_{k-1}u_{k-1} \\ \widehat{y}_k = \sum_{i=0}^{2n} \omega_m^i \widetilde{y}_{i,k-1} \end{cases} \tag{6.9}$$

Acquisition of the observation covariance matrix:

$$P_{yy} = \sum_{i=0}^{2n} \omega_c^i \left[\widetilde{y}_{i,k-1} - \widehat{y}_k\right]\left[\widetilde{y}_{i,k-1} - \widehat{y}_k\right]^T + R_k \tag{6.10}$$

Acquisition of the covariance matrix between observation variables and state variables:

$$P_{xy} = \sum_{i=0}^{2n} \omega_c^i \left[\widetilde{\chi}_{i|k-1} - \widehat{x}_k\right]\left[\widetilde{y}_{i,k-1} - \widehat{y}_k\right]^T \tag{6.11}$$

Kalman filter gain acquisition:

$$K_k = \frac{P_{xy}}{P_{yy}} \tag{6.12}$$

Matrix acquisition of state equations and error correction equations:

$$\begin{cases} x_k = \widehat{x}_k + K_k\left(y_k - \widehat{y}_k\right) \\ P_{xx} = P_{xx}^{-} - K_k P_{yy} {K_k}^T \end{cases} \tag{6.13}$$

The unscented Kalman filtering algorithm employs the UT as its core to process data and construct weights, thereby significantly enhancing estimation accuracy. Moreover, this method requires only the initial values of x and P to be set, enabling it to continuously update state estimates by adapting to operational data, thus achieving effective characterization and analysis of complex nonlinear systems.

6.2.2 Application of the UKF algorithm in SOP estimation

For the application of the UKF algorithm in estimating the SOP of lithium-ion batteries, establishing a battery model that is both accurate and suitable for online computation is fundamental to SOP estimation. As an example, equivalent circuit models are favored in practical BMS applications due to their clear physical interpretation, simple structure, and ease of parameter identification. The following analysis will be based on the second-order RC equivalent circuit model illustrated in Fig. 6.1.

In Fig. 6.1, Uoc denotes the Open-circuit voltage (OCV), R0 represents the ohmic internal resistance, and the two parallel RC networks, R1-C1 and R2-C2, characterize electrochemical polarization and concentration polarization, respectively. The voltages across the two RC branches are denoted by U1 and U2, respectively, while I(t) represents the current at time t.

6.2.2.1 SOC estimation strategy based on the UKF algorithm

Taking the second-order RC equivalent circuit shown in Fig. 6.1 as an example, three key internal states are selected as the state vector:

$$\mathbf{x}_k = \left[\mathrm{SOC}_k, U_{1,k}, U_{2,k}\right]^T \tag{6.14}$$

In Eq. (6.14), SOC_k denotes the state of charge at time step k; $U_{1,k}$ and $U_{2,k}$ represent the polarization voltages across the first and second RC networks at time step *k*, respectively. The battery state-space model can be expressed as:

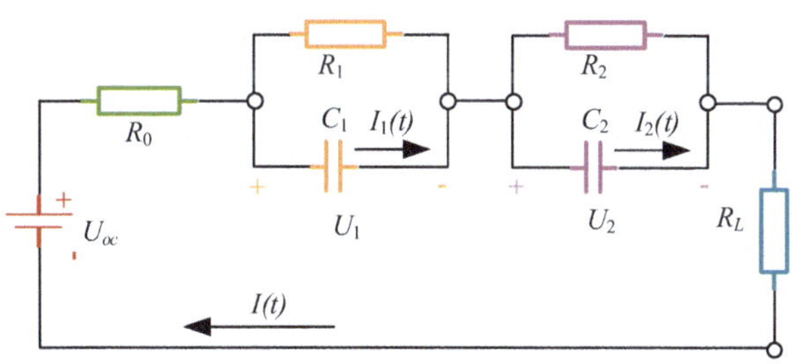

Figure 6.1 Second-order RC equivalent circuit model of a lithium-ion battery.

$$\begin{bmatrix} SOC_{k+1} \\ U_{1,k+1} \\ U_{2,k+1} \end{bmatrix} = \begin{bmatrix} SOC_k - \frac{\eta \Delta t}{Q_n} \cdot I_k \\ U_{1,k} \cdot e^{-\frac{\Delta t}{\tau_1}} + I_k \cdot R_1 \cdot (1 - e^{-\frac{\Delta t}{\tau_1}}) \\ U_{2,k} \cdot e^{-\frac{\Delta t}{\tau_2}} + I_k \cdot R_2 \cdot (1 - e^{-\frac{\Delta t}{\tau_2}}) \end{bmatrix} + \boldsymbol{w}_k \tag{6.15}$$

In Eq. (6.15), w_k denotes process noise, representing the uncertainty in the model. The observation equation can be expressed as:

$$V_{t,k} = OCV(SOC_k) - I_k \cdot R_0 - U_{1,k} - U_{2,k} + v_k \tag{6.16}$$

In Eq. (6.16), v_k denotes the observation noise, representing the measurement error of the voltage sensor.

Subsequently, the SOC is estimated according to the procedures outlined in Section 6.2.1.

6.2.2.2 SOP estimation method

At each estimation cycle, the UKF algorithm provides an accurate estimate of the state of charge $\widehat{SOC}_{1,k}$ and polarization voltage $\widehat{U}_{1,k}$ and $\widehat{U}_{2,k}$ at the current moment, Combined with battery constraints—such as terminal voltage limits—the peak power over a future time horizon Δt_p can be calculated. Taking discharge state of power (SOP) as an example, we need to determine the maximum discharge current such that the terminal voltage reaches its lower bound Vmin exactly after Δt_p seconds.

① Voltage constraint

Assuming the current remains constant within Δt_p, the terminal voltage after Δt_p seconds, as predicted by the model, is:

$$\begin{aligned} V_{t,k+\Delta t_p} = {} & OCV\left(\widehat{SOC}_{k+\Delta t_p}\right) - I_{\text{discharge}} \cdot R_0 - \widehat{U}_{1,k} \cdot e^{-\frac{\Delta t_p}{\tau_1}} \\ & - \widehat{U}_{2,k} \cdot e^{-\frac{\Delta t_p}{\tau_2}} - I_{\text{discharge}} \cdot R_1 \cdot \left(1 - e^{-\frac{\Delta t_p}{\tau_1}}\right) \\ & - I_{\text{discharge}} \cdot R_2 \cdot \left(1 - e^{-\frac{\Delta t_p}{\tau_2}}\right) \end{aligned} \tag{6.17}$$

Among them,

$$\widehat{SOC}_{k+\Delta t_p} = \widehat{SOC}_k - \frac{I_{\text{discharge}} \cdot \Delta t_p}{Q_n} \tag{6.18}$$

② SOC constraints

$$I_{\text{discharge,SOC}}^{\max} = \frac{\left(\widehat{SOC}_k - SOC_{\min}\right) \cdot Q_n}{\Delta t_p} \tag{6.19}$$

③ Current constraint:

$$I_{\text{discharge},I}^{\max} = I_{\max} \tag{6.20}$$

Ultimately, the discharge SOP can be expressed as:

$$SOP_{\text{discharge}} = \min\left(I_{\text{discharge},V}^{\max}, I_{\text{discharge,SOC}}^{\max}, I_{\text{discharge},I}^{\max}\right) \cdot V_{\min} \tag{6.21}$$

Similarly, the SOP under charging conditions can be calculated. This model-based, multi-constrained SOP estimation method can fully exploit the accurate internal state information provided by the UKF, enabling safer and more efficient power prediction.

6.2.3 Improved UKF algorithm and its application in SOP estimation

Although the standard UKF demonstrates excellent performance in state-of-charge (SOP) estimation for lithium-ion batteries, it still faces several challenges in practical engineering applications: the fixed noise statistics assumption struggles to handle time-varying operating conditions, the time-varying nature of model parameters is not adequately considered, and the relatively high computational load limits its application on low-cost hardware. To address these issues, researchers have proposed a series of improved algorithms aimed at enhancing the robustness, adaptability, and practicality of the UKF.

6.2.3.1 *Adaptive Unscented Kalman filter*

The core improvement of this method lies in online real-time estimation and adjustment of the process noise covariance matrix Q and the observation noise covariance matrix R, enabling the filter to adaptively handle time-varying noise, thereby suppressing estimation bias and preventing filter divergence.

The standard UKF assumes that noise statistical characteristics remain constant; however, the battery operates in a complex environment, and both model errors and sensor noise vary with temperature, aging level, and current rate. The core idea of Adaptive unscented Kalman filter (AUKF) is to reverse-adjust the noise covariance matrices using the statistical properties of the observation residual sequence (also known as the "innovation sequence"). When the filter is optimal, the residual sequence should be a zero-mean white noise sequence. By monitoring the discrepancy between the actual covariance of the residual sequence and its theoretical value, Q and R can be dynamically updated. In the AUKF algorithm proposed by a research team, the key formula for the adaptive update mechanism of the process noise covariance matrix Q and observation noise covariance matrix R is:

$$\begin{cases} \mathrm{H}_k = \frac{1}{L} \sum_{i=k-L+1}^{k} \varepsilon_k \varepsilon_k^{\mathrm{T}} \\ \mathrm{Q}_{k+1} = \mathrm{K}_k \mathrm{H}_k \mathrm{K}_k^{\mathrm{T}} \\ \mathrm{R}_{k+1} = \mathrm{H}_k + \sum_{i=1}^{2N} w_i^{(c)} \left(y_i^-(k) - y_k \right) \left(y_i^-(k) - y_k \right)^{\mathrm{T}} \end{cases} \tag{6.22}$$

In Eq. (6.22), ε_k represents the observation residual; $\boldsymbol{H}_k$ denotes the observation residual covariance matrix; L stands for the window length.

In SOC estimation, the AUKF can effectively handle model mismatches caused by battery aging, severe temperature fluctuations, or sensor performance degradation. Through adaptive tuning, it consistently maintains high-precision SOC and internal state estimation, thereby ensuring that the resulting SOP predictions based on these states are more reliable and robust.

6.2.3.2 Multi-time scale/dual Kalman filter

The core improvement of DKF lies in employing two filters with different update frequencies to handle state variables and model parameters with varying rates of change, thereby achieving a balance between rapid response in state estimation and stable tracking in parameter estimation.

In battery systems, state variables vary rapidly with load current, while model parameters change slowly with temperature and aging. The DKF typically employs a state filter (e.g., UKF) and a parameter filter (which may be either UKF or EKF) operating in parallel. The state UKF estimates state variables x at millisecond or tens-of-millisecond intervals, as shown in Eq. (6.14); meanwhile, the parameter UKF/EKF estimates model parameters θ at second- or minute-level intervals. During operation, the two filters are coupled: the state filter uses the most recent parameter estimates for state prediction and update, while the parameter filter utilizes the state estimates generated by the state filter as known quantities to identify model parameters.

In SOP estimation, the DKF enables online, real-time updating of model parameters used for SOP calculation, allowing the SOP prediction values to adaptively respond to battery aging degradation and temperature variations. This avoids the low estimation accuracy caused by fixed parameter sets, thereby enabling early warnings of declining battery power capability and ensuring operational safety.

6.2.3.3 Square-root unscented Kalman filter

The core idea of the SR-UKF algorithm is to directly recursively propagate the square-root factor of the state covariance matrix rather than the covariance matrix itself, thereby ensuring the symmetry and positive definiteness of the covariance matrix during numerical computation, which significantly enhances the algorithm's numerical stability and robustness.

Standard UKF may suffer from filter divergence due to the accumulation of rounding errors during computation, which can cause the covariance matrix P to lose positive semi-definiteness. In contrast, SR-UKF employs Cholesky decomposition such that $P = SS^T$, and updates only the triangular matrix S throughout the algorithm. All core steps are implemented using a series of efficient QR decompositions and Cholesky factor updates, thereby avoiding explicit computation of the covariance matrix. The flowchart of the SR-UKF algorithm is illustrated in Fig. 6.2.

In SOP estimation, the SR-UKF outperforms the standard UKF in terms of computational accuracy and stability, making it particularly suitable for BMS requiring long-term continuous operation and extremely high reliability. It provides a more robust mathematical foundation for estimating SOC and internal states, thereby indirectly enhancing the credibility of SOP predictions based on these states. Moreover, its computational complexity is

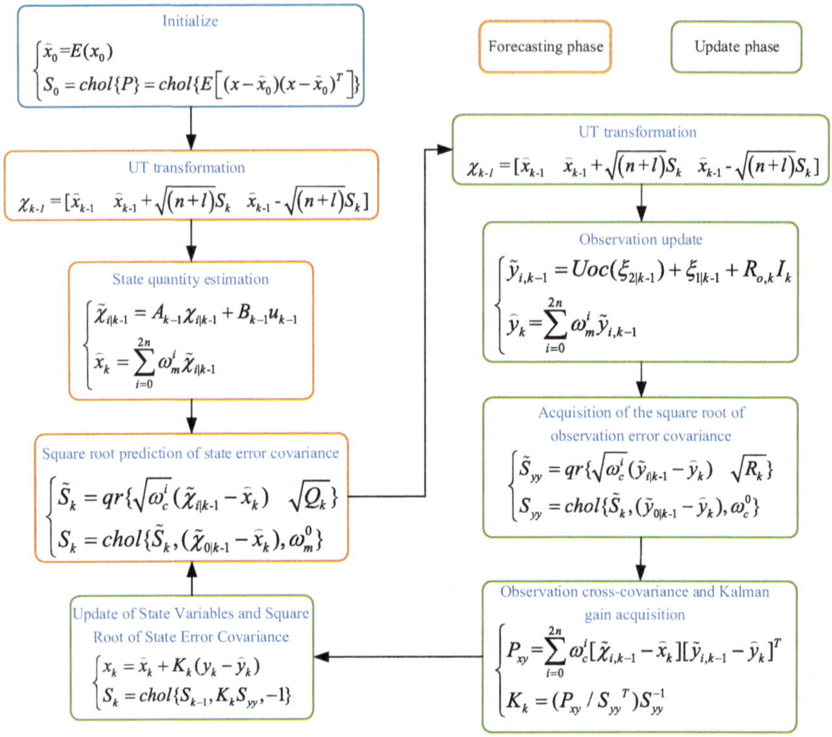

Figure 6.2 Flowchart of the SR-UKF algorithm.

comparable to that of the standard UKF, rendering it a highly practical and effective engineering improvement algorithm.

6.2.3.4 Fusion of the conventional UKF with other algorithms

Combining the conventional UKF with other algorithms is also a common approach for performance improvement, such as integrating the UKF with H∞ filtering. Standard UKF is sensitive to the statistical characteristics of noise, and its estimation performance may deteriorate under external disturbances. In contrast, H∞ filtering aims to minimize the estimation error under the worst-case scenario, without making any assumptions about noise characteristics, thus exhibiting exceptional robustness.

The combined approach integrates the unscented transformation concept from the UKF into the *H*∞ filtering framework, resulting in the *H*∞ unscented Kalman filter (*H*∞-UKF). This algorithm propagates sigma points during state prediction and update stages—similar to UKF—to capture nonlinearities, while also employing the minimax criterion of *H*∞

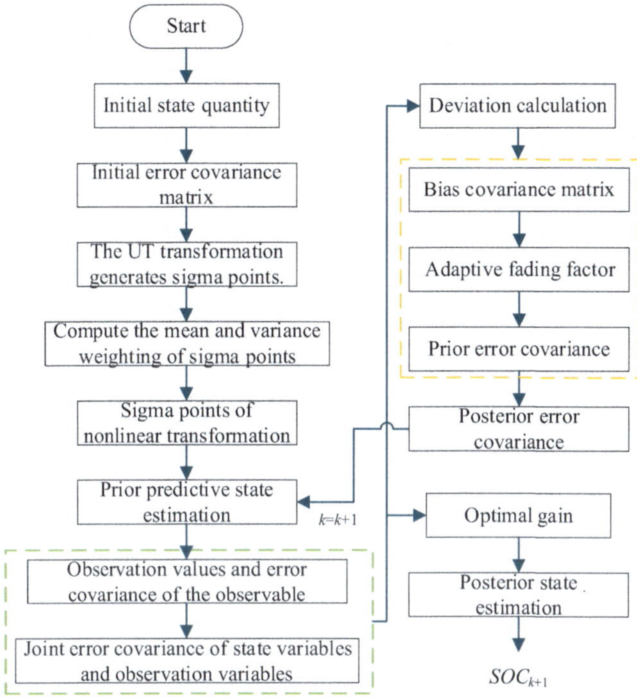

Figure 6.3 Flowchart of the adaptive unscented $H\infty$ filtering algorithm.

filtering to design the gain matrix Kk, ensuring stable estimation performance even under worst-case disturbances. An adaptive unscented H∞ filtering algorithm has been proposed in the literature [1], with its procedure illustrated in Fig. 6.3.

In SOP estimation, this hybrid algorithm is particularly suitable for scenarios involving high battery model uncertainty and strong environmental disturbances. It effectively mitigates adverse effects arising from model simplification and parameter identification errors, ensuring reliable and conservative SOP estimates even under complex operating conditions, thereby enhancing the safety margin of the BMS.

6.3 SOP estimation based on LSTM and improved algorithms

As a core state parameter in lithium-ion battery management systems, SOP directly determines the maximum instantaneous power output available from the battery at a given moment. The estimation accuracy of SOP critically influences the dynamic response timeliness of electric vehicles, the safety of power scheduling in energy storage systems, and the operational stability of portable devices. Although conventional SOP estimation methods can achieve dynamic state tracking, they tend to accumulate estimation errors when handling long-sequence, multi-dimensional coupled battery data due to insufficient modeling of temporal dependencies. The LSTM network, an enhanced architecture of Recurrent neural networks (RNN), addresses the vanishing gradient problem inherent in traditional RNNs through its gated mechanisms and demonstrates strong capability in capturing long-term temporal dependencies. Consequently, LSTM effectively meets the demand for multi-dimensional temporal feature representation in SOP estimation and has thus emerged as one of the leading technological directions for intelligent SOP estimation.

6.3.1 Adaptability of LSTM principles to SOP estimation

6.3.1.1 Basic structure of the LSTM network

The core advantage of LSTM lies in its ability to selectively retain and forget temporal information through the "cell state" and gating mechanisms, enabling effective preservation of critical information within long sequences. As illustrated in Fig. 6.4, a typical LSTM cell structure comprises four major components:

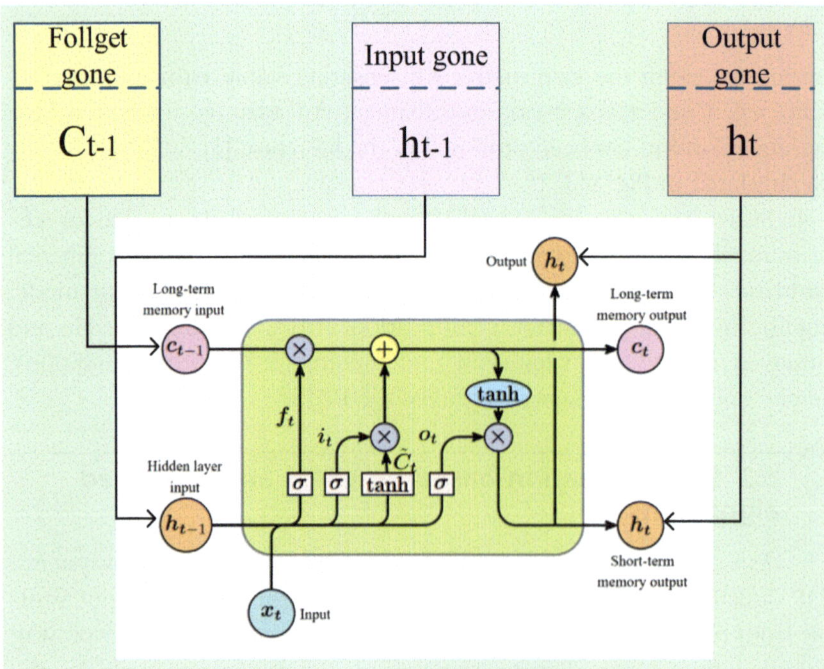

Figure 6.4 LSTM cell structure.

1. Forget Gate: selects information from the previous time step's cell state that should be retained. The output value is mapped to the [0,1] interval through a sigmoid function, where 0 indicates complete forgetting and one indicates complete retention.
2. Input Gate: Controls the proportion of effective updates from the current input information, implemented by combining the sigmoid and tanh functions;
3. Cell state update: perform linear update of the cell state based on the forget gate output and the candidate state from the input gate;
4. Output Gate: Determines the information from the cell state that will be passed to the next time step's hidden state, generated through the collaborative operation of the sigmoid and tanh functions.

The mathematical expressions of the components of LSTM are as follows:

$$f_t = \sigma\left(W_f \cdot [h_{t-1}, x_t] + b_f\right) \tag{6.23}$$

In Eq. (6.23), f_t denotes the output of the forget gate; σ represents the sigmoid activation function, expressed as $\sigma(z) = \frac{1}{1+e^{-z}}$; W_f is the weight matrix of the forget gate; $[h_{t-1},x_t]$ denotes the concatenation vector of the previous hidden state h_{t-1} and current input x_t; b_f is the bias vector of the forget gate.

$$i_t = \sigma(W_i \cdot [h_{t-1}, x_t] + b_i) \tag{6.24}$$

In Eq. (6.24), i_t denotes the output of the input gate; W_i represents the weight matrix of the input gate; b_i denotes the bias vector of the input gate; other parameters are defined similarly as in formula (6.23).

$$\widetilde{c}_t = \tanh(W_c \cdot [h_{t-1}, x_t] + b_c) \tag{6.25}$$

In Eq. (6.25), $\widetilde{c}_t$ represents the candidate cell state; tanh denotes the hyperbolic tangent activation function with a range of [-1, 1]; W_C is the weight matrix for the candidate state; b_C is the bias vector for the candidate state.

$$C_t = f_t \odot C_{t-1} + i_t \odot \widetilde{C}_t \tag{6.26}$$

In Eq. (6.26), C_t denotes the cell state at the current time step, C_{t-1} denotes the cell state at the previous time step, and $\odot$ represents element-wise multiplication.

$$o_t = \sigma(W_o \cdot [h_{t-1}, x_t] + b_o) \tag{6.27}$$

In Eq. (6.27), o_t is the output of the output gate; W_O is the weight matrix of the output gate; b_O is the bias vector of the output gate.

$$h_t = o_t \odot \tanh(C_t) \tag{6.28}$$

In Eq. (6.28), h_t denotes the hidden state at the current time step, serving as the core output of the LSTM and used for subsequent SOP calculations.

6.3.1.2 Temporal features of SOP data

The essence of SOP is the maximum available power of the battery at time t, whose value depends on the battery's state parameters at time t and historical moments, exhibiting significant temporal dependence. Its core characteristics include:

1. Continuity: During battery charging and discharging processes, input parameters such as voltage, current, and temperature fluctuate continuously over time, requiring the model to capture these dynamic variation patterns.
2. Long-term dependency: SOP is influenced by the battery's historical number of cycles and long-term aging degree. Traditional sequential models struggle to capture long-term dependencies exceeding 10 steps, whereas LSTM can effectively store such information over extended periods through its cell state.
3. Multidimensional coupling: The SOP input features encompass multiple dimensions such as U, I, T, SOC, and SOH. The temporal coupling among these parameters necessitates that the model simultaneously handle the interaction relationships across multiple temporal streams.

6.3.1.3 Advantages of LSTM in processing SOP data

1. Compared to conventional SOP estimation methods, the advantages of LSTM in handling SOP temporal features are primarily reflected in:
2. Gradient Stability: During backpropagation in traditional RNNs, gradients decay exponentially over time steps, making it difficult to learn long-term dependencies; in contrast, LSTM maintains stable gradients over long sequences through its linear update pathway of the cell state, enabling effective learning of long-term dependencies spanning over 100 time steps.
3. Feature adaptive extraction: Methods such as UKF require manual design of the mapping relationship between input parameters and SOP, whereas LSTM automatically learns the nonlinear mapping from U–I-T-SOC to SOP through its gating mechanism, without

relying on prior physical assumptions, effectively adapting to the SOP characteristics of different battery types, including ternary lithium and lithium iron phosphate batteries.

6.3.2 Limitations of traditional LSTM in SOP estimation

Although traditional LSTM possesses advantages in temporal processing, its direct application to SOP estimation still faces four core limitations that require overcoming through improved algorithms.

6.3.2.1 Insufficient short-term temporal attention

Conventional LSTM assigns identical weights to input features across all time steps, whereas in SOP estimation, the U and I data from neighboring time points significantly influence the SOP at time t compared to earlier moments. Conventional LSTM does not adequately focus on such short-term critical information, resulting in estimation delay.

6.3.2.2 Multi-feature weight imbalance

In the multi-feature weighted imbalance SOP input features, SOC has a significantly higher influence weight on SOP compared to temperature; however, traditional LSTM fails to differentiate feature importance and over-learns low-weight features, leading to increased estimation bias.

6.3.2.3 Insufficient model generalization ability

Conventional LSTM training relies on data from specific operating conditions, and when applied to unseen operating conditions, its accuracy significantly degrades, failing to meet practical requirements.

6.3.2.4 Poor real-time performance

Conventional LSTM requires processing a large volume of sequential data, with a single-step inference time of approximately 20 ms, which fails to meet the 10 ms per update requirement for SOP in electric vehicle BMS. Consequently, the real-time capability of conventional LSTM is inadequate for highly dynamic operating conditions such as sudden braking or rapid acceleration.

6.3.3 SOP estimation algorithm based on improved LSTM

To address the limitations of conventional LSTM, this section proposes three improvement directions: optimizing temporal dynamics, optimizing feature weights, and incorporating physical constraints, thereby enhancing SOP estimation performance from different dimensions.

6.3.3.1 Optimizing the temporal direction (BiLSTM-SOP model)

① Principle: Bidirectional long short-term memory (BiLSTM) networks learn forward and backward temporal dependencies of sequential data through forward and backward LSTM units, respectively. The concatenation of the bidirectional hidden states is used as the output, enabling the model to simultaneously capture both short-term forward and backward dependencies within SOPs.

② Model structure: The BiLSTM-SOP model consists of an input layer, a BiLSTM layer, a fully connected layer, and an output layer, as illustrated in Fig. 6.5. The input layer is an N×D temporal matrix (where N denotes the number of time steps and D = 5, corresponding to U, I, T, SOC, and SOH); the BiLSTM layer comprises forward and backward LSTMs, producing a bidirectional hidden state of size N×256; the fully connected layer maps the 256-dimensional features to a single dimension; the output layer provides the SOP estimate at time step *t*.

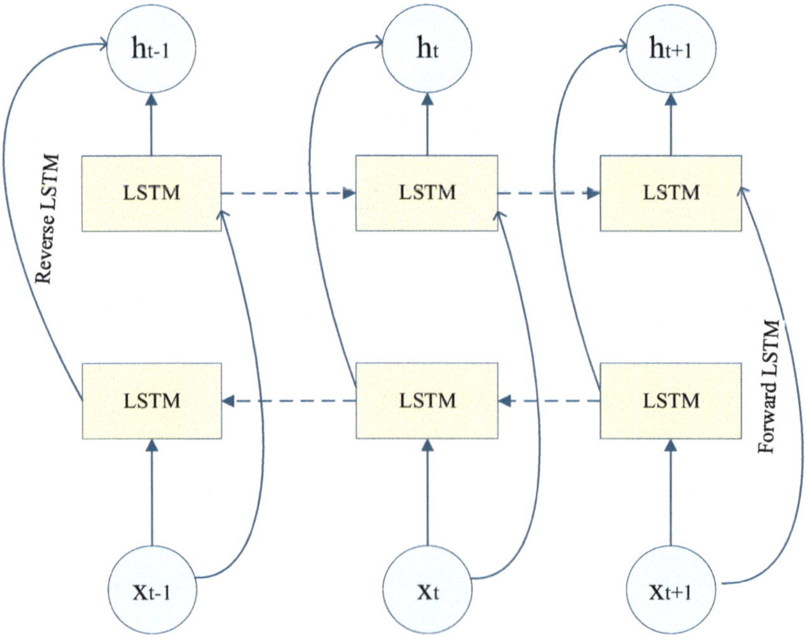

Figure 6.5 Schematic diagram of the BiLSTM-SOP model structure.

6.3.3.2 LSTM-attention model

① Principle: Introducing the attention mechanism to assign dynamic weights to LSTM hidden states, as shown in Fig. 6.6, thereby focusing on time steps and feature dimensions that significantly impact SOP, thus addressing the issue of weight imbalance. Attention weights are categorized into temporal attention (Optimizing time step weights) and feature attention (Optimizing feature dimension weights).

② Attention weight calculation:

$$\alpha_t = \frac{\exp(e_t)}{\sum\limits_{K=1}^{N} \exp(e_k)} \$, \$e_t = v^T \tanh(W_h h_t + b_h) \tag{6.29}$$

In Eq. (6.29), α_t is the temporal attention weight $\sum_{K=1}^{N} \alpha_t = 1$ at time t; e_t is the energy value at time t; v is the attention vector; W_h is the weight matrix; b_h is the bias vector.

$$\beta_d = \frac{\exp(f_d)}{\sum\limits_{k=1}^{D} \exp(f_k)}, f_d = u^T \tanh(W_x x_d + b_x) \tag{6.30}$$

In Eq. (6.30), β_d is the weight of the d-th feature $\sum_{K=1}^{D} \beta_d = 1$; f_d is the energy value of feature d; u is the feature attention vector; W_x is the feature weight matrix; bx is the feature bias vector.

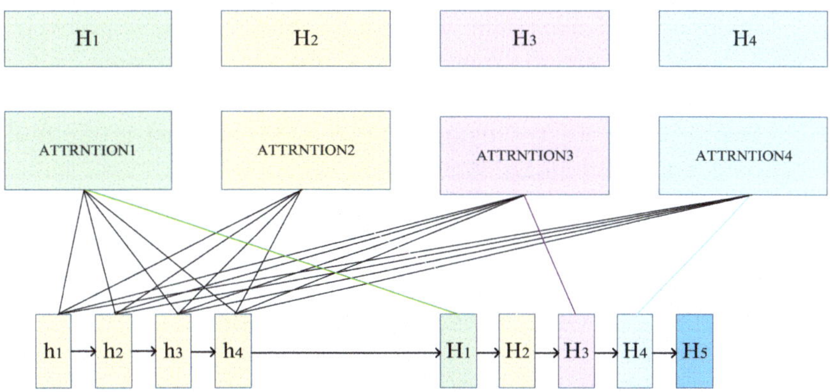

Figure 6.6 Schematic diagram of the LSTM-Attention model architecture.

③ Model output: By combining bidirectional attention weights, the final output of SOP is given by Eq. (6.31):

$$\mathrm{SOP}_t = W_0\left[\sum_{K=1}^{N} \alpha_t(h_t \odot \beta)\right] + b_0 \tag{6.31}$$

In Eq. (6.31), β is the feature weight vector $\left([\beta_1, \beta_2, \cdots, \beta_D]^T\right)$; $h_t \odot \beta$ represents assigning weights to each feature dimension of h_t; W_0 is the output layer weight matrix; b_0 is the output layer bias.

6.3.3.3 LSTM-physics fusion model

① Principle: Conventional LSTM lacks physical constraints and is prone to producing state-of-power (SOP) estimates that exceed the battery's safe power limits. Therefore, by incorporating power constraints derived from an Equivalent circuit model (ECM), the LSTM output is corrected to ensure physical plausibility of the estimates.

② Fusion mechanism:

Step 1: The LSTM outputs an initial estimate of the SOP, denoted as SOPLSTM;

Step 2: Calculate the maximum safe power $\mathrm{SOP}_{physics}$ at time t based on the ECM model, as shown in Eq. (6.32):

$$\mathrm{SOP}_{\mathrm{physics}} = \min\left(\frac{(U_{oc} - U_{\min}) \times I_{\max}}{\eta}, \frac{(U_{\max} - U_{oc}) \times I_{\min}}{\eta}\right) \tag{6.32}$$

In Eq. (6.32), U_{oc} is the open-circuit voltage; $U_{\min}$ and $U_{\max}$ are the minimum and maximum allowable battery voltages, respectively; $I_{\min}$ and $I_{\max}$ are the minimum and maximum discharge/charge currents, respectively; and η is the charge-discharge efficiency.

Step 3: Physically constrain the SOPLSTM, resulting in the final output as shown in Eq. (6.33):

$$\mathrm{SOP}_{\mathrm{final}} = \mathrm{clamp}\left(\mathrm{SOP}_{\mathrm{LSTM}}, 0, \mathrm{SOP}_{\mathrm{physics}}\right) \tag{6.33}$$

In Eq. (6.33), the function *clamp* restricts $\mathrm{SOP}_{\mathrm{LSTM}}$ to the range $[0, \mathrm{SOP}_{\mathrm{physics}}]$.

6.4 Multi-factor coupled SOP strategies

Facing the practical operating conditions characterized by temperature fluctuations, aging evolution, and cell-to-cell inconsistency, this section presents a unified engineering framework and key implementation principles for SOP estimation under multi-factor coupling. In contrast to Sections 6.2 and 6.3, which respectively emphasize model-based filtering and data-driven approaches along single pathways, Section 6.4 focuses on establishing an end-to-end chain within the same time window Δt and under a consistent constraint set: from measurement acquisition and data conditioning, through joint state and parameter estimation, to constraint prediction and final SOP@Δt output. Moreover, it incorporates dominant constraint annotation and confidence level management into the estimation results to enhance interpretability and diagnostic capability. Section 6.4.1 outlines the overall workflow of multi-factor coupled estimation and extends the state space accordingly; Section 6.4.2 addresses coupled electrothermal estimation under temperature influence and consistency in $\Delta \mathrm{t}$ constraint prediction; Section 6.4.3 discusses SOH-driven parameter drift induced by aging and consistency coupling, along with low-dimensional representation of pack-level bottlenecks; finally, Section 6.4.4 concludes with multi-source information fusion, unifying model-consistent and data-driven candidate estimates into a deployable SOP state estimation output via constraint-consistency verification and confidence-based arbitration mechanisms.

6.4.1 Overview of multi-factor coupling SOP state estimation

The multi-factor coupled SOP strategy aims to transform SOP estimation results into a power upper limit that remains compliant with constraints across a wide operating range and is directly executable by the BMS. In contrast to single-path SOP research that focuses solely on the internal mechanisms of estimation algorithms, the multi-factor coupled perspective emphasizes the end-to-end chain from measurement to power limit delivery, treating temperature evolution, aging, consistency deviations, and algorithm robustness as core constraints determining power boundaries rather than secondary perturbations. This section presents a unified framework to standardize how multiple factors enter the SOP process and how strategy outputs should be defined and structured during engineering deployment. Fig. 6.7 illustrates the overall architecture, consisting of six main stages

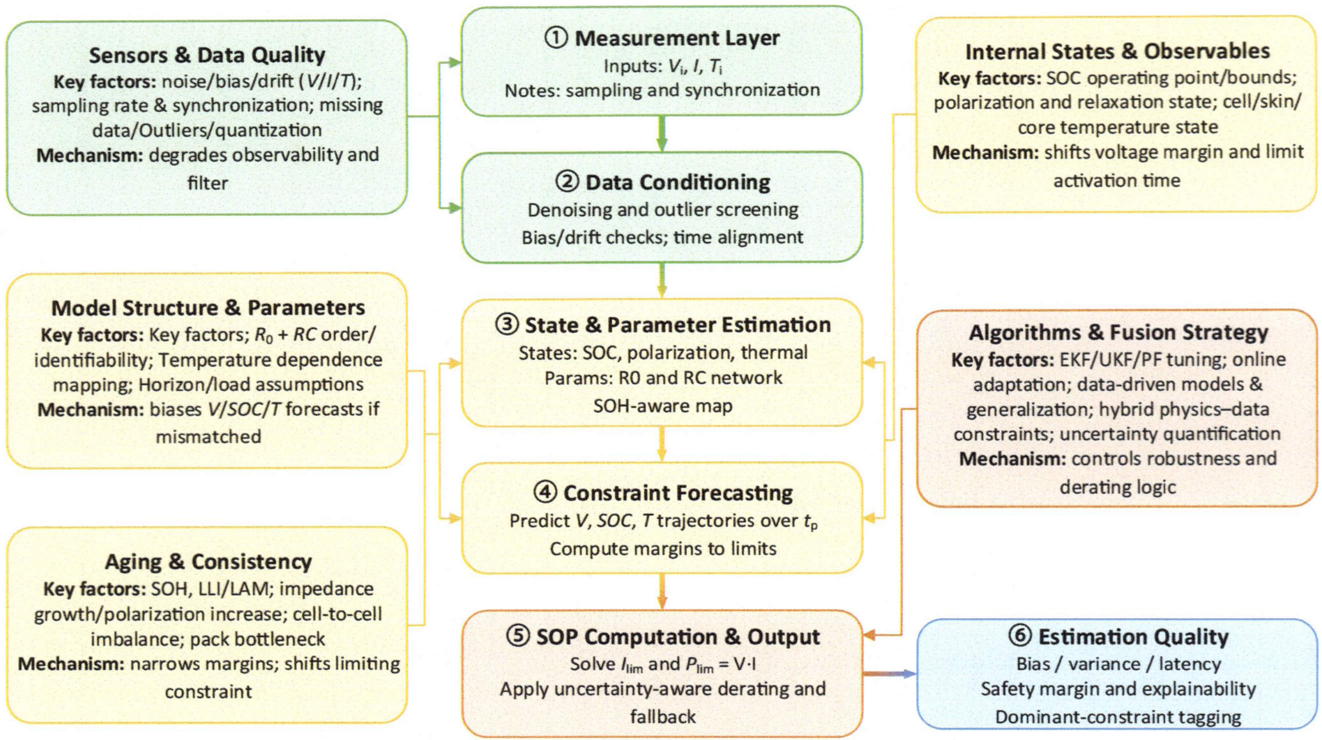

Figure 6.7 End-to-end framework for multi-factor coupled SOP estimation with power limitation output.

labeled ①–⑥, complemented by five types of auxiliary modules that collectively capture key coupling sources influencing the SOP envelope.

As illustrated in Fig. 6.7, the multi-factor SOP pipeline begins at the measurement layer, where the terminal voltage V_i, current I, and temperature T_i are acquired at the cell or battery pack level. In engineering implementation, sampling rate, channel synchronization, and temporal consistency critically determine whether subsequent estimators can operate on a consistent signal basis—particularly critical when large-current transients coexist with thermal lags. The second stage is data conditioning, encompassing noise reduction and outlier removal, bias and drift detection, and temporal alignment of multi-channel sequences. These preprocessing steps are not ancillary; they directly influence the observability of internal states and the stability of downstream filtering or learning modules.

The third stage is state and parameter estimation, whose objective is to provide essential internal variables for constraint activation discrimination. Typical states include SOC, polarization/relaxation states, and thermal states; typical parameters include ohmic resistance R_0, resistor-capacitor network parameters, and mapping relationships associated with State of health (SOH). It should be emphasized that ③ provides a consistent state basis rather than directly delivering SOP results. Subsequently, constraint forecasting is performed, predicting the evolution trajectories of voltage, SOC, and temperature over the prediction horizon t_p (or equivalently, the SOP time window Δt), and computing the margin to constraint boundaries. In other words, ④ transforms estimated states into actionable information—namely, the remaining margins to constraint boundaries and the expected times of constraint violation under candidate current conditions.

On the basis of available margin, SOP computation and output determines the maximum allowable current I_{lim}, and corresponding power limit P_{lim}, typically calculated as $P_{lim} = V \cdot I$, with separate definitions for charging and discharging directions. In multi-factor coupled operation, ⑤ should additionally incorporate a derating mechanism oriented toward uncertainty, defining conservative degraded outputs when measurement quality degrades, parameters drift, or prediction reliability is insufficient—thereby preventing constraint violations rather than merely triggering protection after violations occur. Finally, estimation quality evaluates the bias, variance, and latency of the end-to-end output and explicitly specifies the interpretability requirements necessary for engineering deployment.

In addition to the main workflow, Fig. 6.7 also identifies five categories of lateral coupling sources to explain why the SOP envelope contracts and constraints migrate under varying conditions. The first category, sensor and data quality, includes noise, bias, and drift in voltage/current/temperature measurements, as well as missing data, outliers, and quantization errors. Its underlying mechanism lies in degrading observability and introducing estimation bias, primarily affecting quality assessment at stages ②–③ and ⑥. The second category, model structure and parameters, encompasses model order, identifiability issues, temperature-dependent parameter mapping, and assumptions regarding prediction horizon. Model mismatch systematically biases predictions of voltage, SOC, and temperature, thereby influencing margin computation and power limits across stages ③–⑤. The third category, aging and consistency, involves SOH evolution, Loss of lithium inventory (LLI) and Loss of active material (LAM), impedance growth, and cell-to-cell inconsistency. These factors lead to reduced voltage/thermal margins, migration of dominant constraints, and result in the battery pack's SOP being predominantly governed by the weakest individual cell. The fourth category, internal states and observables, emphasizes that SOC operating points, relaxation effects, polarization phenomena, as well as core-shell temperature differences can alter voltage margins and timing of constraint activation—rendering SOP highly sensitive to operational conditions. The fifth category—algorithms and fusion strategy—summarizes tuning of estimators, online adaptive mechanisms, data-driven generalization capabilities, physics-data fusion constraints, and uncertainty quantification. This module determines the robustness of end-to-end output performance as well as derating behavior while preserving a consistent definition of unified SOP across all contexts.

To ensure result comparability and safe deploy ability, the multi-factor coupled SOP framework requires standardized definitions centered on constraint sets and the time window Δt. Constraint sets typically include voltage bounds (V_{min}, V_{max}), SOC bounds (SOC_{min}, SOC_{max}), maximum current capability I_{max} and power device limitations, as well as maximum temperature T_{max} and thermal constraints influenced by cooling capacity. A key intermediate variable within the framework is margin—the predicted distance to each constraint boundary over Δt. Margin enables the identification of dominant constraints under a unified metric, thereby determining the bottleneck categories of I_{lim} and P_{lim}. Given that temperature, aging, and consistency factors may cause switching among dominant constraints across different operating conditions, labeling dominant constraints holds direct engineering value for calibration and maintenance.

From an engineering interface perspective, the SOP output should be formulated in a manner directly mapable to BMS decision logic. At the control layer, SOP should be translated into traction and regenerative power upper limits ($P_{trac,max}$, $P_{regen,max}$), or current upper limits in discharge and charge directions ($I_{dis,max}$, $I_{chg,max}$). At the protection layer, the same output should include boundary margins for voltage, current, and temperature to support a proactive protection strategy based on derating followed by trip activation. At the scheduling layer, SOP defines the dispatchable power domain and enables power allocation under battery pack constraints. Regardless of application context, strategy outputs must simultaneously provide limit values and dominant constraint labels to maximize system availability while maintaining safety margins.

6.4.2 State estimation of SOP under temperature coupling

Estimating the SOP under temperature-coupled scenarios fundamentally constitutes an online inference problem of power boundaries determined jointly by electrical and thermal states and parameters. Temperature not only alters the battery's ohmic internal resistance and polarization dynamics, thereby affecting terminal voltage headroom and the timing of constraint activation, but also introduces additional uncertainty through thermal inertia and non-uniform temperature distribution. This makes SOP computation based solely on voltage and current responses prone to systematic bias across different operating conditions. Therefore, rather than focusing primarily on power limitation curves, this section centers on the systematic estimation of SOP under temperature coupling, emphasizing joint estimation of electrical-thermal states, online updating of temperature-dependent parameters, and a constraint prediction mechanism tailored to a time window Δt. The objective is to establish a consistent and reliable state foundation for subsequent SOP computation.

Within the estimation framework, temperature-coupled SOP estimation should follow the primary chain of measurement-conditioning-state/parameter estimation-constraint prediction-SOP solution as described in Section 6.4.1, with emphasis placed on the temperature-coupling treatment at stages ③–④. First, state variables should not be limited to SOC and polarization state alone, but must explicitly incorporate thermal state. Second, model parameters must be expressed with temperature dependence—for example, ohmic internal resistance $R_0(T)$, polarization branch parameters $R_p(T)$ and $C_p(T)$, or mapping relationships between equivalent time constants and temperature. If temperature-dependent parameters are still treated as constants, the estimator will often absorb temperature effects into state biases, leading to drift in SOC and polarization state estimates, which subsequently results in erroneous voltage margin assessment during SOP computation.

The key advancement in thermal coupling lies in the joint estimation of electro-thermal states. At the level of engineering deployability, this joint estimation does not require complex thermal field solving; instead, the thermal state is treated as a low-order dynamic process coupled with the electrical state. The irreversible and reversible heat generated by current alters the temperature, which in turn affects voltage response and parameters. Based on this framework, the estimator's objective can be formulated as simultaneously updating the SOC, polarization state, and thermal state under measured $V/I/T$ sequences, while online correcting temperature-dependent parameters to ensure model consistency with short-term voltage responses and medium-to-long-term thermal evolution. This joint estimation mechanism directly serves the core requirement of SOP estimation by providing a foundation for predicting V, SOC, and T evolution within a given time window Δt, rather than merely fitting the terminal voltage at the current instant.

Under conditions of limited temperature measurement points or non-uniform temperature fields, the primary challenge in temperature-coupled SOP estimation lies in thermal state observability and the selection of representative temperature metrics. Since BMS typically possess only a few surface-mounted sensors, core and hotspot temperatures cannot be directly measured. Using a single-point temperature as the sole input for thermal state estimation may lead to underestimation of hotspot temperatures and delay the triggering of thermal constraints. A more robust approach involves modeling the thermal state as an equivalent temperature or an envelope temperature at hotspots, while incorporating conservative bias terms or upper-bound corrections during estimation to ensure predictions remain on the safer side. Concurrently, temperature measurement noise, bias drift, and sampling asynchrony should be explicitly treated as sources of uncertainty. Through data conditioning and adaptive filter gain or confidence level management, their adverse effects on joint estimation stability can be mitigated, thereby preventing amplification of temperature noise into power boundary fluctuations during SOP computation.

Following the completion of temperature-coupled state/parameter estimation, the next critical step in SOP estimation is constraint prediction with respect to Δt. Unlike the static determination of whether current boundaries are exceeded, SOP estimation requires predicting the trajectories of terminal voltage, SOC, and temperature over the future time horizon Δt, and subsequently computing margins to constraint boundaries as well as identifying the dominant constraint that will be triggered first. The role of temperature coupling is particularly evident here: as Δt increases, thermal accumulation and heat dissipation capabilities significantly influence temperature trajectories, thereby altering the sustainable current upper limit. Conversely, for shorter Δt values, voltage margin and current capability may become more dominant; however, temperature still affects the accuracy of voltage transient predictions through parameter mapping. Therefore, temperature-coupled SOP estimation must explicitly distinguish between dominant sources of uncertainty in constraint prediction across different Δt values and ensure that the prediction model remains closed-loop consistent with the estimated states from joint estimation.

SOP estimation under temperature coupling should not be simplified to treating temperature as a mere input variable; instead, it should be implemented as co-estimation of electrical and thermal states and parameters at the state and parameter levels, prediction of voltage, temperature, and SOC trajectories with respect to $\Delta \mathrm{t}$ at the forecasting level, and output of SOP@Δt along with dominant constraint explanations under a unified constraint framework. This organizational approach enables a natural integration between Section 6.4.2 and Sections 6.2 (model- and filter-based estimation approaches) and 6.3 (sequence learning-based estimation approaches), where the former emphasizes that temperature coupling expands the state space and parameter mapping, while the latter highlights that temperature coupling necessitates learning mappings covering thermal dynamics and cross-temperature generalization.

6.4.3 SOP state estimation under aging and consistency coupling

In multi-factor coupled SOP estimation, aging and inconsistency issues are often the primary sources of cross-lifetime and cross-cell estimation distortions. Unlike random disturbances such as measurement noise, aging typically manifests as slow-time-scale systematic variations, continuously altering voltage responses and boundary margins, thereby progressively reducing the sustainable power ceiling under identical operating conditions. Meanwhile, cell-to-cell variations within a battery pack cause the pack-level available power to be dominated by the weakest cell. If estimation is still conducted under assumptions of cell mean values or ideal consistency, SOP estimates in practical engineering scenarios tend to become overly optimistic or difficult to interpret. Therefore, this section transforms aging and inconsistency from noise terms into state variables and parameter drifts. By focusing on online parameter updates driven by SOH and low-dimensional representation of pack-level limiting factors, a consistent SOP estimation framework for cross-lifetime and cross-cell applications is established.

The impact of aging on SOC estimation can be summarized into two primary pathways, both of which result in reduced constraint margins and a shift in the dominant constraints. The first pathway involves increased impedance and exacerbated polarization: as cycling and calendar aging progress, ohmic and polarization resistances rise, leading to more pronounced voltage drops under the same current. Consequently, the terminal voltage reaches the lower bound $V_{\min}$ earlier or approaches the upper bound $V_{\max}$ more readily during charging. This directly compresses voltage margins, reducing the sustainable current upper limit based on a Δt time window, thereby manifesting as a systematic downward adjustment of SOP. The second pathway stems from capacity fade and changes in lithium inventory, which alter the SOC boundary triggering behavior. Reduced usable capacity modifies the baseline and sensitivity near SOC boundaries, making it easier for the same power request to trigger either SOC or voltage constraints within the SOC extremities. Since SOP@Δt is determined by the first constraint to be violated within the time window, aging not only diminishes SOP values but also induces a migration of dominant constraints across different life stages—for instance, shifting from voltage-limited dominance toward thermal or component capability limitations. This provides essential physical insights for subsequent interpretation of output behavior.

To maintain consistent estimation of SOP under aging conditions, the key lies in explicitly incorporating SOH into the state and parameter estimation process, enabling online updates of parameters as they evolve with life. The parameters most closely related to SOP typically include the ohmic internal resistance R_0, RC network parameters, and equivalent capacity. In terms of deployable implementation, two complementary approaches may be adopted. The first is a parameter drift model, wherein parameters are treated as slowly varying processes and updated gradually at a constrained rate during online estimation to prevent excessive fluctuations due to short-term noise. The second approach involves hierarchical or piecewise SOH mapping, where the life cycle is segmented and temperature-dependent parameter mappings or lookup tables are maintained for each segment. Regardless of the chosen method, it is essential that the updated parameter set simultaneously supports both state estimation and subsequent constraint prediction, thereby avoiding inconsistencies such as using outdated parameters for prediction despite an updated SOH. Failure to ensure consistency may induce structural bias in margin calculations, which would be directly propagated into the SOP output.

At the battery pack level, cell-to-cell variations inherently introduce a bottleneck effect in SOP estimation. The pack-level available power is typically determined by the weakest individual cell rather than the average cell performance. Therefore, pack-level SOP estimation should first adhere to the weakest-cell principle: during discharge, the cell with the minimum voltage margin is most likely to trigger $V_{\min}$ first; during charging, the cell closest to $V_{\max}$ becomes the primary constraint; under thermal constraints, the cell with the highest temperature limits sustained power output. Meanwhile, incorporating the full state vector of each individual cell into the estimator would result in dimensionality growing linearly with cell count, making it computationally prohibitive and difficult to maintain in real-time applications. A more feasible approach involves adopting low-dimensional consistency representations that substitute full-dimensional cell states with cluster-level statistics—such as maximum single-cell temperature, minimum voltage margin, maximum internal resistance estimate, or polarization intensity indicators—and optionally including metrics like SOC dispersion to assess consistency. These low-dimensional features can effectively capture pack-level bottlenecks while maintaining deploy ability and directly feed into constraint prediction and dominant constraint identification within a Δt time window, thereby enabling consistent SOP estimation across cells.

Aging and temperature often simultaneously affect the R_0 and RC parameters, leading to challenges in online identification such as parameter collinearity and reduced identifiability. For instance, impedance changes induced by temperature variations may exhibit similar voltage response characteristics to impedance growth caused by aging at the observation level. Without proper distinction, estimators may erroneously attribute temperature effects to aging or absorb aging effects into polarization state bias, ultimately resulting in SOC drift, parameter divergence, or fluctuation in SOP output. To ensure estimation stability, engineering practices typically impose structural constraints on parameter updates. On one hand, parameter update rates are limited to allow life-related drifts to evolve on a slow timescale, thereby preventing overfitting driven by short-term noise. On the other hand, prior knowledge or piecewise mapping is introduced to hierarchically separate temperature-dependent and life-dependent effects, reducing the risk of misinterpretation due to collinearity. Furthermore, residual consistency or innovation statistics can be employed to gate the update process—freezing parameter updates when observational information is insufficient or when fitting reliability is questionable—thus maintaining continuity and maintainability of estimator outputs. These strategies are independent of specific algorithmic formulations but play a decisive role in ensuring the reliability of SOP state estimation across different lifetimes and temperatures.

Under aging and consistency-coupled conditions, the output of SOP estimation should not merely provide numerical values of SOP@Δt, but also include explanatory information related to lifetime evolution, to support calibration iteration and engineering diagnostics. It is recommended that, alongside the output of maximum allowable current ($I_{dis,max}$, $I_{chg,max}$) or maximum sustainable power ($P_{dis,max}$, $P_{chg,max}$), dominant constraint labels be appended to indicate whether the current SOP is primarily governed by under-voltage, over-temperature, device capability, or other constraints. Furthermore, key margin metrics—such as voltage margins relative to V_{min}/V_{max}, temperature margins relative to T_{max}, and consistency-related bottleneck indicators—should be simultaneously provided. These explanatory metrics enable a clear mapping between SOP variations and the battery's life stage as well as its consistency status, thereby allowing unambiguous attribution of SOP degradation rather than misinterpreting it as estimator instability or random operational fluctuations. The diagnostic feedback loop established through interpretable outputs also provides triggering signals and arbitration criteria for the subsequent section on multi-source information fusion and confidence management, thereby ensuring long-term consistency and engineering usability of SOP state estimation under multi-factor coupled conditions.

6.4.4 Coupled SOP estimation through multi-source information fusion

This section does not introduce new estimation algorithms, but instead addresses the practical engineering scenario characterized by coexisting temperature coupling, aging evolution, and consistency discrepancies. It discusses how to integrate multi-source information—such as voltage/current/temperature, thermal states, SOH indicators, consistency statistics, model priors, and learning outputs—into a single, consistent, interpretable, and diagnosable state estimate for the SOP. In contrast to the model-consistent and data-driven estimation pathways presented in Sections 6.2 and 6.3, multi-source fusion places greater emphasis on evidence complementarity and unified interpretation frameworks. Under inevitable conditions such as measurement noise, parameter drift, and distribution shifts, a stable SOP@Δt estimate is achieved through parallel candidate estimation, confidence arbitration, and constraint-consistent projection, while simultaneously providing information on dominant constraints and confidence levels.

Under multi-factor coupling conditions, a single pathway often fails to simultaneously satisfy requirements for accuracy, generalization, and interpretability. The model-consistent path offers clear state semantics and constraint interpretability; however, when parameter mismatches arise due to temperature variations and SOH evolution, systematic biases may emerge and propagate into the Δt margin prediction. The data-driven path can capture complex coupling relationships and latent patterns, yet its generalization error and interpretability risks significantly increase under out-of-domain operating conditions or distributional shifts. Therefore, the core value of multi-source fusion lies in enhancing observability and robustness through information complementarity. Temperature-coupling information supports electro-thermal joint estimation and thermal-constraint-triggered timing, while SOH and consistency evidence help explain parameter drifts and pack-level performance bottlenecks. Meanwhile, the two candidate estimation results provide comparative references and redundancy for fusion arbitration. This logic directly aligns with Sections 6.4.2 and 6.4.3: temperature and aging correspond to slow-time-scale parameter evolution, while consistency determines the dominant pack-level bottleneck—both necessitating evidence fusion to establish a stable state-of-pack (SOP) estimation closed-loop.

To ensure comparability in the fusion process, the multi-source inputs must first be standardized and organized into evidence items that can be fed into the estimator and arbitrator. It is recommended to categorize inputs into three types: the first type comprises measurement evidence, including synchronized, denoised, and drift-checked $V/I/T$ sequences along with their health scores, which support state updates and confidence assessment; the second type consists of state and parameter evidence, including SOC, polarization, thermal equivalent states, SOH parameter sets, and low-dimensional consistency statistics of pack-level weaknesses; the third type refers to candidate SOP evidence—candidate SOP@Δt values derived from model paths and data paths, or equivalently, candidate estimates of constraint margins or triggering times. More critically, consistency in terminology must be emphasized: all evidence must be bound to the same time window Δt and the same constraint set; otherwise, outputs from different sources become mathematically incomparable, rendering fusion meaningless.

On the basis of unified evidence standards, multi-source fusion can be generalized into three deployable estimation templates. The first is model-driven with data correction. It takes the state interpretation, constraint predictions, and interpretable boundaries provided by a consistent model path as the core, and corrects system-level biases caused by temperature or aging mismatches using data-driven outputs or statistical bias terms. This approach enhances cross-condition accuracy while preserving physical consistency. The second is data-driven with constraint projection. It employs a learned model to directly generate SOP@Δt or margin candidates, which are then projected into a unified feasible constraint domain to ensure outputs comply with V_{min}/V_{max}, T_{max}, I_{max}, and SOC boundaries; dominant constraint annotations are additionally used to enhance interpretability. The third is dual-channel parallelism with confidence arbitration. Both model-based and data-based paths generate SOP candidates and quality scores in parallel, with confidence assessment determining either weighted fusion or path switching; this template is particularly suitable for scenarios involving rapid temperature changes, lifecycle transitions, or fluctuating data quality, enabling more robust coupled estimation without introducing additional complexity. All three templates share the common requirement that fusion and output must adhere to the same Δt and constraint set, while retaining intermediate variables such as candidate estimates and constraint margins to support diagnostic closed-loop operations.

To prevent uncontrolled outputs under abnormal measurements, parameter drift, or out-of-distribution operating conditions, an explicit confidence management mechanism must be introduced. Confidence sources can be categorized into four types: residual/innovation statistics reflect current fitting consistency; consistency between voltage and temperature trajectories indicates prediction plausibility; distribution shift indicators assess whether inputs deviate from the training/calibration domain; and sensor health status evaluates measurement chain reliability. Based on confidence levels, arbitration logic can be implemented: when one evidence source becomes unreliable, its weight is reduced or its update is frozen, and in critical cases, the system switches to a more trustworthy alternative output. When both sources are deemed unreliable, a degraded estimation mode is triggered, producing a conservative safety-critical candidate—such as a safe lower bound for SOP@Δt derived from worst-case statistics and piecewise mapping. It should be emphasized that this degradation pertains to confidence management at the estimation layer, aiming to maintain output continuity and safety-consistent behavior rather than implementing specific control derating strategies. The final output must simultaneously include the SOP candidate value, dominant constraint label, confidence level, and key margin information—enabling the system to answer: what constraints are currently affecting SOP. How much margin remains? Is this result trustworthy—thereby establishing a diagnosable closed-loop system.

Multi-source fusion provides a unified estimation interface and evaluation framework, which does not replace Sections 6.2 or 6.3, but rather incorporates both as candidate estimators and evidence sources within a single fusion framework. Through confidence arbitration and constrained consistent projection, deployable SOP@Δt results are achieved. Under multifactor coupled conditions, consistent state estimation of SOP across temperature, lifetime, and individual cells should be ensured by adhering to constrained consistency as the output baseline, confidence controllability as the engineering safeguard, and explainable outputs to support calibration iteration and diagnostic closed-loop processes.

Reference

[1] W. Qian, D. Zhao, X. Guo, et al., Research on adaptive unscented H∞ filtering for state of charge estimation of lithium batteries, Energy Storage Sci Technol 13 (11) (2024) 4078–4088.

CHAPTER SEVEN

Joint estimation of battery states

7.1 Theoretical background

7.1.1 Coupled characteristics of battery states

The core states of a battery primarily include the State of charge (SOC), State of health (SOH), Temperature (T), and State of power (SOP), with their coupling representing the central challenge in Battery management system (BMS) control.

7.1.1.1 Coupling between state of charge and state of health

The coupling between SOC and SOH represents the most fundamental state relationship throughout a battery's entire life cycle. These two parameters interact dynamically and directly determine the accuracy of battery SOC management and the battery's service life [1].

The influence of SOH on SOC is realized based on the principle of reference offset. The essence of SOC lies in the ratio of remaining capacity to the current maximum available capacity. As SOH degrades, the "current maximum available capacity" serving as the reference baseline, diminishes directly, thereby affecting the accuracy of SOC estimation. The mechanism underlying the reference offset principle varies according to battery aging level: for a new battery (SOH = 100%), the initial rated capacity equals the current maximum available capacity; at this stage, an SOC of 50% corresponds to an actual remaining charge of 25 Ah, with estimation error maintained within 5%. For an aged battery (SOH = 80%), the current maximum available capacity decreases proportionally due to SOH degradation, reaching 40 Ah [2]. If estimation still relies on the initial 50 Ah reference, an SOC reading of 50% would be incorrectly interpreted as 25 Ah; however, the actual remaining charge is only 20 Ah, resulting in a false positive deviation exceeding 25%. When severe aging occurs (SOH = 60%), the current maximum available capacity further declines to 30 Ah. Under identical estimation logic, an SOC value of 50% corresponds to only 15

Modelling, Estimation and AI Applications for Lithium-Ion Battery Management Systems
ISBN: 978-0-443-45347-2
https://doi.org/10.1016/B978-0-443-45347-2.00007-8

Ah of actual remaining charge, leading to a deviation exceeding 40%. Therefore, the impact of SOH on SOC estimation is positively correlated with SOH degradation magnitude, ultimately rendering BMS incapable of accurately determining residual charge.

The influence of SOC on SOH is mediated through the long-term maintenance of SOC within specific ranges, which disrupts electrode structure and accelerates or decelerates side reaction rates, thereby directly affecting the rate of SOH degradation. When SOC is maintained above 90% for prolonged periods, lithium ions continuously occupy interstitial sites in the cathode lattice, inducing lattice distortion and active material detachment. This leads to "lattice collapse" in the cathode material due to excessive lithiation, directly reducing the number of lithium ions available for electrochemical reactions [3]. Concurrently, at high charge states, electrolyte decomposition occurs continuously on the cathode surface, generating impurities that block lithium ion transport pathways and further diminish the battery's actual capacity. As a result, SOH degrades approximately 5% to 8% more per year compared to normal operating conditions. Conversely, when SOC is maintained below 10% for extended durations, the graphite interlayer spacing shrinks due to insufficient lithium ion support, impairing subsequent lithium ion intercalation during charging. This results in structural disordering of the anode material due to lithium depletion. More critically, under low charge conditions, lithium ions fail to intercalate into the anode in a timely manner and instead deposit as metallic lithium on its surface. Lithium dendrites are hard and poorly conductive; they not only consume active lithium ions but also progressively penetrate the separator over time, triggering internal microshort circuits and causing SOH degradation exceeding 10% within 3–6 months.

The coupling relationship between SOC and SOH does not exist in isolation but rather induces a "chain reaction" that affects battery charge management, lifespan, and safety, with their core impacts being far from negligible. This SOC and SOH coupling further amplifies the adverse effects of temperature and power states, resulting in a "multistate coupling risk accumulation." SOC serves as the fundamental basis for BMS control over charging and discharging; coupling induced SOC overestimation can cause irreversible damage to both charging and discharging ends. In practical applications, this leads to insufficient actual energy storage capacity. When SOC overestimation results in "misjudgment of charge level," continued discharge may trigger excessive delithiation at the negative electrode, leading to "over discharge", which causes permanent structural damage. Under such conditions, SOH may degrade abruptly by 5% to 10%, and severe cases may induce internal short circuits. Moreover, the coupling between SOC and SOH can initiate a vicious cycle of "mutual deterioration," significantly shortening battery lifetime, as illustrated in Fig. 7.1.

7.1.1.2 Coupling of SOP with SOC, SOH, and temperature

SOP refers to the maximum instantaneous power that a battery can currently deliver or absorb. Its performance does not exist independently but is strongly coupled with temperature, SOH, and SOC. By influencing the rate of electrochemical reactions within the battery, ion transport efficiency, and structural stability, these factors collectively determine the battery's power status, serving as a comprehensive manifestation of multiple states.

As shown in Fig. 7.2, temperature is a key variable determining the rate of electrochemical reactions in batteries, directly influencing the SOP through the coupled effect of "ion migration velocity + charge transfer resistance," exhibiting a significantly nonlinear relationship. When the temperature falls below 0°C, electrolyte viscosity increases markedly, resulting in a substantial reduction in lithium ion migration speed within the electrolyte. Concurrently, charge transfer resistance at the electrode surface rises sharply, impairing the battery's ability to rapidly complete ion deintercalation and charge exchange. Under such conditions, even with sufficient SOC and SOH, SOP deteriorates dramatically. At ambient temperature, battery power performance reaches its optimum: electrolyte viscosity is moderate, lithium ion migration resistance is low, and interfacial charge transfer efficiency between electrodes and electrolyte is high, enabling stable output/input of the maximum designed instantaneous power. At this point, temperature exerts minimal constraint on SOP, while SOH and SOC become more dominant factors. When temperature exceeds 45°C, ion migration resistance further decreases, potentially leading to a slight increase in SOP. However, when temperature surpasses 60°C, internal side reactions, such as electrolyte decomposition and cathode material structural collapse intensify rapidly, causing accelerated SOH degradation. Moreover, to prevent thermal runaway, the battery management system actively restricts SOP. In this scenario, the decline in SOP is not due to insufficient capability but rather an intentional constraint imposed for safety reasons.

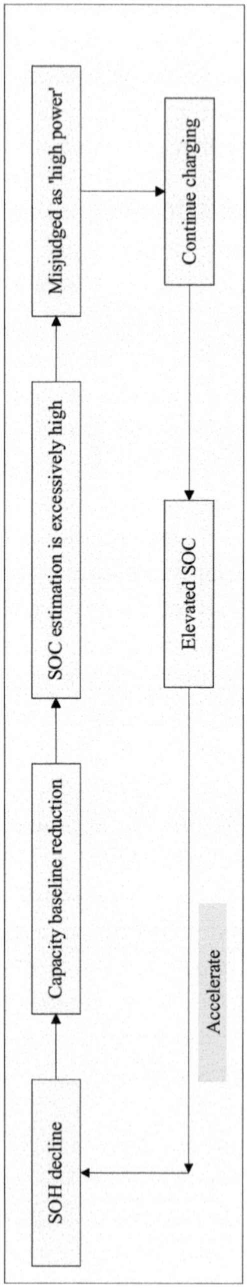

Figure 7.1 SOH and SOC coupling vicious cycle.

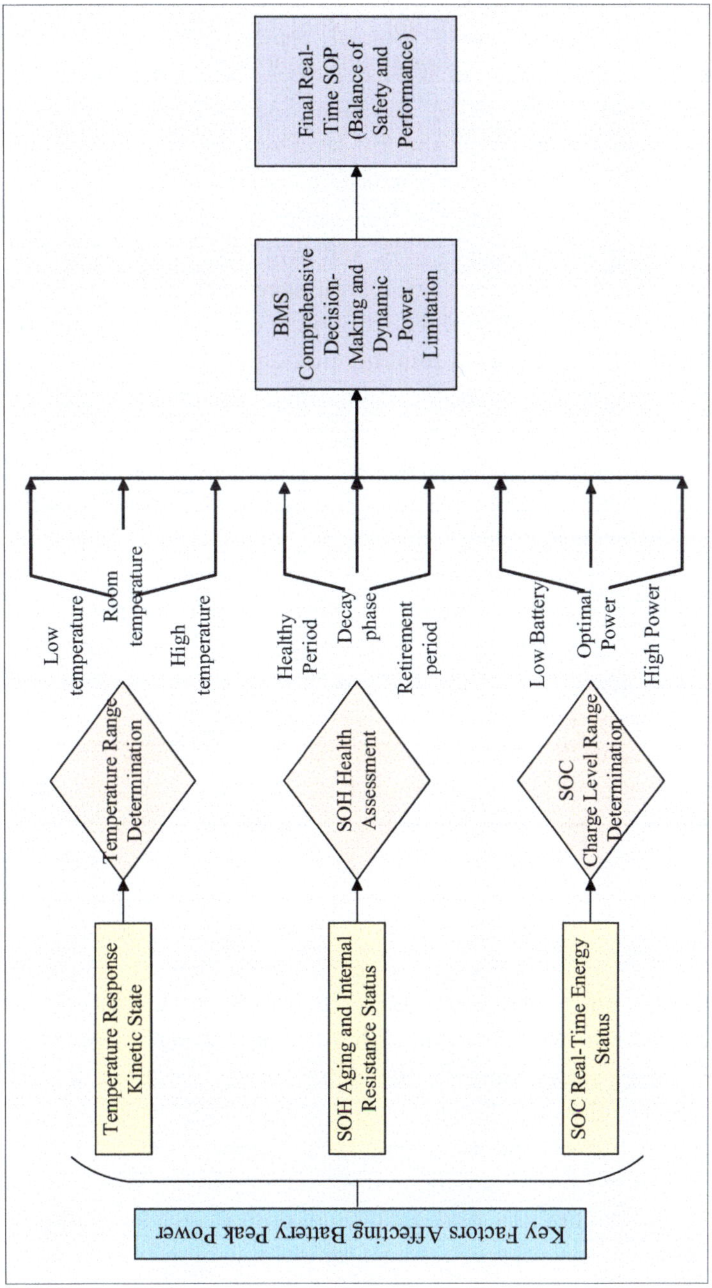

Figure 7.2 Coupling relationships among SOP, temperature, SOH, and SOC.

SOH reflects the degree of aging in core battery performance such as capacity and internal resistance. Its degradation weakens SOP through a dual mechanism of "capacity loss + internal resistance growth," with the coupling effect accelerating as SOH decreases. When the battery is in a healthy state (SOH above 80%), the loss of active materials in both the anode and cathode is minimal, the SEI layer remains thin and stable, and both ohmic and polarization resistances increase slowly. Consequently, SOP experiences only a slight decline, imperceptible during daily use, with SOP primarily regulated by SOC and temperature. When SOH drops below 80% but remains above 50%, prolonged charge/discharge cycling leads to cathode lattice collapse, reduction in anode active sites, thickening and instability of the SEI layer, resulting in a significant increase in polarization resistance. Under conditions requiring instantaneous high power output, voltage drop due to internal resistance rises sharply, causing the battery terminal voltage to rapidly fall below or exceed its limits. As a result, the BMS is forced to limit power output, leading to a marked acceleration in SOP degradation. When battery damage becomes severe, capacity reduced to less than half of its initial value and internal resistance increased to two to three times its original level, even under normal temperature and full SOC conditions, SOP may drop to only 30%–40% of that of a new battery, failing to meet user requirements. This condition thus becomes one of the primary criteria for battery retirement.

SOC represents the percentage of the battery's remaining capacity. The available range of SOP is determined by the "ion concentration gradient + electrode potential constraints," exhibiting a characteristic "U shaped" coupling relationship. When SOC is below 20%, SOP is constrained by "ion deficiency + risk of lithium plating/structural degradation." At this stage, the amount of lithium ions intercalated into the anode is low, resulting in a small concentration gradient for reversible ion insertion/extraction, which prevents rapid fulfillment of high power discharge demands. Moreover, if high current charging is forced under such conditions, lithium ions on the anode surface are prone to metallic lithium plating due to "insertion rate < deposition rate," potentially damaging the battery and triggering safety hazards. Consequently, the BMS strictly limits both charging and discharging SOP. When SOC exceeds 80%, the lithium ion concentrations in both electrodes are moderate, ensuring a stable concentration gradient for ion migration and electrode potentials within safe ranges. The battery can stably deliver its designed maximum instantaneous power output, this constitutes the "golden operating zone" for SOP. Under these conditions, temperature and SOH become the primary influencing factors. However, when SOC surpasses 80%, the cathode approaches full delithiation. Continuing high current discharge may cause excessively high cathode potential, accelerating active material dissolution. Conversely, during high current charging, the cathode cannot rapidly accept lithium ions, increasing the likelihood of electrolyte oxidation and decomposition. Therefore, the BMS progressively reduces both charging and discharging SOP under such conditions.

7.1.1.3 Impact of battery state coupling characteristics on BMS

The coupling characteristics of battery state directly determine the design logic of the core functions of the BMS, compelling the BMS to shift from "single state monitoring" to "multistate collaborative control," and thereby posing critical challenges in four dimensions: algorithmic complexity, control accuracy, safety redundancy, and dynamic response speed, as shown in Fig. 7.3.

7.1.1.3.1 The BMS adopts a "multistate coupling algorithm," significantly increasing algorithmic complexity

Since SOP is jointly determined by SOC, SOH, and temperature, and any change in one state induces cascading effects on the others, the BMS cannot

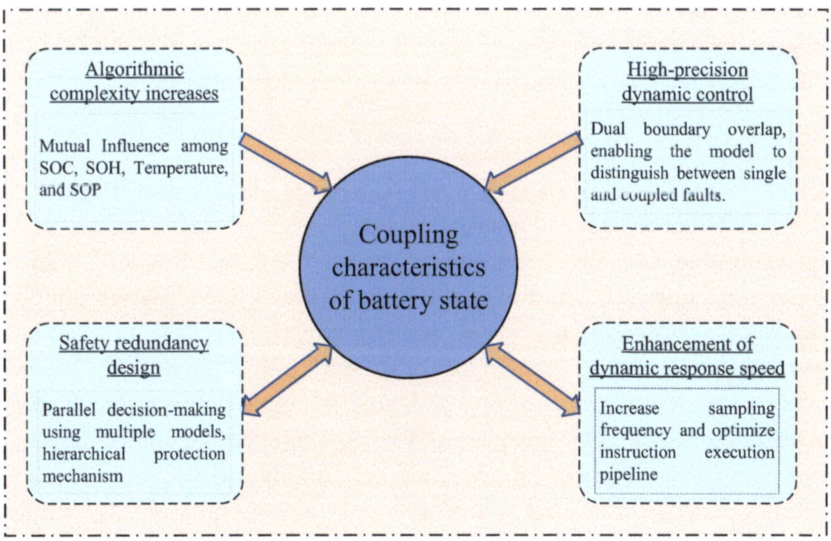

Figure 7.3 Coupling characteristics of battery state in the multidimensional challenges of BMS.

independently calculate any single state. Instead, it must establish a "multi-variable coupled model." First, the BMS accumulates SOC cycling data across multiple temperatures; then, by applying filtering algorithms to eliminate temperature induced interference, SOH can be precisely recalibrated. Subsequently, SOH baseline parameters and temperature compensation factors are incorporated to ensure high accuracy in SOC estimation. Moreover, the BMS must not compute SOP based solely on fixed formulas. Rather, it must first determine the "power base upper limit" according to the current temperature, then adjust the "degradation coefficient" based on SOH aging level. Finally, by considering the current SOC range, it outputs an SOP upper limit that achieves a balanced trade off between safety and performance. This algorithm requires realtime processing of cross variable computations involving multiple parameters.

7.1.1.3.2 The BMS possesses "high precision dynamic control capability," balancing safety and performance

The state coupling characteristics cause a high overlap between the "safety boundary" and "performance boundary" of batteries, requiring the BMS to identify an optimal balance point amid dynamically changing multistate conditions. Upon encountering fault warnings, the BMS must determine whether the anomaly stems from "coupled state failure," as a single state abnormality may actually result from multistate coupling. The BMS must distinguish between "single faults" and "coupled faults" to avoid false triggering of protective measures or missed detection risks that could lead to battery damage. This necessitates the BMS establishing a "multistate anomaly correlation model."

7.1.1.3.3 Promote BMS enhancement of "safety redundancy design" to address extreme coupling scenarios

When multiple states simultaneously fall within the "critical region" of low temperature, low SOC, and low SOH, the battery's safety margin rapidly diminishes, necessitating robust risk mitigation by the BMS through redundant design across hardware, algorithms, and control strategies. On the hardware side, "multisensor cross validation" is employed to monitor critical parameters such as temperature and voltage, preventing erroneous state estimation due to failure of a single sensor. On the algorithmic front, parallel multimodel decision making is implemented, running multiple algorithms concurrently to compute battery states; if the upper bound deviation among outputs from different algorithms exceeds 10%, redundancy verification is triggered to ensure power limitation remains effective under extreme coupled conditions. Additionally, a tiered protection mechanism is established, dividing protective logic into three levels: "warning, current limiting, and power disconnection." This tiered response prevents overreaction and avoids "all or nothing" protection triggered by abrupt changes in a single parameter.

7.1.1.3.4 The BMS enhances "dynamic response speed" to match realtime state variations

The coupling effect of battery state exhibits "instantaneity"; if the BMS response is delayed, it may lead to state parameters exceeding safety boundaries within a short period. Therefore, in addition to conventional voltage and temperature acquisition, integrating coupling dynamics with increased sampling frequency is essential to meet realtime requirements under coupling scenarios, ensuring timely capture of variations in temperature, SOC, and voltage. Furthermore, after the BMS computes the upper limit of SOP, a high speed communication protocol should be adopted along with optimization of the instruction execution path to rapidly transmit commands to the charger or motor controller.

In summary, the coupled characteristics of battery state constitute the "core constraint" and "technical orientation" in BMS design. This necessitates a transition from "passive monitoring" to "active collaborative control" in BMS, achieving a balanced trade off among safety, performance, and lifespan through more sophisticated algorithms, higher precision, enhanced redundancy, and faster response times. This evolution represents the core driving force behind current iterations of BMS technology.

7.1.2 Limitations of independent estimation

Against the backdrop of rapid development in the new energy industry, batteries, as core energy components, directly determine the safety, reliability, and economic efficiency of critical applications such as new energy vehicles and energy storage systems. Battery state estimation, a central function of BMS, encompasses key parameters including SOC, SOH, temperature, and SOP. The accuracy of estimating these parameters constitutes the foundation for ensuring stable system operation. Currently, some BMSs still employ a "decoupled estimation" approach, ignoring the inherent coupling characteristics among various state parameters, and instead perform isolated, noncooperative modeling and computation for SOC, SOH, SOP, and other parameters using separate algorithms or models. However, battery operation involves a complex integration of electrochemical reactions and thermodynamic processes; the coupling relationships among state parameters are intrinsic physical properties. The decoupled estimation paradigm fundamentally contradicts this underlying physical reality.

This section will systematically analyze the fundamental limitations of independent estimation when battery state coupling characteristics are neglected, conducting a comprehensive examination from multiple perspectives including inherent technical principle deficiencies, practical application risks, failure in full life cycle management, and insufficient adaptability to diverse operating scenarios. The analysis aims to provide theoretical foundations and practical references for subsequent optimization of battery state estimation models and enhancement of BMS overall performance.

7.1.2.1 The inherent limitations of independent estimation when coupling characteristics are not considered

7.1.2.1.1 Estimation accuracy is severely distorted, with deviations in key parameters triggering a "chain reaction" effect

The core parameters of a battery, such as SOC, SOH, and SOP, are not isolated but dynamically coupled through electrochemical reactions and thermodynamic processes. Estimating these parameters independently, without accounting for their coupling characteristics, causes individual parameter estimations to deviate significantly from the actual physical and chemical operating conditions of the battery. This deviation not only compromises the accuracy of each parameter's estimation but also propagates through the coupling relationships to other parameters, resulting in a "cascading error effect" that severely undermines the accuracy of battery state assessment by the BMS.

7.1.2.1.2 Prominent safety and reliability concerns, with a significant increase in the risk of thermal runaway and failures

The safety boundary of batteries is jointly determined by the coupled relationship among "SOC -T -SOH." Under different parameter combinations, significant variations exist in the critical conditions for thermal runaway, lithium plating risk, and internal resistance heating characteristics. Independently estimating these parameters severs this interdependence, leading to failure of safety warnings and malfunction of protective mechanisms, thereby introducing serious risks into the system.

7.1.2.1.3 Failure of life cycle management leads to a significant reduction in economic efficiency and resource utilization

Battery full life cycle management must be based on the coupled relationships among SOC, T, SOH, and SOE to optimize charging strategies, replacement cycles, and residual value assessment. Independent estimation leads to a disconnect between management strategies and actual battery states, resulting in economic losses and resource waste.

7.1.2.1.4 Lack of scene generalization capability, unable to adapt to diverse application requirements

Under extreme environmental conditions such as cold, high temperature, and high altitude, the coupled characteristics of batteries become more pronounced. Independent estimation methods fail to reflect the influence of temperature on battery states such as SOC and SOH, resulting in an inaccurate assessment of battery operating conditions. Consequently, timely protective measures cannot be implemented, causing batteries to remain in an overloading state and leading to irreversible damage. Furthermore, certain devices must operate across multiple scenarios; independent estimation cannot dynamically adjust coupling parameters in real time, resulting in functional failure and inability to meet user requirements, thereby delivering a poor user experience.

7.1.2.2 Comparative analysis of core characteristics between independent estimation and coupled estimation

To clearly reveal the limitations of independent estimation, the following table provides a comprehensive comparison between independent and coupled estimation from perspectives including core concepts, model design, parameter accuracy, and so forth, as shown in Table 7.1.

In summary, the independent estimation approach that disregards battery coupling characteristics is an outdated technological scheme based on oversimplified assumptions. Its fundamental flaw lies in violating the objective coupling laws of battery parameters, resulting in low estimation accuracy, high safety risks, and poor management efficiency, rendering it entirely incapable of meeting the stringent requirements of modern new energy systems regarding safety, reliability, and lifespan. As the new energy industry transitions toward high quality development, the application of collaborative estimation techniques enables precise sensing and intelligent management of battery states, providing a critical foundation for the safe and efficient operation of new energy systems and driving industry advancement toward greater reliability, economic efficiency, and environmental sustainability.

Table 7.1 Comparison between independent estimation and coupled estimation.

Characteristics	Independent estimation	Coupling estimation
Core idea	Estimate SOC, SOH, and SOP separately and independently.	Collaboratively estimate all states within a unified framework or model
Model	Fixed parameter model	Adaptive model with parameters online identified according to SOH and operating conditions
SOC accuracy	Low, prone to accumulating errors	High, capacity is calibrated via SOH, and adapts to aging through model parameter updates
SOH accuracy	Lag, inaccuracy	Real time/quasi realtime, with accuracy enhanced through collaboration with SOC
System complexity	Low	High (requiring more powerful processors and more complex algorithms, such as dual Kalman filtering, particle filtering, machine learning, etc.)
Result	Inaccurate state estimation, shortened lifespan, and high safety risks	Accurate state estimation, life optimization, safe and reliable operation

7.1.3 Full life cycle health management

7.1.3.1 Overview of battery full life cycle health management

The full life cycle of a battery encompasses the entire process from design and manufacturing to recycling, including stages such as raw material acquisition, production and manufacturing, transportation and storage, application and usage, and final recycling treatment. Each stage significantly influences the battery's performance, lifespan, environmental impact, and economic viability. Battery health management refers to the realtime monitoring, prediction, management, and optimization of a battery's health status throughout its entire life cycle. Its core objective is to extend battery lifespan, optimize performance, and ensure safety. Battery health management spans the entire life cycle of the battery, with health management strategies at each stage interacting synergistically. Properly managing the battery's full life cycle not only maximizes its utilization value but also reduces environmental pollution and enhances resource efficiency, thereby promoting sustainable development. Fig. 7.4 illustrates the various stages of battery life cycle health management.

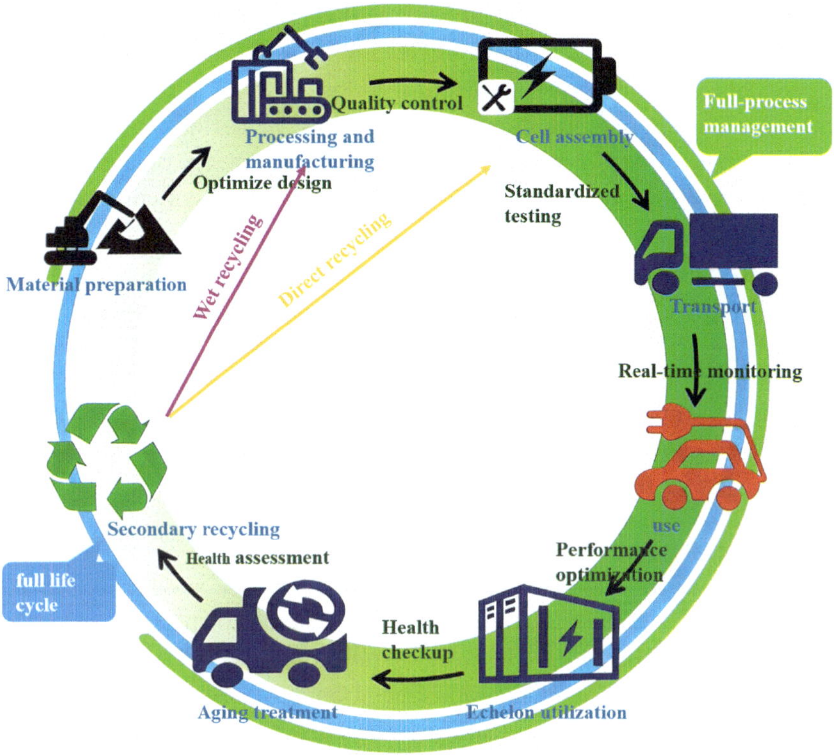

Figure 7.4 Life cycle management process diagram.

7.1.3.1.1 Design phase

The battery design phase marks the beginning of the entire life cycle, directly determining the battery's fundamental performance, safety level, and potential lifespan, while also laying the foundation for subsequent manufacturing, usage, decommissioning, and recycling. A meticulously designed battery ensures efficient energy storage, extended service life, safe operating conditions, and feasibility of future recycling. Consequently, the battery design phase profoundly influences performance, longevity, costeffectiveness, and environmental impact; therefore, it must be carefully planned and implemented through comprehensive consideration.

7.1.3.1.2 Manufacturing phase

The manufacturing phase represents the stage at which design specifications are transformed into tangible products, during which the battery's performance, quality, and reliability begin to take definitive form. The core objective of battery health management is to ensure that each individual cell meets established quality standards, thereby preventing performance issues arising from manufacturing defects. This phase entails not only rigorous quality control but also careful selection and processing of raw materials to guarantee consistency among cells and stability in performance. Battery health management during manufacturing typically relies on well established production processes and advanced testing equipment, enabling comprehensive evaluation of multiple parameters for each batch to ensure compliance with design specifications.

At this stage, battery health monitoring primarily focuses on physical and chemical characteristics. In particular, initial capacity, cycle performance, and safety testing are the key aspects of health management during this phase. By monitoring these critical indicators, potential manufacturing defects can be effectively identified, thereby reducing the risk of premature failure.

7.1.3.1.3 Initial usage phase

During the initial phase of battery usage, performance is typically stable; however, precise battery management is still required to ensure continuous performance optimization. The primary objective of battery management during this phase is to maintain stable operation during early usage and prevent performance degradation or safety hazards caused by environmental factors such as temperature, humidity, and charge/discharge rates. Estimation of the SOC and SOH constitutes a key task during this stage. Fig. 7.5 illustrates the curves representing different battery stages as SOH gradually declines.

During the initial usage phase, the battery's state of health is typically good, and the primary management tasks involve preventing adverse operating conditions such as overcharging, overdischarging, and overheating. Based on realtime battery status data, the BMS automatically adjusts the charging strategy, controlling charging current and voltage to prevent battery overcharging. Simultaneously, realtime monitoring of battery temperature helps avoid thermal runaway caused by excessive heat. Table 7.2 outlines the management objectives and strategies during the early stage of battery usage.

7.1.3.1.4 Intermediate utilization phase

As battery usage time increases, entering the mid life stage, battery capacity and internal resistance gradually change, leading to performance degradation. This phase represents a critical period in the lithium ion battery life

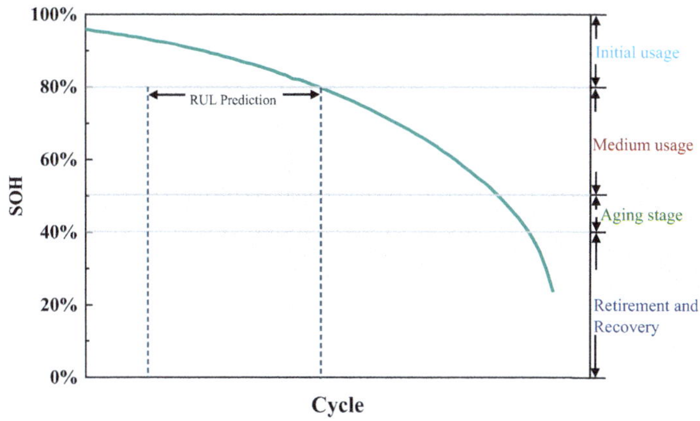

Figure 7.5 Schematic diagram of battery stages divided according to the decline curve of SOH.

Table 7.2 Key points for health management during the initial usage phase.

Management objectives	Key parameters	Management policy	Major risks
Ensure performance stability	SOC, SOH	Accurate SOC estimation and SOH monitoring	Overcharge, overdischarge, high temperature
Enhancing security	Battery temperature, voltage	Temperature control and voltage regulation	Thermal runaway, short circuit
Prolong lifespan	Number of cycles, current fluctuation	Limit high-power charging and discharging	Battery degradation

cycle, posing greater challenges for battery health management. The primary management objective is to accurately predict battery health status by monitoring states such as SOC and SOH, and to establish corresponding early warning strategies.

During this stage, the battery's SOH begins to decline, and both its capacity and internal resistance gradually change with increasing cycle counts. By continuously monitoring the battery's charge/discharge data, temperature, internal resistance, and other parameters, the SOH can be estimated and potential failures can be predicted in advance. To extend the battery's service life, aging can be slowed by limiting excessively high charge and discharge rates and optimizing charging strategies.

7.1.3.1.5 Aging stage

When a battery enters the aging phase, its capacity and internal resistance decline significantly, rendering its performance unable to meet initial requirements; however, the residual energy it retains can still power certain low demand applications. The primary objective of battery health management during this phase is to maximize the remaining value of the battery while ensuring no safety incidents occur prior to its retirement. At this stage, the focus shifts toward establishing retirement criteria and determining whether to continue using, recycling, or repurposing the battery based on its state of health.

During this phase, battery health monitoring should focus on changes in parameters such as residual capacity, internal resistance, and leakage current. By analyzing the battery's degradation trends, the usage patterns of the battery can be reasonably planned. Prior to the end of its service life, battery charging and discharging strategies can be optimized based on health data, thereby extending the effective operational lifespan while ensuring safety and recyclability.

7.1.3.1.6 Retirement and recovery phase

The primary task of health management during this stage is to evaluate whether the battery possesses reuse potential or whether it needs to be disposed of in accordance with environmental protection standards. As battery performance degrades, its reuse value gradually diminishes; however, under certain conditions, retired batteries that have undergone appropriate treatment can still be utilized in low power applications or energy storage systems.

For the management of retired batteries, the first step is to evaluate the SOH and Remaining useful life (RUL) of the batteries to determine whether they meet the criteria for reuse. If batteries are no longer suitable for reuse, they must be disassembled and recycled in accordance with environmental protection standards to prevent environmental pollution. At the same time, battery health management during the retirement phase should ensure that data from all recycled batteries are properly managed and traceable. Table 7.3 outlines the management objectives and strategies for the battery retirement and recycling phase.

7.1.3.2 Objectives of life cycle health management

The objective of full life cycle health management is to optimize battery usage, maintenance, and retirement strategies at every stage of battery

Table 7.3 Key points for health management during the retirement and recovery phase.

Management objectives	Key parameters	Management policy	Major risks
Establishing the value of reuse	SOH, internal resistance, capacity	Assessing battery remaining performance and reuse criteria	Environmental pollution, resource waste
Safe handling	Battery chemistry	Dispose of batteries in accordance with environmental protection requirements	Hazardous material spill

operation by monitoring SOH variations, thereby maximizing performance, extending service life, and ensuring safety.

1) Extended service life: By implementing precise health assessment and management strategies, the effective service life of batteries can be prolonged, delaying their retirement and reducing replacement frequency.
2) Enhanced safety: Ensure the battery remains in a safe and controllable state throughout its entire life cycle, preventing hazardous conditions such as overcharging, overdischarging, short circuits, and overheating.
3) Enhancing efficiency and reliability: By enabling realtime monitoring and data analysis, the battery's operating mode can be dynamically adjusted, thereby improving operational efficiency and reliability, and preventing unnecessary losses caused by failures.
4) Reduce environmental impact: By optimizing usage and recycling management, minimize waste generated after battery disposal and promote the sustainable development of lithium ion batteries.

7.1.3.3 Application of joint estimation in whole life cycle health management

The joint estimation method, by comprehensively considering multiple battery state variables, enables a more thorough understanding of battery health. This approach can identify coupling relationships among different states and provide more accurate information for battery management systems, thereby effectively overcoming the limitations of conventional single state estimation methods. In contrast to traditional methods that rely solely on the estimation of a single state, joint estimation allows for integrated analysis of multidimensional battery health data, resulting in more precise and comprehensive health assessment. This is particularly crucial for the full life cycle management of batteries, especially during their usage, maintenance, and end of life phases, ensuring adequate protection of battery performance, lifespan, and safety.

7.1.3.3.1 Multistate comprehensive evaluation

The health condition of a battery is not determined by a single state variable, but rather influenced by the interaction of multiple factors. Traditional battery state estimation often relies on a single state variable for prediction and decision making. However, battery health is jointly affected by several factors, including internal resistance, temperature, and capacity. Relying solely on one estimation method may lead to misjudgment of the battery's health status. Joint estimation not only avoids the risks associated with misleading single indicators but also enables more detailed analysis of battery state based on multidimensional health data. Table 7.4 provides a comparison between joint and individual assessment methods.

7.1.3.3.2 Real-time monitoring and dynamic adjustment

During battery operation, the battery's health condition deteriorates to varying degrees over time and under different operating conditions. To ensure long term performance and safety, realtime monitoring and dynamic adjustment have become indispensable functions of the BMS. These capabilities enable the BMS to respond promptly to changes in battery state, preventing the battery from entering hazardous or suboptimal operating conditions. Through joint estimation, the BMS can continuously monitor critical parameters during battery operation, leveraging historical data and predictive models for accurate state forecasting. Simultaneously, it can dynamically adjust the battery's operational mode. Battery performance is influenced by environmental conditions during use, and variations in these

Table 7.4 Comparison of advantages in multistate comprehensive evaluation.

State estimation methods	Advantages	Limitations
Single estimation	Simple, fast	Ignoring other influencing factors, the accuracy is poor
Joint estimation	Providing a comprehensive health assessment	Requires more computational resources, high algorithmic complexity

factors can lead to changes in battery health. By jointly estimating multiple state variables, the BMS can detect early signs of health degradation and accordingly adjust the battery's operating mode. Table 7.5 lists several real-time adjustment strategies.

7.1.3.3.3 Health management optimization

The joint estimation method provides a scientific basis and accurate prediction for whole life cycle health management, enabling optimization of battery management strategies and thereby enhancing battery utilization efficiency. During the early usage phase, joint estimation facilitates assessment of battery performance degradation and allows adjustment of charging and discharging strategies to improve battery efficiency. In the mid and late life phases, joint estimation enables optimization of battery usage strategies according to the battery's health condition, thereby maximizing its service life. During the decommissioning phase, joint estimation enables comprehensive evaluation based on multiple states, including SOC, SOH, and RUL to determine whether batteries are suitable for secondary use or require recycling. Accurate estimation of battery health status ensures timely decommissioning or recycling when batteries can no longer be used, thus preventing adverse environmental impacts. Table 7.6 summarizes applications of battery health management across different stages of the battery's life cycle.

By optimizing battery usage patterns and resource recovery strategies, joint estimation can contribute to more efficient battery management and promote the sustainable development of battery life cycles. Battery health management is no longer limited to extending battery lifespan; it also involves the proper handling of spent batteries, reducing environmental

Table 7.5 Battery adjustment strategies under different states.

Battery status	Adjustment measures	Objectives and outcomes
SOH decreases.	Limit the charging rate and reduce the charging depth.	Slow battery degradation and reduce the risk of overcharging.
Increased internal resistance.	Limit the charging rate and reduce the charging depth.	Reduce energy losses caused by internal resistance and minimize the risk of thermal runaway.
The temperature is too high.	Initiate the cooling system to reduce charging or load stress.	Prevent battery faults or failures caused by excessively high temperatures.

Table 7.6 Application of joint estimation in different stages of battery health management.

Phase	Applications of joint estimation	Advantages
Design and manufacturing phase	By predicting battery performance, optimizing manufacturing processes, and ensuring quality control	Enhance production quality, reduce defects, and optimize design solutions
Usage phase	Real-time monitoring of SOC, SOH, internal resistance, temperature, etc., dynamically adjusting the battery's operating state	Enhance battery utilization efficiency, extend battery lifespan, and reduce failure occurrences
Aging stage	Assessing the battery degradation process, optimizing charging and discharging strategies, and delaying aging	Provides real-time health assessment, supporting battery maintenance and replacement decision making
Retirement and recovery phase	Assessing the recycling value and applicability of batteries to support resource reuse	Enhance recycling efficiency, reduce environmental pollution, and promote sustainable development

burden, improving resource utilization efficiency, and advancing the development of a circular economy.

7.2 Traditional joint estimation methods

7.2.1 Extended Kalman filter based methods

7.2.1.1 Analysis of iterative computation in extended Kalman filtering

The Extended Kalman filter (EKF) algorithm is a nonlinear extension of the classical Kalman filter. Its core lies in addressing nonlinear problems through local linearization. Specifically, the state transition function and observation function of the system are approximated via first order Taylor expansion around the current state estimate, and their Jacobian matrices are computed to construct a linearized model [4]. This linear approximation enables the application of the standard framework of linear Kalman filtering for state estimation.

The dynamic characteristics of lithium-ion batteries exhibit strong nonlinearity; for instance, parameters such as open circuit voltage and internal resistance demonstrate nonlinear relationships with SOC, temperature, and operating current [5]. By employing the aforementioned linearization mechanism, the EKF can effectively perform realtime online estimation of the internal states of lithium ion batteries, thereby achieving widespread application in battery management systems. When applying the EKF to model lithium ion battery systems, the system's state space equations are typically expressed as shown in Eq. (7.1).

$$\begin{cases} x_k = f(x_{k-1}, u_{k-1}) + w_k \\ y_k = h(x_k, u_k) + v_k \end{cases} \tag{7.1}$$

In Eq. (7.1), x_k denotes the system state variable at time k, y_k represents the system observation variable at time k, and u_k is the system input; ω_k is the noise in the system state equation, i.e., process noise; ν_k is the noise in the measurement equation, observation noise; the covariance matrices of w and v are denoted by Q and R, respectively. After performing a first order Taylor series expansion to linearize the equation, the resulting expression is obtained.

$$\begin{cases} x_k = A_{k-1}x_{k-1} + B_{k-1}u_{k-1} + w_k \\ y_k = C_{k-1}x_k + D_k u_k + v_k \end{cases} \tag{7.2}$$

Eq. (7.2) represents the linearized state space equation of the system, where the specific values of each component are obtained via first order Taylor expansion. Here, A_k denotes the state matrix, B_k the input matrix, C_k the measurement matrix, and D_k the feedforward matrix. Kalman filtering is applied based on the aforementioned linearized state space equation.

Initialization must first be performed, which involves setting the initial values for the state variables x and covariance P. After initialization, time update is carried out, including the time update of the state variables and covariance. Based on the optimal estimate from the previous time step, a preliminary prediction is made, as shown in Eq. (7.3).

$$\begin{cases} \widehat{x}_{k+1|k} = A\widehat{x}_{k|k} + BI_k \\ P_{k+1|k} = AP_{k|k}A^T + Q_k \end{cases}$$

In this Eq. (7.3), $\widehat{x}_{k|k}$ is the optimal estimate value at the previous time step, $\widehat{x}_{k+1|k}$ is the prior estimate obtained from the previous time step's estimate, and $P_{k+1|k}$ is the predicted value calculated based on the covariance from the preceding time step. After performing time update, the first step prior estimate can be obtained. However, to further derive the optimal estimate, Kalman gain coefficient update and system measurement update are required, as shown in Eq. (7.4).

$$\begin{cases} K_{k+1} = P_{k+1|k}C^T\left(CP_{k+1|k}C^T + R_{k+1}\right) \\ \widehat{x}_{k+1|k+1} = \widehat{x}_{k+1|k} + K_{k+1}\left[y_{k+1} - h\left(\widehat{x}_{k+1|k}\right)\right] \\ P_{k+1|k+1} = (I - K_{k+1}C)P_{k+1|k} \end{cases}$$

In this Eq. (7.4), K_{k+1} denotes the Kalman gain coefficient at the current time step, which $\widehat{x}_{k+1|k+1}$ represents the optimal estimate of the state, $P_{k+1|k+1}$ variables based on $\widehat{x}_{k+1|k+1}$ the updated covariance. By utilizing the most recent state estimate obtained through time update, an estimate of the current observation can be derived. Combining this with the predicted covariance and the Kalman gain coefficient yields the optimal estimate at the current time step. To enable subsequent iterative computation, further updating of the covariance matrix is required.

7.2.1.2 Joint estimation of state of charge and state of health based on the dual extended Kalman algorithm

The SOH is estimated using the Dual extended Kalman filter (DEKF) algorithm, which also estimates the SOC, as accurate SOH estimation of lithium ion batteries relies fundamentally on accurate SOC estimation. This algorithm enables the joint estimation of both states. Prior to state estimation, a state space model must be established based on the equivalent circuit model and relevant definitions, as shown in the following equation [6].

The DEKF is an advanced estimation algorithm that employs a parallel dualfilter architecture to enable simultaneous and coordinated estimation of the SOC and SOH of lithium ion batteries. Its fundamental principle lies in recognizing the strong coupling between SOH, typically associated with slow varying parameters such as capacity degradation and increasing internal resistance, and SOC, which represents the fast varying state of residual battery charge. Accurate SOH estimation critically depends on the ability to track SOC variations in real time.

The DEKF algorithm achieves joint state and parameter estimation by deploying two cooperating EKF.

1) State filter: Primarily responsible for estimating the battery's instantaneous state, specifically the SOC. Based on the current SOH parameter estimate and realtime measured battery voltage and current data, this filter rapidly updates the SOC.
2) Parameter filter: Primarily responsible for estimating the battery model parameters that characterize the battery's SOH, such as battery capacity and internal resistance. This filter slowly adjusts and tracks SOH related parameters using the SOC estimation results from the state filter and the observed data.

These two filters operate alternately at each time step, providing each other with the most recent estimation information and forming a closed loop feedback system, thereby enabling realtime collaborative estimation and updating of SOC and SOH.

Prior to applying the DEKF algorithm, state space equations must be established based on the equivalent circuit model and relevant definitions, as shown in Eq. (7.5).

$$\begin{cases} \begin{bmatrix} SOC(k) \\ U_1(k) \end{bmatrix} = \begin{bmatrix} 1 & 0 \\ 0 & e^{-\frac{\Delta t}{\tau}} \end{bmatrix} \begin{bmatrix} SOC(k-1) \\ U_1(k-1) \end{bmatrix} \\ \qquad + \begin{bmatrix} -\frac{\Delta t}{Q} \\ R_1\left(1-e^{-\frac{\Delta t}{\tau}}\right) \end{bmatrix} I(k-1) + w(k) \\ U_L(k) = U_{oc}(k-1) - U_1(k-1) - I(k-1)R_0 + v(k) \\ Q_n(k) = Q_n(k-1) + c(k) \\ SOC(k) = SOC(k-1) + \int \frac{I}{NQ_n(k-1)} dt + d(k) \end{cases} \tag{7.5}$$

In Eq. (7.5), the first two equations represent the state space equations for estimating the SOC, while the last two equations correspond to the state space equations for estimating capacity. Based on the state space equations in Eq. (7.5) and considering the meanings of the respective coefficient matrices, the specific expressions for estimating SOC and capacity are given by Eq. (7.6).

$$\begin{cases} A_k^x = \begin{pmatrix} 1 & 0 \\ 0 & e^{-t/\tau} \end{pmatrix} & A_k^Q = 1 \\ B_k^x = \begin{pmatrix} -\dfrac{t}{Q_n} \\ R_1\left(1 - e^{-t/\tau}\right) \end{pmatrix} & \\ C_k^x = \left(\dfrac{\partial U_{oc}}{\partial \mathrm{SOC}} \quad -1\right)\Bigg|_{x_k=\hat{x}_k} & C_k^Q = -\dfrac{1}{{Q_n}^2}\displaystyle\int_N Idt \end{cases} \tag{7.6}$$

In Eq. (7.6), the superscript x denotes the variables and parameters associated with the estimated SOC, while the superscript Q denotes the variables and parameters related to the estimated capacity.

Capacity degradation of lithium ion batteries is a slow process, whereas the SOC changes rapidly. Based on this difference, the dual time scale DEKF algorithm employs different update frequencies: SOC estimation uses a small time scale to achieve rapid tracking, while capacity estimation adopts a large time scale to reduce computational load and avoid interference from SOC fluctuations. This approach enhances algorithmic efficiency without compromising accuracy. According to the coefficient matrices in Eq. (7.6), the iterative procedure of the dual time scale DEKF is as follows.

7.2.1.2.1 Initialization

$$\begin{cases} \hat{x}_0 = E[x_0] \\ \hat{Q}_{n,0} = E\left[Q_{n,0}\right] \\ P_{x,0} = E\left[(x_0 - \hat{x}_0)(x_0 - \hat{x}_0)^T\right] \\ P_{Q,0} = E\left[\left(Q_{n,0} - \hat{Q}_{n,0}\right)\left(Q_{n,0} - \hat{Q}_{n,0}\right)^T\right] \end{cases} \tag{7.7}$$

7.2.1.2.2 The EKF algorithm estimates SOC at a short time scale

Time update

$$\begin{aligned} \hat{x}(k+1|k) &= A_x\hat{x}(k|k) + B_xI(k) \\ P_x(k+1|k) &= A_x(k)P_x(k|k)A_x^T(k) + W(k) \end{aligned} \tag{7.8}$$

In Eq. (7.8), $\hat{x}(k+1|k)$ is the preliminary estimate of the state variable at time step $k+1$ for EKF, $\hat{x}(k|k)$ is the estimate of the state variable at time step k, $P_x(k+1|k)$ is the error covariance matrix at time step $k+1$, $P_x(k|k)$ is the error covariance matrix at time step k, and $W(k)$ is the expectation of w_k.

Compute the Kalman gain K_x

$$K_x(k+1)=P_x(k+1|k)C_x^T(k+1)\left[C_x(k+1)P_x(k+1|k)C_x^T(k+1)+V_x\right]^{-1} \tag{7.9}$$

In the equation, $K_x(k+1)$ denotes the Kalman gain at time step k+1 in the small time scale EKF, and V_x represents the expectation of v_k. SOC state update and covariance update

$$\begin{aligned}\hat{x}(k+1|k+1) &= \hat{x}(k+1|k)+K_x(k+1)[y(k+1)-\hat{g}(k+1|k)]\\ P_x(k+1|k+1) &= [H-K_x(k+1)C_x(k+1)]P_x(k+1|k)\end{aligned} \tag{7.10}$$

In Eq. (7.10), $y(k+1)$ denotes the realtime observation value at time step k+1 in EKF, $\hat{g}(k+1|k)$ represents the estimated observation value at time step k+1, $\hat{x}(k+1|k+1)$ denotes the updated state variable value at time step k+1 in EKF, $P_x(k+1|k+1)$ denotes the updated covariance value at time step k+1 in EKF, and H represents the identity matrix.

7.2.1.2.3 Estimating capacity using the EKF algorithm under large time scales

Capacity state prediction and covariance prediction at large time scales

$$\begin{aligned}\hat{Q}(k+1|k) &= \hat{Q}(k|k)\\ P_Q(k+1|k) &= P_Q(k|k)+R(k)\end{aligned} \tag{7.11}$$

In Eq. (7.11), $\hat{Q}(k+1|k)$ represents the prior estimate of the state variable at time instant k+1 in the large time scale EKF, $\hat{Q}(k|k)$ denotes the estimated state variable at time instant k, $P_Q(k+1|k)$ is the error covariance matrix at time instant k+1 in the large time scale EKF, $P_Q(k|k)$ is the error covariance matrix at time instant k, and $R(k)$ is the expectation of r.

Compute the Kalman gain K_Q

$$K_Q(k+1)=P_Q(k+1|k)C_Q^T(k+1)\left[C_Q(k+1)P_Q(k+1|k)C_Q^T(k+1)+V_Q\right]^{-1} \tag{7.12}$$

In Eq. (7.12), $K_Q(k+1)$ denotes the Kalman gain at time step k+1 in the large time scale EKF, and V_Q represents the expectation of ξ;

Capacity update and covariance update

$$\widehat{Q}(k+1|k+1) = \widehat{Q}(k+1|k) + K_Q(k+1)\left[y_Q(k+1) - \widehat{g}_Q(k+1|k)\right]$$
$$P_Q(k+1|k+1) = \left[H - K_Q(k+1)C_Q(k+1)\right]P_Q(k+1|k) \tag{7.13}$$

In Eq. (7.13), $y_Q(k+1)$ denotes the realtime observation at time step $k+1$ within the large time scale EKF, $\widehat{g}_Q(k+1|k)$ represents the estimated observation at time step $k+1$, $\widehat{Q}(k+1|k+1)$ signifies the updated value of the state variable at time step $k+1$ in the large time scale EKF, and $P_Q(k+1|k+1)$ denotes the updated value of the covariance at time step $k+1$ in the large time scale EKF.

As shown in Fig. 7.6, the flowchart of the dualtime scale DEKF algorithm is illustrated according to the aforementioned steps.

In Fig. 7.6, which illustrates the flowchart of the dualtime scale DEKF algorithm, N denotes the number of SOC estimation cycles required before performing one capacity estimation. The dualtime scale ADEKF algorithm

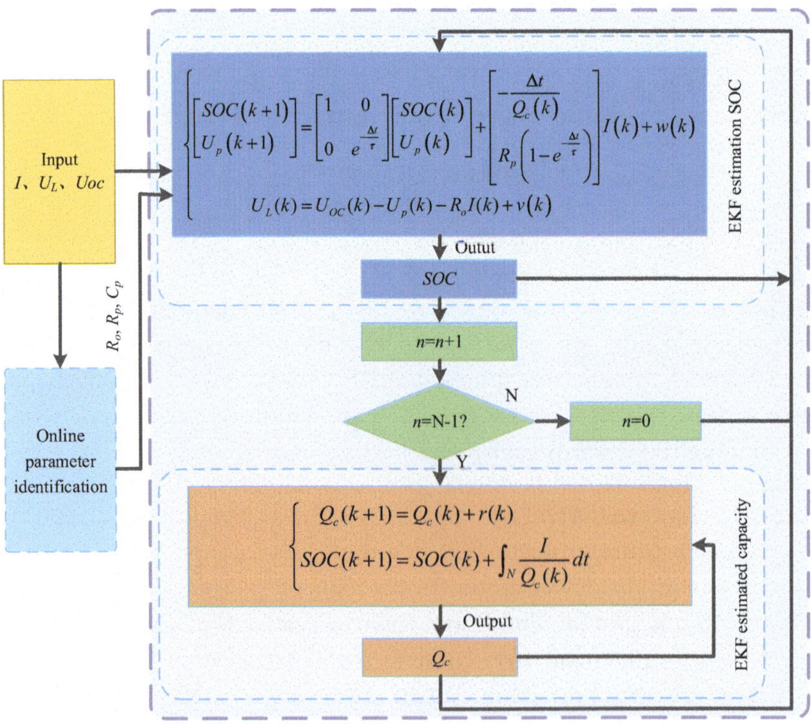

Figure 7.6 Flowchart of the dual-time-scale DEKF algorithm.

consists of two AEKF algorithms operating at different time scales. The small time scale AEKF uses SOC as the state variable and estimates SOC at a smaller time scale; in this study, a sampling interval of 0.1 s is selected as the time scale for SOC estimation. The large time scale adaptive extended Kalman filter takes the SOC estimated by the small time scale algorithm as input, with battery capacity as the state variable, and predicts the battery capacity through iterative computation at a larger time scale, whose specific duration is determined according to actual conditions, thereby obtaining the current SOH.

7.2.2 Particle filter-based methods

Particle filtering (PF) is a nonlinear filtering algorithm based on Monte Carlo methods and Bayesian estimation, primarily used to address state estimation problems in nonlinear and non Gaussian systems. Its core idea is to approximate the posterior probability distribution of system states using a set of weighted random samples, referred to as "particles," thereby enabling recursive estimation of unknown states. Traditional particle filtering often requires a large number of particles to achieve satisfactory performance when the state transition or observation models are highly nonlinear; otherwise, the particle degeneracy problem may occur. To mitigate the impact of particle degeneracy, improvements can be achieved by selecting an appropriate proposal density function. Within the particle filtering framework, the Unscented Kalman filter (UKF) is employed to replace the sampling process based on importance density in conventional PF.

7.2.2.1 Construction of a joint estimation model for SOC and SOH

Considering the different scales of SOC and SOH, the AUKFUPF algorithm with two sampling intervals is adopted to achieve collaborative estimation of SOC and SOH. Key parameters derived from this process enable accurate prediction of remaining power and cycle life. The estimation model for SOC and SOH of lithium ion batteries is illustrated in Fig. 7.7.

As illustrated in Fig. 7.7, during the SOC estimation, realtime acquisition of battery current and voltage is performed based on equivalent circuit characterization and state description, with current data serving as model input and voltage data as observation output. State space models are established separately using SOC and SOH as state variables, and realtime estimation is conducted through two coordinated iterative pathways. The estimated SOC result is then fed as an input variable into the SOH iterative process. During SOH estimation, the battery's realtime maximum available

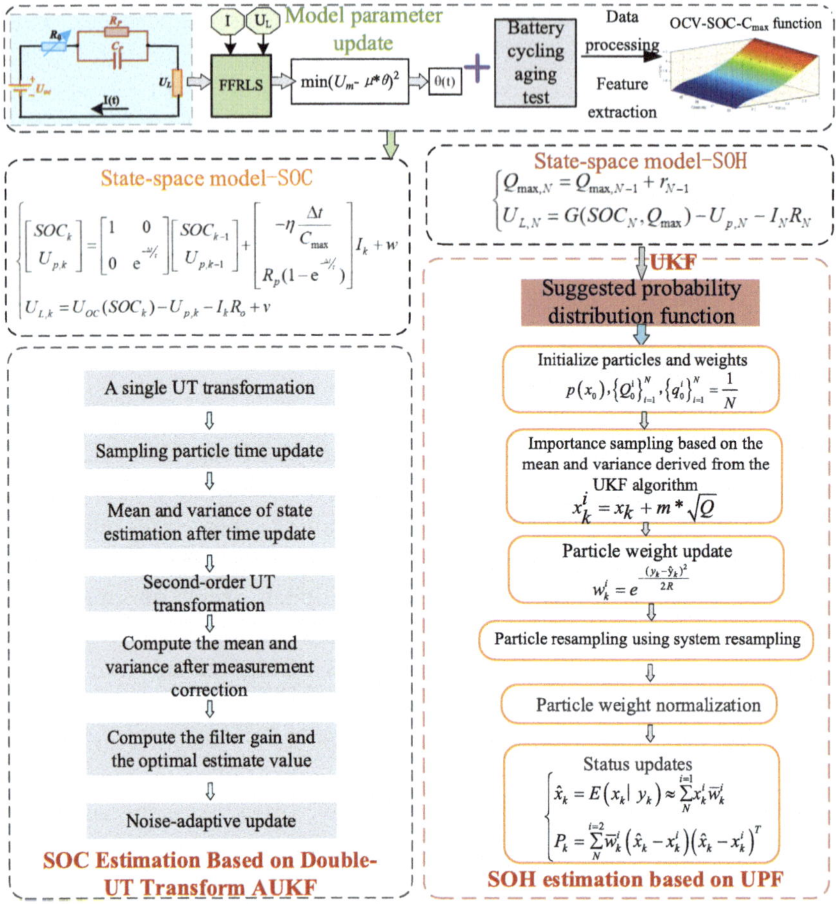

Figure 7.7 Flowchart of the cooperative estimation based on the AUKF-UPF algorithm.

capacity is treated as the state variable to be estimated. Battery current and the estimated SOC values are used as inputs to the estimation model, and an unscented particle filter based estimation model is constructed to achieve realtime estimation of the battery's maximum available capacity. During iteration, the Kirchhoff's voltage law (KVL) relationship within the equivalent circuit model, combined with the output voltage, forms a correction mechanism following initial estimation. Consequently, the established OCV-SOC-Q_{max} functional relationship reduces modeling errors caused by battery aging compared to conventional SOC-OCV curves, thereby indirectly enhancing estimation accuracy. The state space equations are expressed in Eq. (7.14).

$$\begin{cases} X_k = \begin{pmatrix} \frac{R_P C_P}{R_P C_P + 1} & 0 \\ 0 & 1 \end{pmatrix} X_{k-1} + \begin{pmatrix} \frac{R_P}{R_P C_P + 1} \\ -\frac{\eta^* T}{Q_N} \end{pmatrix} * I_{k-1} \\ + \begin{pmatrix} w_{0,k} \\ w_{1,k} \end{pmatrix}, X_{k=} \begin{pmatrix} U_{p,k} \\ SOC_k \end{pmatrix} \\ U_{L,k} = G\left(SOC_{k-1}, Q_{\max,N-1}\right) - U_{p,k} - I_k R_{0,k} \end{cases} \tag{7.14}$$

In Eq. (7.14), $G(SOC_{k-1}, Q_{\max}, N-1)$ represents the coupled functional relationship among the Open-circuit voltage (OCV), SOC, and maximum available capacity ($Q_{\max}$) of a lithium ion battery. This functional relationship quantitatively characterizes the time varying evolution of the SOC-OCV mapping as well as the battery model parameters during battery aging.

7.2.2.2 SOC-SOH coiterative estimation procedure

Establishing a collaborative online estimation model for battery SOC and SOH, the entire iterative process comprises two distinct update cycles. Initially, system initialization is performed, setting the initial estimates for SOC and SOH based on prior information within the filtering models that conduct iterative estimation at two different scales. First, a state space model for SOC is established, and SOC is estimated using an improved Adaptive unscented Kalman filter (AUKF) algorithm, as illustrated in the following procedure.

Apply the UT transformation to the state value $X = [SOC\ U_p]$ to obtain 2n+1 sampling points, as shown in Eq. (7.15).

$$X^i(k-1) = \begin{bmatrix} X(k-1) \\ X(k-1) + \sqrt{(n+\lambda)P(k-1)} \\ X(k-1) - \sqrt{(n+\lambda)P(k-1)} \end{bmatrix} \tag{7.15}$$

Obtain the weights of the state variables and the corresponding variance weights. Perform time updates on the sampled particles using the state equation, as shown in Eq. (7.16).

$$\widehat{X}^{i}(k) = f\left(X^{i}(k-1), Q_{\max}\right) = \begin{pmatrix} \frac{R_c C_P}{R_c C_P + 1} & 0 \\ 0 & 1 \end{pmatrix} X^{i}(k-1) + \begin{pmatrix} \frac{R_c}{R_c C_P + 1} \\ -\frac{\eta * T}{Q_N} \end{pmatrix} * I_{k-1} \tag{7.16}$$

The one step prediction of the state matrix and the prediction variance are obtained from the one step predicted values of the sampled particles, as shown in Eq. (7.17).

$$\begin{cases} X(k|k-1) = \sum_{i=0}^{2n} \omega^{(i)} \widehat{X}^{(i)}(k|k-1) \\ P(k|k-1) = \sum_{i=0}^{2n} \omega^{(i)} \left[X(k|k-1) - \widehat{X}^{(i)}(k)\right] \\ * \left[X(k|k-1) - \widehat{X}^{(i)}(k)\right]^{T} + Q \end{cases} \tag{7.17}$$

After obtaining the updated state values and covariance, and considering the strong nonlinearity of the measurement equation, a second Unscented transform (UT) is performed on the state variables to generate particles consistent with the predicted state mean and covariance. The UT transformation process and weight calculation are shown in Eq. (7.15). Upon obtaining the new sampling particles, the estimated output voltage for each particle is computed based on the measurement equation, as expressed in Eq. (7.18).

$$\begin{aligned} \overset{\wedge}{UL}{}^{i}(k) &= g\left(X^{i}(k-1), Q_{\max}\right) \\ &= G\left(X^{i}(k-1,1), Q_{\max}\right) - X^{i}(k-1,2) \quad - R_c * I_{k-1} \end{aligned} \tag{7.18}$$

Subsequently, the output voltage value under the predicted state can be obtained from the estimated output voltage under sampled particles. The required variance value can then be derived from the difference between the predicted output voltage and the actual voltage output, as shown in Eq. (7.19).

$$\begin{cases} U_L(k|k-1) = \sum_{i=0}^{2n} \omega^{(i)} \widehat{U}_L{}^{(i)}(k) \\ P_{U_L(k)U_L(k)} = \sum_{i=0}^{2n} \omega^{(i)} \left[\widehat{U}_L{}^{(i)}(k) - U_L(k|k-1) \right] \\ \left[\widehat{U}_L{}^{(i)}(k) - U_L(k|k-1) \right]^T + RP_{X(k)U_L(k)} \\ = \sum_{i=0}^{2n} \omega^{(i)} \left[\left[X(k|k-1) - \widehat{X}^{(i)}(k) \right] \right] \left[\widehat{U}_L{}^{(i)}(k) - U_L(k|k-1) \right]^T \end{cases} \tag{7.19}$$

Through the aforementioned recursive steps, the Kalman gain K, the optimal state estimate, and the estimation error covariance can be obtained, as shown in Eq. (7.20).

$$\begin{cases} K(k) = P_{X(k)U_L(k)} P^{-1}_{U_L(k)U_L(k)} \\ X(k) = X(k|k-1) + K(k)[U_L(k) - U_L(k|k-1)] \\ P(k) = P(k|k-1) - K(k) P_{U_L(k)U_L(k)} K^T(k) \end{cases} \tag{7.20}$$

After predicting and correcting the SOC, the estimated SOC is used as an input condition to initiate the iterative process for estimating the SOH. The SOH estimation process employs the maximum available capacity, Q_{max}, as the state variable, and its state space model is described by Eq. (7.21).

$$\begin{cases} Q_{max,N} = \dfrac{SOC_{N-1} - SOC_N}{\int I dt} + r_{N-1} \\ U_{L,N} = G(SOC_N, Q_{max}) - U_{p,N} - I_N R_N \end{cases} \tag{7.21}$$

In Eq. (7.21), the subscript N denotes the sampling period under the scale of SOH update, which is 20 s in this study. Based on the state space equation derived from Q_{max} and the estimated SOC results, the UPF algorithm is employed to iteratively update the lithium ion battery SOH. Initially, the initial state value is obtained from the prior probability.

In the subsequent importance sampling step, the state mean and covariance are first obtained using the UKF algorithm, and particles are sampled according to the corresponding probability distribution, as shown in Eq. (7.22).

$$x_k^i = x_k + m^* \sqrt{Q} \tag{7.22}$$

In equation (7.22), m is a random number between [0,1]. After performing importance sampling on the particle swarm, the corresponding weights of the particles are computed. Assuming the state variables follow a Gaussian distribution, the particle weights of the swarm are given by Eq. (7.23).

$$w_k^i = e^{-\frac{(y_k - \hat{y}_k)^2}{2R}} \tag{7.23}$$

The particle swarm is resampled using the system's resampling mechanism to prevent particle degeneracy. After resampling, a new set of particles is obtained, each assigned a weight of 1/N. The filtered optimal state estimate at this stage is given by Eq. (7.24).

$$\widehat{X}_k = E(X_{\text{set}}) = \frac{1}{N} \sum_{N}^{i=1} w_i x_i \tag{7.24}$$

After each iterative computation at the micromacro time scale, the battery's state of charge and maximum available capacity can be estimated in real time, while the battery's health status is simultaneously estimated in real time based on its defined health status criteria. The closed loop logic formed by coordinated state estimation enables mutual feedback and correction among state variables, achieving highly reliable realtime monitoring of multiple critical state variables in high energy density lithium ion batteries.

7.3 AI based joint estimation approaches

7.3.1 Support vector machines

The complex coupling relationship between SOC and SOH, along with their differing time scale characteristics. Traditional step by step estimation methods, unable to effectively handle this bidirectional coupling, often lead to error propagation and amplification across estimation stages, ultimately degrading the overall accuracy of state estimation [7]. Support vector machines (SVMs) demonstrate unique advantages in addressing this complex problem, with their applicability primarily reflected at the following levels:

From a methodological perspective, SVMs, as a fully data driven approach, effectively circumvent the reliance of conventional model based methods on precise battery mathematical models. Batteries constitute complex electrochemical systems involving multiphysics coupling effects internally, posing significant challenges in establishing accurate mechanistic models [8]. In contrast, SVMs directly learn the input and output relationships of the system from historical operational data, thereby effectively avoiding system errors arising from model inaccuracies. This provides a novel solution pathway for addressing state estimation problems in such complex systems.

The kernel function is the core mathematical tool employed by SVMs to handle nonlinear problems. It maps sample data from the original space into a high dimensional feature space, rendering data that is linearly inseparable in the original space linearly separable in the higher dimensional space. The choice of an appropriate kernel function decisively influences the performance of SVMs. Specific types of kernel functions are listed in Table 7.7.

In terms of nonlinear processing capability, SVMs demonstrate outstanding performance through the kernel trick. Battery systems exhibit pronounced nonlinear dynamic responses, and these nonlinear characteristics evolve continuously as aging progresses. By selecting an appropriate kernel function, SVMs can transform nonlinear problems in the original space into linear ones in a high dimensional feature space, thereby enabling accurate modeling of the complex dynamic behaviors of battery systems. This powerful nonlinear mapping ability allows SVMs to effectively capture the intricate coupling relationships between SOC and SOH.

From the perspective of few shot learning capability, SVMs, based on the principle of minimizing structural risk, maintain strong generalization performance even under limited training samples. This characteristic is particularly crucial for battery state estimation, as acquiring experimental data across the entire battery life cycle typically involves substantial time and economic costs. SVMs can establish effective estimation models with limited sample sizes, significantly reducing the cost and difficulty associated with data collection.

In terms of robustness, Support vector regression (SVR) employs an ε insensitive loss function, which inherently confers robustness against measurement noise and outliers. In practical battery systems, various sources of uncertainty exist, including sensor measurement noise, modeling errors, and variations in operating conditions. By introducing slack variables and a penalty parameter, SVR effectively balances model complexity against fitting accuracy, thereby preventing overfitting to noisy data and enhancing the reliability of the model in real world applications.

Table 7.7 Comparison of different kernel function types.

Kernel function name	Mathematical expressions	Main characteristics	Use cases
Polynomial kernel function	$K(x, y) = (\langle x, y \rangle + 1)^d$	Strong global characteristics, with fitting ability increasing as d increases; however, it involves a large number of parameters and computational complexity rises with increasing d.	Suitable for classification problems with large data volumes and moderately high feature dimensions.
Sigmoid kernel function	$K(x, y) = \tanh(\alpha \langle x, y \rangle + r)$	Can be regarded as a multilayer perceptron, which does not satisfy the Mercer condition under certain parameters and is therefore not always a valid kernel function.	Its applicability is somewhat limited and it is commonly used for simulation in specific scenarios.
Power-exponent kernel function	$K(x_i, x) = \exp\left\{\frac{-\lVert x_i - x \rVert}{2\delta^2}\right\} = \exp\{-\gamma \lVert x - y \rVert\}$	The function is nonstationary and highly sensitive to changes in the input data.	Sample problems suitable for handling discontinuities
Tensor product kernel function	$K(x, y) = \prod_i K_i(x_i, y_i)$	Multidimensional kernel functions are constructed via the tensor product of univariate kernel functions.	Particularly suitable for constructing multidimensional B spline functions
Gaussian kernel function	$K(x_i, x) = \exp\left\{\frac{-\lVert x_i - x \rVert^2}{2\delta^2}\right\}$	Data can be mapped into an infinite dimensional feature space; the model has simple parameters and exhibits strong robustness against noise.	Highly universal

Moreover, SVMs also possess the capability of multioutput estimation. By employing multioutput support vector regression techniques, a unified framework can be established to simultaneously estimate SOC and SOH. This approach not only effectively exploits the correlation between the two state parameters but also prevents the propagation and accumulation of errors. Such an end to end joint estimation method provides an effective technical pathway for decoupling SOC from SOH.

Battery state joint estimation based on SVMs is a systematic engineering endeavor, whose complete technical roadmap comprises three key stages: feature engineering, model construction, and model training and validation. Fig. 7.8 illustrates the technical roadmap for battery state joint estimation

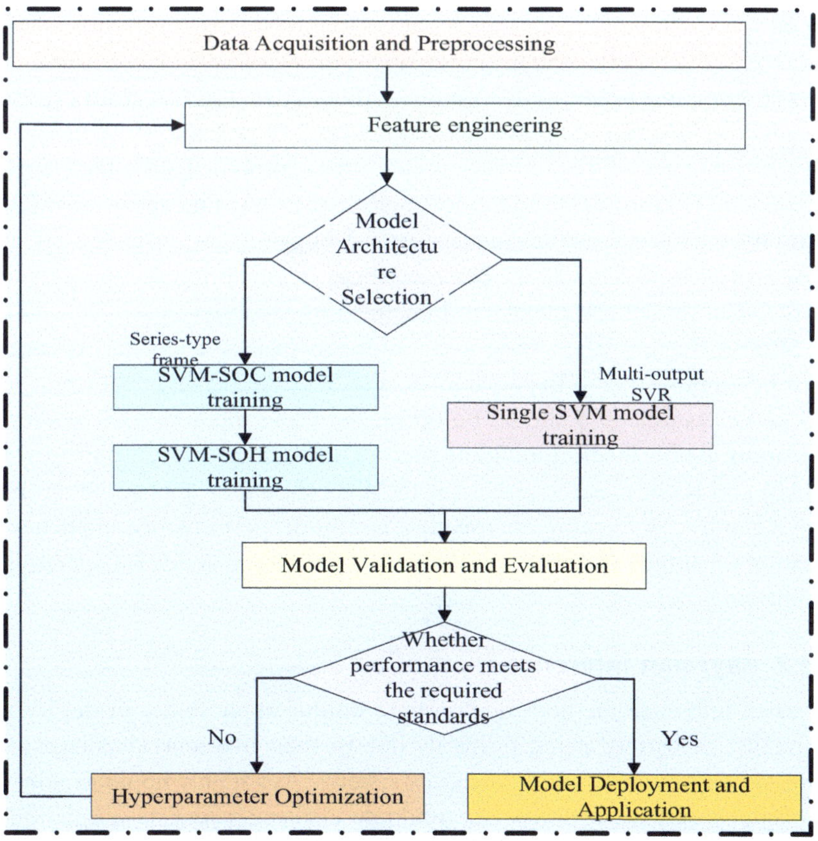

Figure 7.8 Roadmap of the joint state estimation technique for batteries based on SVM.

based on SVM, clearly depicting the entire process from data acquisition to model application.

Feature engineering is fundamental to model success and requires the construction of feature vectors that simultaneously reflect both the short term dynamics and long term degradation characteristics of batteries. These features primarily fall into three categories: first, instantaneous features, which are direct measurements at the current time point; second, short term historical features, derived from statistical measures over a time window to capture battery dynamic behavior, mainly associated with SOC estimation; and third, long term cumulative features, used to characterize battery aging and closely related to SOH estimation. The quality of these features directly determines the upper performance limit of the model.

In terms of model construction, there are primarily two architectures. Multioutput support vector regression (multioutput SVR) adopts an end to end approach, employing a single model to directly map from mixed features to [SOC, SOH], enabling automatic learning of intrinsic relationships among states and resulting in a compact model structure. In contrast, the cascaded architecture employs two sequentially connected SVM models: the first estimates SOC, and its output is then combined with long-term features and fed into a second model to estimate SOH. This approach offers clear physical interpretability and facilitates debugging, but carries the risk of error propagation. The choice between these architectures requires a trade off between model complexity and engineering feasibility.

During the model training and validation phases, annotated datasets must be employed, and hyperparameters such as the penalty coefficient C and kernel parameter γ should be optimized using methods such as cross validation. For evaluation, multiple metrics, including Mean absolute error (MAE) and Root mean square error (RMSE) should be applied on an independent test set to comprehensively assess model performance and generalization capability, ensuring stability and reliability across diverse operating conditions.

7.3.2 Bayesian inference methods

Bayesian inference methods demonstrate unique applicability in the joint estimation of battery states, primarily due to their inherent advantages in handling uncertainties within complex systems. As a highly nonlinear electrochemical system, batteries face multiple challenges in state estimation: first, intricate electrochemical reaction processes inside the battery make it difficult to establish accurate mathematical models; second, the system is influenced by various sources of uncertainty; third, strong coupling exists among different state parameters. These characteristics often render conventional deterministic estimation methods ineffective, whereas Bayesian inference provides a systematic solution to these challenges.

First, the battery system suffers from significant model uncertainties and measurement noise. Due to the complex electrochemical reaction processes occurring within batteries, establishing an accurate mathematical model is extremely challenging. Meanwhile, during actual operation, measured signals such as voltage and current are inevitably affected by noise. Bayesian inference explicitly handles these uncertainties through probability distributions, incorporating both model errors and measurement noise into the estimation framework. This probabilistic approach yields more reliable estimation results and provides a crucial foundation for battery safety management.

Second, SOC and SOH exhibit distinctly different temporal scale characteristics, imposing specific requirements on joint estimation methods. The Bayesian framework, through the design of rational estimation algorithms, can naturally address this multitime scale estimation challenge. For instance, a dualtime scale estimation strategy can be employed: high frequency updates of SOC on a fast time scale and low frequency updates of SOH on a slow time scale, thereby enabling coordinated estimation of both state parameters.

Moreover, there exists a strong coupling relationship between SOC and SOH, making it difficult for decoupled independent estimation methods to achieve satisfactory results. On one hand, SOH degradation directly affects the battery's maximum available capacity and internal resistance characteristics, thereby altering the OCV and SOC relationship and introducing bias into SOC estimation. On the other hand, accurate SOC estimation is a prerequisite for SOH calculation. Bayesian methods naturally describe this coupling relationship through probabilistic frameworks, simultaneously accounting for the mutual influence of both parameters during the state estimation process. In practical implementation, SOC and SOH can be treated as joint state variables, their dynamic evolution relationships modeled via a state-space framework, and joint estimation achieved using Bayesian filtering algorithms. This approach effectively mitigates error propagation and accumulation issues, thereby enhancing overall estimation accuracy.

It is noteworthy that Bayesian methods also demonstrate good adaptability in addressing the joint estimation of battery states. As batteries age, their parameter characteristics evolve; however, the Bayesian framework can adapt to these changes through online parameter updating algorithms. Moreover, this approach allows for the convenient integration of multiple types of observational data, such as voltage, current, and temperature, thereby enhancing estimation accuracy via information fusion. Furthermore, with advancements in computational technologies, various improved Bayesian filtering algorithms have continuously emerged, offering greater flexibility and options for battery state estimation across diverse application scenarios.

In summary, Bayesian inference demonstrates significant advantages in joint state estimation of batteries. Its probabilistic framework is particularly well suited for handling system uncertainties, effectively reconciling multi-time scale estimation requirements, and naturally capturing the coupling relationships among state parameters. These characteristics render Bayesian methods an ideal choice for joint battery state estimation, exhibiting robust performance in practical applications.

The joint estimation of battery state based on Bayesian inference adopts a systematic technical approach, whose complete workflow encompasses multiple critical stages from fundamental theory to practical implementation. The specific procedure is illustrated in Fig. 7.9. The method begins with an in depth analysis of the battery system, followed by the establishment of an accurate state space model, including the state equations that describe the dynamic evolution of SOC and SOH, as well as the observation equations that define the relationship between terminal voltage and state variables. After model construction, it is necessary to evaluate the degree of nonlinearity inherent in the system, a step that decisively influences the subsequent selection of filtering algorithms.

For systems with varying degrees of nonlinearity, appropriate filtering algorithms must be adopted. For weakly nonlinear systems, the EKF employs first order Taylor expansion of nonlinear functions, achieving a balance between acceptable accuracy and low computational cost. When the system exhibits general nonlinear characteristics, the UKF provides more accurate approximation of the state distribution through unscented transformation and sigma point sampling. For strongly nonlinear or non Gaussian systems, PF utilizes Monte Carlo methods to approximate the posterior probability distribution using a set of weighted particles; although computationally intensive, PF offers broader applicability.

On the basis of algorithm selection, a dualtime scale estimation strategy must be designed to accommodate the differing dynamic characteristics of

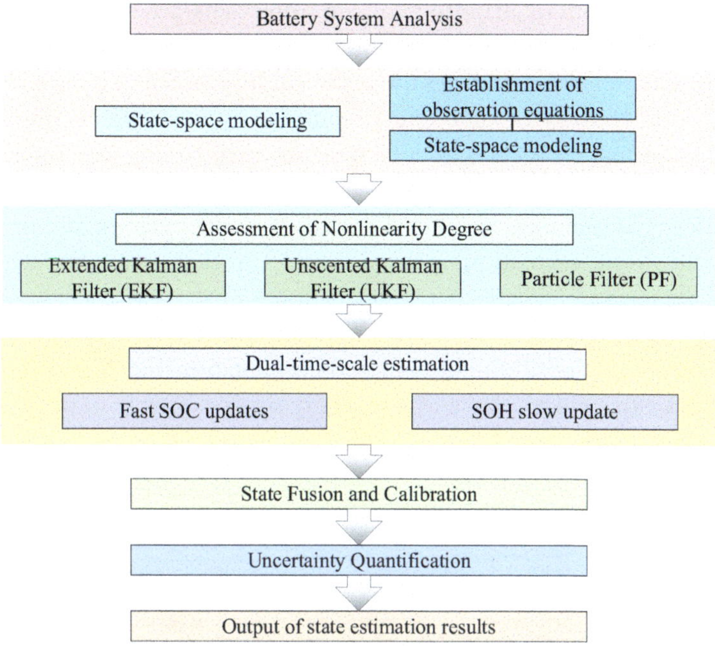

Figure 7.9 Implementation methods of Bayesian inference approaches.

SOC and SOH. SOC estimation is performed on a fast time scale, with state updates conducted at each sampling interval to track instantaneous state variations of the battery. In contrast, SOH estimation occurs on a slow time scale and is typically updated after a complete charge/discharge cycle, enabling accurate representation of long term battery aging trends. The two estimation processes are coordinated through an information exchange mechanism: fast estimation provides accurate instantaneous state information for slow estimation, while slow estimation supplies updated model parameters to enhance the accuracy of fast estimation, thereby establishing a collaborative optimization framework.

In practical implementation, uncertainty quantification represents a significant advantage of Bayesian methods. Through posterior probability distributions, point estimates of the state can be obtained alongside confidence intervals, providing crucial information for safe decision making in battery management systems. Furthermore, practical issues such as determining noise statistical characteristics, ensuring numerical stability, and managing computational complexity must be addressed. Particularly in real world applications, it is necessary to design parameter adaptive update strategies that enable the algorithm to adapt to evolving battery characteristics during aging, thereby maintaining long term estimation accuracy.

The implementation of the entire technical approach requires a comprehensive consideration of theoretical requirements and engineering constraints, ensuring estimation accuracy while maintaining computational efficiency, thereby guaranteeing the feasibility of the method within practical battery management systems. Through systematic design and optimization, the joint estimation method based on Bayesian inference can provide accurate and reliable state information for battery management systems, supporting the safe and efficient operation of batteries. With ongoing advancements in algorithmic theory and computational technologies, this approach is expected to play an increasingly significant role in the field of battery state estimation.

References

[1] T. Junjie, S. Wang, C. Wen, et al., A comprehensive review of state-of-charge and state-of-health estimation for lithium-ion battery energy storage systems, Ionics 30 (10) (2024) 5903–5927.

[2] J. Zhang, L. Kang, State-of-health estimation for lithium-ion batteries in hybrid electric vehicles—A review, Energies 17 (22) (2024) 5753.

[3] Q. Wan, Q. Jie, J. Xu, A feature optimization framework for improving state of health estimation of lithium-ion batteries based on electrochemical impedance spectroscopy, J. Energy Storage 144 (2026) 119713.

[4] S. Peng, D. Zhang, G. Dai, et al., State of charge estimation for LiFePO4 batteries joint by PID observer and improved EKF in various OCV ranges, Appl. Energy 377 (PA) (2025) 124435.

[5] X. Zhao, B. Sun, W. Zhang, et al., Error theory study on EKF-based SOC and effective error estimation strategy for Li-ion batteries, Appl. Energy (2024) 353 (PA).

[6] C. Wang, S. Wang, J. Zhou, et al., A novel back propagation neural network-dual extended Kalman filter method for state-of-charge and state-of-health co-estimation of lithium-ion batteries based on limited memory least square algorithm, J. Energy Storage 59 (2023).

[7] D. Yang, Y. Wang, R. Pan, et al., State-of-health estimation for the lithium-ion battery based on support vector regression, Appl. Energy 227 (2017) 273–283.

[8] T. Raoofi, M. Yildiz, Comprehensive review of battery state estimation strategies using machine learning for battery management systems of aircraft propulsion batteries, J. Energy Storage (2023) 59.

CHAPTER EIGHT

Experimental design and validation

8.1 Development of experimental platform

8.1.1 Electrochemical impedance spectroscopy instruments

Electrochemical impedance spectroscopy (EIS) is a powerful technique for characterizing the frequency dependent properties of electrochemical systems by applying small alternating perturbations and measuring the system's response. The fundamental principle involves applying a small amplitude sinusoidal voltage or current perturbation to an electrochemical system under steady state or quasi steady state conditions, and recording the amplitude and phase of the resulting current or voltage across a range of frequencies, thereby determining the variation of complex impedance with angular frequency [1]. As a "frequency domain resolved" technique, EIS has become an indispensable methodological foundation for studying and diagnosing batteries and other electrochemical systems due to its nondestructive nature, high sensitivity, and ability to resolve multiscale processes.

As illustrated in Fig. 8.1, the fundamental procedure of electrochemical impedance spectroscopy analysis can be divided into three main stages: signal excitation, signal acquisition, and data processing. A sinusoidal alternating current disturbance signal is generated by a waveform generator, typically superimposed as a small amplitude alternating voltage onto a direct current bias potential to ensure the system remains within an approximately linear response regime. This signal is subsequently applied to the electrochemical system or battery model via a potentiostat, thereby imposing an external perturbation on the system under test. During signal transmission, the system's current response varies over time. To enable precise detection of the current response, a current to voltage converter is employed to transform the current signal generated by the system into a corresponding

Modelling, Estimation and AI Applications for Lithium-Ion Battery Management Systems
ISBN: 978-0-443-45347-2
https://doi.org/10.1016/B978-0-443-45347-2.00008-X

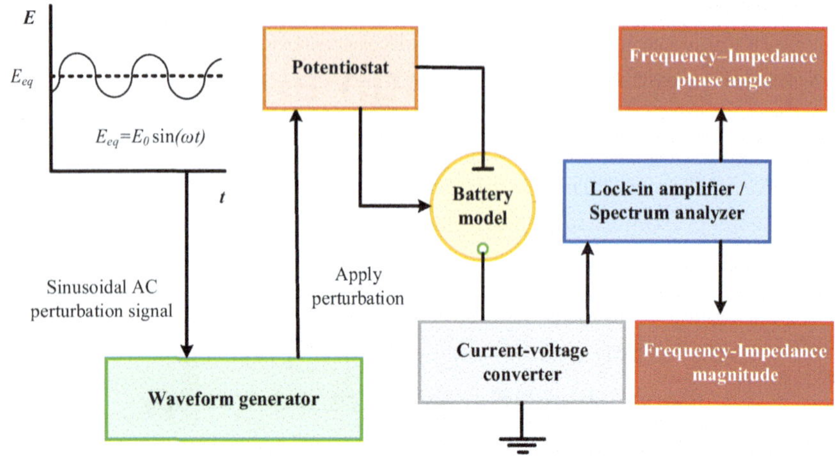

Figure 8.1 EIS measurement system and workflow diagram.

voltage signal suitable for subsequent processing. Simultaneously, the voltage response of the electrochemical system is directly fed back into the detection circuit, enabling dual channel data acquisition of both voltage and current signals.

The converted voltage signal is fed into high-sensitivity detection instruments such as a lockin amplifier or a spectrum analyzer. The lockin amplifier extracts the response component at the same frequency as the applied AC perturbation signal, thereby effectively filtering out noise and enhancing measurement accuracy. The spectrum analyzer records the amplitude ratio and phase difference between the voltage and current signals over a broad frequency range, enabling the calculation of the frequency dependent variation of complex impedance.

The magnitude and phase distribution of impedance are obtained through the data processing module, and Nyquist and Bode plots are subsequently constructed. The Nyquist plot clearly reveals features such as charge transfer, electrolyte resistance, and diffusion processes, while the Bode plot illustrates the frequency dependent variations in impedance magnitude and phase angle, facilitating the separation and identification of processes with different time constants [2]. Based on these analyses, further interpretation of experimental data can be achieved through equivalent circuit fitting, distributed relaxation time domain analysis, or physically based modeling approaches, enabling the extraction of key parameters and kinetic information from the electrochemical system.

The aforementioned content analyzes the experimental procedure and data processing of electrochemical impedance spectroscopy. To systematically investigate the electrochemical impedance characteristics and their application in battery performance evaluation, a complete electrochemical impedance spectroscopy experimental platform must be established. The core design philosophy of the platform emphasizes high precision, modularity, and scalability, enabling it to meet various experimental requirements for impedance measurement, charge and discharge testing, and temperature monitoring. The overall architecture comprises three main components: the electrochemical testing unit, the data acquisition and control module, and the temperature control system.

8.1.1.1 Electrochemical testing unit

The electrochemical testing unit is the core component of the entire experimental platform, primarily responsible for providing high-precision electrochemical measurement conditions and enabling comprehensive analysis of battery charge–discharge behavior and impedance characteristics. This unit mainly comprises a multichannel electrochemical working platform, a battery interface system, an excitation signal source, and protective circuits.

8.1.1.1.1 Multichannel electrochemical experimental platform

As shown in Fig. 8.2, high-precision electrochemical impedance spectroscopy requires a multichannel electrochemical testing platform to simultaneously perform charging/discharging and impedance measurements on multiple battery cells. The experimental platform must support excitation signals across a wide frequency range, from millihertz to above 100 kHz, and be capable of delivering constant and step current signals at various levels, ensuring measurements can be conducted on batteries with different capacities and types.

8.1.1.1.2 Battery interface system

As shown in Fig. 8.3, the electrochemical impedance spectroscopy experimental platform requires appropriate battery mounting and connection methods, and the battery interface system must be adaptable to lithium ion batteries of various sizes and capacities. The interface should ensure

Figure 8.2 Multichannel electrochemical experimental platform.

low contact resistance and stable electrical contact while enabling rapid battery replacement to support highthroughput experiments. Mounting methods may employ spring clamps or custom designed slots to prevent loosening and thereby minimize measurement errors.

Figure 8.3 Different battery interfaces.

8.1.1.1.3 Excitation signal source

Impedance measurements require the experimental platform to provide various programmable excitation signals, the signal source interface is shown in Fig. 8.4 including sinusoidal waves, step currents, and random noise signals. The signal amplitude should be adjustable to ensure measurements are conducted within the battery's linear response range, thereby minimizing nonlinear errors induced by large signals. The frequency range should span from low to high frequencies to enable separation of time constants associated with different dynamical processes.

8.1.1.1.4 Protective circuits and safety design

As shown in Fig. 8.5, safety considerations for batteries and equipment must be taken into account during experimentation; therefore, the electrochemical impedance spectroscopy testing unit requires multilevel protection circuits, including overcurrent, overvoltage, and short circuit protection. The experimental platform should be equipped with an automatic disconnection mechanism to rapidly cut off the circuit under abnormal conditions, thereby preventing equipment damage or battery failure. Meanwhile, the protection mechanism should not interfere with normal measurements, ensuring the authenticity and reliability of the acquired data.

8.1.1.1.5 Ensuring experimental accuracy and reproducibility

High precision impedance measurement requires an experimental platform equipped with a low noise data acquisition system and synchronous multichannel acquisition capability to minimize the impact of timing delays and interference on measurements. Proper shielding and grounding design effectively suppress external electromagnetic interference, ensuring consistent results across repeated measurements. Amplifiers and the data

Figure 8.4 Excitation signal source.

Figure 8.5 Protection circuit design.

acquisition system must be capable of detecting minute voltage variations, providing a reliable foundation for subsequent data analysis and equivalent circuit fitting.

8.1.1.2 Data acquisition and control module

8.1.1.2.1 Data acquisition functionality

In the electrochemical impedance spectroscopy experimental platform, the data acquisition component must achieve high precision recording of voltage and current responses under alternating current perturbation signals, while maintaining high resolution and low noise characteristics to ensure the accuracy and reproducibility of impedance data. The system should support multichannel synchronous sampling to enable real time measurements at various frequency points, minimizing signal delay and phase errors, thereby providing a necessary foundation for obtaining complete and reliable Nyquist and Bode plots.

8.1.1.2.2 Signal processing functions

In response to the stringent requirements for data quality in EIS experimental platforms, the data acquisition end must possess real-time signal preprocessing capabilities. Common preprocessing steps include bandpass filtering, amplitude normalization, and frequency-domain analysis based on the Fast Fourier transform (FFT), which are employed to extract the phase difference and amplitude ratio between voltage and current signals, thereby enabling impedance calculation. By performing partial frequency domain transformations during the acquisition stage, not only is the efficiency of impedance spectrum construction significantly improved, but also the computational burden on subsequent fitting and modeling processes is markedly reduced.

8.1.1.2.3 Experimental control functions

In terms of experimental control, an EIS system must provide programmable management of excitation and measurement procedures. Its core functionality lies in the high precision application of small signal sinusoidal current or voltage, combined with point-by-point scanning or rapid frequency sweeps across a broad frequency range. By integrating controllable constant current charge/discharge operations with dynamic step response measurements, the platform enables flexible design of experimental sequences and parameter ranges. The automated control mechanism effectively minimizes human intervention effects on EIS experimental outcomes, significantly enhancing the comparability and reproducibility of experiments.

8.1.1.2.4 Extension and compatibility

To meet the research demands of complex electrochemical systems, the data acquisition and control module must be compatible with external sensors and advanced analytical instruments. Through a modular interface design, the electrochemical impedance spectroscopy experimental platform not only extends the applicability of impedance analysis but also enhances the capability for studying multiphysics coupling behaviors.

8.1.1.3 Temperature control system

8.1.1.3.1 Temperature control system

In EIS experiments, temperature is a critical factor influencing battery impedance characteristics and kinetic behavior. The environmental control system must provide wide range, high precision temperature regulation capabilities to cover actual operating conditions from low to high temperatures. By implementing constant temperature and temperature gradient control, the platform can effectively investigate the effects of temperature on impedance spectrum features and performance degradation patterns.

8.1.1.3.2 Temperature monitoring and feedback function

The system must integrate high-precision sensors to enable real time monitoring of temperature, humidity, and atmospheric composition in the experimental environment, and achieve closed loop control through a feedback mechanism. This ensures stability and consistency of experimental conditions, reduces uncertainties induced by external disturbances, and thereby enhances the reliability of impedance testing and performance evaluation.

After completing the hardware assembly and basic functional configuration of each module, the EIS experimental platform requires comprehensive optimization from the perspectives of system integration, measurement stability, and longterm operational reliability. Since impedance measurements often involve wide-frequency scanning and prolonged experimental durations, performance fluctuations in any single module may be amplified in the frequency domain results. Therefore, the platform's software design must emphasize coordinated operation among modules and parameter consistency, as illustrated in Fig. 8.6.

At the system level coordination, the electrochemical testing unit, data acquisition and control module, and temperature control system do not operate independently but are instead coordinated through a unified control logic and temporal reference. The application of excitation signals, acquisition of response signals, and adjustment of environmental parameters must be synchronized within a consistent timing framework to ensure that impedance data collected at the same frequency point correspond to identical experimental conditions. A centralized control architecture effectively prevents impedance errors caused by time drift or parameter asynchronization, which is particularly significant in low frequency impedance measurements.

From the perspective of measurement stability, an EIS experimental platform must possess robust longterm operational reliability. Low frequency impedance measurements typically require extended acquisition times, imposing stricter demands on system drift, temperature fluctuations, and power supply stability. Consequently, platform designs commonly incorporate high stability reference sources, low drift amplification circuits,

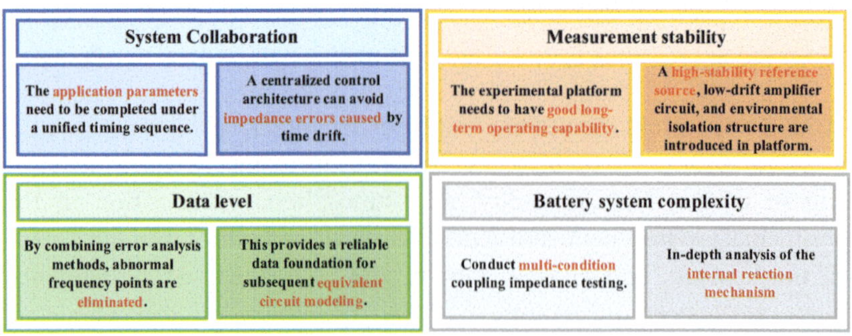

Figure 8.6 Platform software design.

and environmental isolation structures to minimize the influence of nonideal factors on measurement outcomes. Such design strategies not only enhance the accuracy of individual tests but also significantly improve consistency across repeated experiments.

At the data level, the raw impedance data acquired by the experimental platform typically require multistage processing and validation. In addition to basic impedance spectroscopy plotting, outlier frequency points can be identified and eliminated through error analysis and consistency testing methods. By conducting statistical analysis on repeated measurement results, the measurement uncertainty of the experimental platform across different frequency ranges can be evaluated, thereby providing a reliable data foundation for subsequent equivalent circuit modeling and parameter extraction.

Moreover, as the complexity of battery systems continues to increase, traditional impedance testing under single conditions is no longer sufficient to comprehensively reflect their electrochemical behavior. Based on a modular experimental platform, impedance tests under multiple coupled operating conditions can be further conducted, such as simultaneous acquisition of impedance data under varying states of charge, different temperature gradients, and distinct aging stages. The acquisition of multidimensional impedance information provides richer experimental support for indepth analysis of internal reaction mechanisms and degradation processes within batteries.

In summary, EIS instruments are not only fundamental equipment for conducting EIS tests, but their hardware architecture, performance specifications, and operational modes also significantly influence experimental outcomes. The rational selection and configuration of EIS instruments constitute a critical prerequisite for establishing a high precision electrochemical impedance spectroscopy experimental platform and for conducting research on battery impedance characteristics and performance evaluation.

8.1.2 High-precision charge–discharge testing systems

In the comprehensive lifecycle management system encompassing R&D, mass production quality control, and inservice operation and maintenance of power batteries, accurately quantifying electrochemical performance, cycle life, and state of health represents a core technical requirement. The CE-6016n-5V300A-H highrate power battery charge-discharge tester, as a dedicated testing apparatus based on electrochemical testing principles, precisely replicates typical charge/discharge conditions encountered in realworld applications of power batteries through closed loop control technology. This provides reliable data support for the quantitative assessment of the aforementioned performance metrics. The device plays a critical enabling role across the entire power battery industry chain, with its technical characteristics and application scenarios covering the full spectrum of requirements, from validation of novel battery materials to health monitoring of inservice batteries.

As shown in Fig. 8.7,the tester is designed based on a universal adaptability concept, enabling compatibility with mainstream battery chemistries such as lithium ion, lead acid, nickel cadmium, and nickel metal hydride batteries. Considering the distinct electrochemical characteristics of

Figure 8.7 Actual photograph of CE-6016n-5V300A-H.

different battery types and varying test scenario requirements, the tester incorporates four core charging modes, each implemented through independent closed-loop control algorithms to achieve precise regulation. These modes exhibit significant differences in technical features and application scenarios:

Constant-current charging mode stabilizes the charging current at a preset value through an internal current feedback control circuit within the device, enabling constant current charging of the battery. The technical advantages of this mode lie in the low polarization level of the battery during the initial stage of charging, resulting in high energy absorption efficiency and rapid replenishment of battery capacity. Additionally, a constant current ensures a stable rate of electrochemical reactions within the battery, preventing localized overcharging. However, as the charging process progresses, the battery terminal voltage gradually increases. When the voltage approaches the full charge threshold, sustained constant current charging may lead to overvoltage, accelerating electrode material degradation and potentially triggering electrolyte decomposition. Therefore, this mode is suitable only for rapid recharging during low states of charge and is not appropriate for the final stage of full charge topping-off.

The constant voltage charging mode employs a voltage closed loop control strategy to maintain a stable output voltage throughout the charging process. Once the battery terminal voltage reaches the preset constant-voltage level, the battery internal resistance gradually increases with rising State of charge (SOC). According to Ohm's law, the charging current decreases exponentially toward zero. This mode effectively limits the battery's maximum terminal voltage, preventing overcharging and ensuring operation within a safe voltage range; however, during the later stages of charging, the current becomes excessively small, resulting in significantly reduced charging efficiency and prolonged time required to fully charge the battery. Therefore, this mode is primarily applied during the final stage of charging for full charge compensation and must be combined with other charging modes to enhance overall charging efficiency.

The constant current/constant voltage composite charging mode integrates the technical advantages of both constant current and constant voltage charging, employing a "two stage" control logic. During the initial phase, the system operates in constant current mode, utilizing a high charging current to rapidly replenish battery capacity and enhance charging efficiency. When the battery terminal voltage reaches a predefined threshold, a voltage sensor triggers a switching mechanism, automatically transitioning the system to constant-voltage mode. In this stage, the charging current gradually decreases as the SOC increases, until it drops to the cutoff current for constant voltage charging, at which point the charging process terminates. This approach effectively mitigates the risk of overvoltage during the later stages of constant current charging while overcoming the low efficiency characteristic of constant voltage charging in its initial phase. Consequently, it reduces thermal losses during charging and slows down structural degradation of electrode materials. As such, it is currently the most widely adopted charging method in routine performance testing of power batteries, particularly suitable for critical assessments such as capacity testing and cycle life evaluation of lithium-ion power batteries, and is also employed for factory calibration of battery capacity in mass-produced cells.

Constant power charging mode maintains a steady output power during charging through a power closed-loop control algorithm. According to the power formula, as the battery terminal voltage increases, the device dynamically reduces the charging current via its current regulation module in real time, ensuring that the product of voltage and current consistently equals the preset power value. The core value of this mode lies in simulating actual power demand scenarios encountered by battery systems in practical applications. Through testing, it enables evaluation of a battery's ability to accept charging under varying power loads, as well as its voltage stability and temperature variation characteristics. However, this mode imposes stringent requirements on equipment control accuracy: voltage and current signals must be sampled and dynamically adjusted at millisecond level precision to prevent measurement deviations caused by power fluctuations; meanwhile, strict limits must be enforced on maximum current to avoid exceeding current thresholds due to sudden voltage spikes, which could otherwise lead to irreversible battery damage.

To further ensure the safety of the testing process and the reliability of the data, the CE-6016n-5V300A-H tester also integrates multiple protective mechanisms that work in conjunction with the aforementioned performance parameters to establish a comprehensive testing assurance system:

Overvoltage protection: When the battery terminal voltage exceeds the preset protection threshold, the device automatically disconnects the charging circuit to prevent safety incidents such as electrolyte decomposition and electrode burnout caused by overvoltage.

Overcurrent protection: When the charging/discharging current exceeds 1.2 times the maximum output current of 100 A, the device activates overcurrent protection, halting the charging/discharging process to prevent excessive battery internal resistance heating and potential thermal runaway caused by excessive current.

Overtemperature protection: When the battery temperature exceeds a preset threshold, the device automatically reduces the charging/discharging current or terminates the test to prevent accelerated degradation of battery performance or potential safety hazards caused by high temperatures.

Reverse polarity protection: When the battery's positive and negative terminals are incorrectly connected to the test fixture, the internal protection circuit of the device activates, interrupting the output circuit to prevent reverse current from damaging the equipment or the battery.

The parameters of these protective mechanisms can be flexibly adjusted according to testing requirements, making them suitable not only for conventional performance testing but also for security-related performance testing, thereby further expanding the application scenarios of the equipment and providing critical safeguards for both testers and test specimens.

In the full industrial chain technology of power batteries, the application scenarios of the CE-6016n-5V300A-H high-rate battery charger/discharge tester must be aligned with technical requirements at different stages. Leveraging its high precision parameter control and multimode charge/discharge capabilities, the tester enables comprehensive coverage from research and validation to failure analysis. For battery packs, the tester must be used in conjunction with a voltage dividing module to distribute the total voltage of multicell batteries within the single-channel test range. It supports up to 10 series-connected 5V cells, making it suitable for small scale battery packs such as those used in electric bicycles and drones. During testing, the voltage dividing module enables realtime monitoring of individual cell voltages, preventing pack failure due to overcharging or overdischarging of any single cell. Meanwhile, through coordinated multichannel control, the tester achieves precise regulation of overall charge/discharge current for the battery pack, meeting demands for capacity testing and rate performance evaluation in small scale battery systems. It should be noted that for large scale battery packs with total voltages exceeding 50 V or current requirements surpassing 100 A, multiple testers must be connected in series or parallel configurations and integrated with dedicated battery pack testing software to fulfill complex operational condition testing requirements.

8.1.3 Temperature-controlled environmental chambers

The MGDW-408-40BC Constant Temperature and Humidity Test Chamber is a specialized environmental simulation device primarily designed for experimental applications in lithium ion battery material development, cell performance testing, and battery pack reliability validation, as shown in Fig. 8.8. By precisely controlling temperature and humidity parameters, the equipment simulates the operational conditions of lithium ion batteries under various environmental settings, providing a stable environmental foundation for evaluating battery capacity degradation, cycle life, safety, and electrochemical performance. Its design specifically addresses the unique environmental simulation requirements in lithium ion battery experiments, such as low humidity stability, rapid temperature change response, and long term operational reliability during extended duration tests.

Compared with general purpose constant temperature and humidity equipment, the MGDW-408-40BC has undergone targeted improvements in structural optimization and functional adaptation. In performance testing related to lithium-ion battery management systems, the explosion proof high/low temperature test chamber serves as a core device for simulating battery characteristics under varying temperature conditions. Its superior temperature control accuracy, explosion proof safety, and system

Figure 8.8 Physical image of MGDW-408-40BC.

interconnectivity provide a reliable foundation for environmental adaptability testing of batteries. The test chamber covers a temperature range of -40 to 150°C, enabling precise replication of operational scenarios for lithium ion batteries under extreme low temperature and high temperature conditions. Under noload and thermally stable conditions, the temperature fluctuation is ≤ 0.5°C, and the temperature deviation is maintained within ± 2.0°C, ensuring stability and consistency of the test environment. To meet demands for efficient testing, the equipment demonstrates excellent heating and cooling rates: the time required to heat from $+20$ to $+150$°C and to cool from $+20$ to -40°C is each ≤ 60 min, significantly enhancing testing efficiency.

The equipment's structural design balances practicality and safety, with an internal chamber dimension capable of accommodating various lithium ion battery sizes and testing fixtures, while its external dimensions are compatible with standard laboratory layouts. The thermal insulation enclosure is constructed using high quality materials: the outer wall consists of cold rolled steel plates treated with electrostatic spraying and baking paint, while the inner wall employs corrosion resistant SUS304 stainless steel plates. The chamber's thermal insulation layer features a composite structure of 100 mm thick rigid polyurethane foam and glass wool, with the door insulated using glass wool to effectively minimize heat loss and ensure uniform temperature distribution within the chamber. The airconditioning system includes a centrifugal fan, heater, and evaporator, combined with finned explosionproof heaters and contactless pulse width modulation control for precise temperature regulation within the chamber.

Safety design is one of the core advantages of this equipment. Each chamber is equipped with explosion proof pressure relief functionality, and two explosion proof chains are installed on both sides of the door, effectively mitigating potential safety risks during battery testing. The equipment features a high definition color LCD touchscreen with bilingual human and machine interface, complemented by a controller display and overtemperature protection setpoint regulator, ensuring convenient and safe operation. Standard configuration includes four φ100 mm wire-passing holes and four adjustable feet-mounted casters, facilitating test wiring and equipment mobility. In terms of system integration, the equipment can be hardware-connected to a BTS host computer and cell testing devices via channel cables and data communication lines, enabling coordinated operation with battery testing systems. Additionally, the temperature chamber's internal trays can be customized according to battery dimensions, demonstrating strong adaptability for tailored configurations.

This test chamber employs a mechanical compression cascade refrigeration system. It operates under AC (380 ± 38) V, (50 ± 0.5) Hz three-phase five-wire power supply conditions, with a distribution power of 9 kW and a maximum current of 18 A, enabling stable compatibility with laboratory power systems. Its overall performance meets the critical testing requirements for lithium-ion batteries under various temperature environments, including capacity, cycle life, and rate capability. It provides precise and safe environmental simulation support for modeling of lithium-ion battery management systems, state estimation, and validation of artificial intelligence algorithms.

8.2 Design of test protocols under varying conditions

8.2.1 Hybrid pulse power characterization protocol

To better evaluate battery performance and characterize the pulse charge–discharge behavior of power batteries, the Hybrid pulse power characteristic (HPPC) test procedure is conducted. Initially, the battery is charged to a specific voltage level, typically around 80% of its nominal voltage. After charging is completed, a pulsed load is applied to discharge the battery. Subsequently, the battery is allowed to rest for 40 s to recover its state. Following this, a 10 second charging current pulse is applied, and the battery voltage response is observed and recorded. By analyzing the measured data, both ohmic internal resistance and polarization internal resistance can be determined. A resting period is required between consecutive pulse tests to allow the battery to reestablish electrochemical and thermal equilibrium. This experiment evaluates the internal resistance characteristics of the battery through repeated charge/discharge pulse tests combined with rest periods for recovery, thereby providing a more comprehensive understanding of battery performance. The experimental

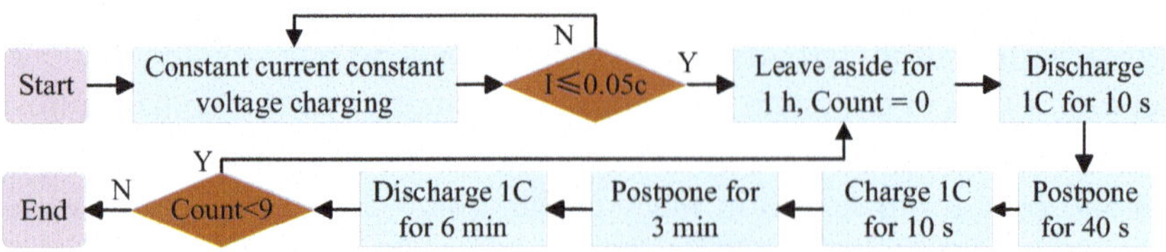

Figure 8.9 Experimental design diagram under HPPC conditions.

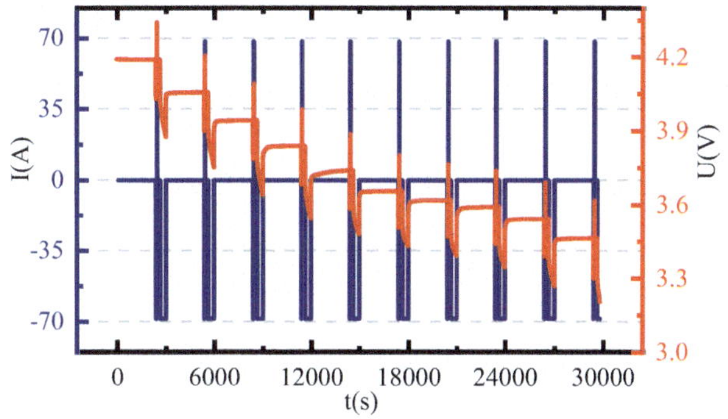

Figure 8.10 Voltage and current data under HPPC operating conditions.

procedure and results of the HPPC test are illustrated in Figs. 8.9 and 8.10, respectively.

As the cell changes, its ohmic resistance and polarization impedance also vary in accordance with certain linear patterns, as shown in Fig. 8.10. State assessment studies can be conducted based on the voltage-current data obtained experimentally.

8.2.2 Bursts of balanced dynamic stress test protocol

To better simulate the complex operating conditions of lithium-ion batteries in practical applications, the Beijing bus dynamic stress test (BBDST) condition was designed. BBDST is a real world operational condition derived from actual data collected from Beijing buses, and its testing protocol comprises two components: static performance testing and dynamic performance testing of the battery pack. The test procedure includes the following steps: startup, acceleration, coasting, braking, constant speed, sudden acceleration, and sudden braking. The specific running times and power calculations for each step are illustrated in Fig. 8.11, while the experimental voltage and current data obtained are shown in Fig. 8.12.

As shown in Fig. 8.11, the duration of a complete BBDST condition is 300 s. Since the BBDST condition involves power discharge, Fig. 8.12 indicates that the battery's discharge capability improves with increasing cycle count. The voltage and current data obtained from the experiment can be used for subsequent model parameter identification validation and algorithm research.

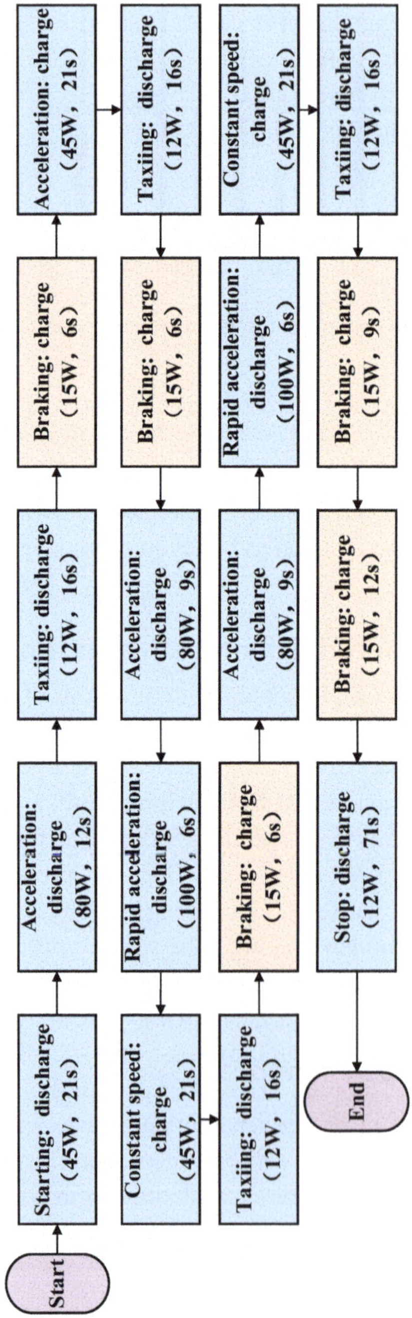

Figure 8.11 Experimental design diagram of the BBDST condition.

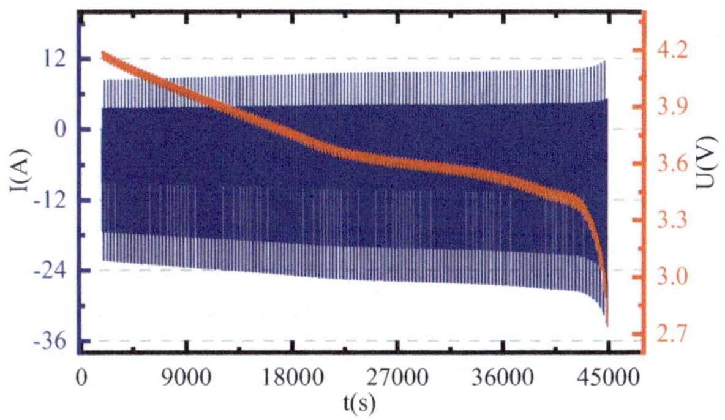

Figure 8.12 BBDST operating condition voltage and current data.

8.2.3 Dynamic stress test protocol

In practical applications, the realtime current of lithium-ion batteries is complex and highly variable. Under different operating conditions, abrupt current switching and stopping frequently occur, necessitating stringent requirements for the battery model's dynamic simulation performance. To better evaluate battery performance and provide effective validation for state of energy estimation algorithms, a customized Dynamic stress test (DST) experiment was designed, with specific experimental procedures illustrated in Fig. 8.13.

In Fig. 8.13, the cutoff condition for constant-current constant-voltage charging is when the current drops to 0.05C, and the discharge cutoff voltage is 2.75 V.

The experimental data for voltage and current under the DST condition are shown in Fig. 8.14.

As shown in Fig. 8.14, with increasing cycle numbers, the discharge current varies uniformly, and the battery terminal voltage generally exhibits a downward trend. The voltage and current data obtained from the experiment can be used for subsequent model parameter identification validation and research on state estimation algorithms.

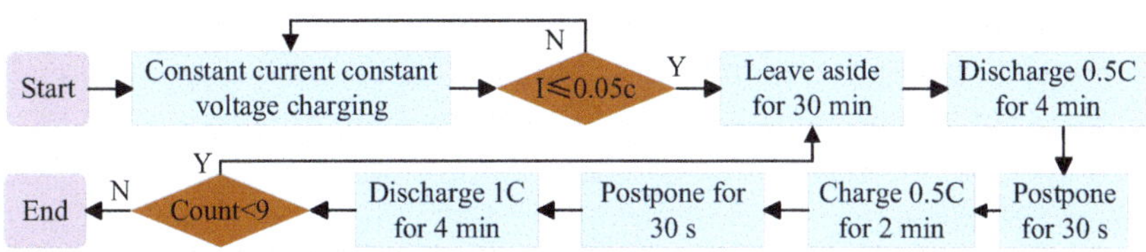

Figure 8.13 Experimental design diagram under DST operating conditions.

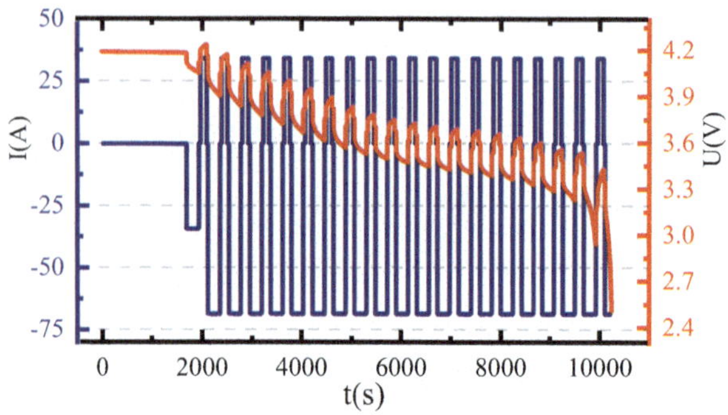

Figure 8.14 Voltage and current data under DST operating conditions.

8.3 Visualization and analysis of experimental data

8.3.1 Data cleaning and preprocessing techniques

In battery experiments and vehicle operation, the raw data collected often contain substantial environmental noise, sensor drift, and abrupt anomalies induced by electromagnetic interference. Adhering to the principle of "garbage in, garbage out," high quality data preprocessing is a prerequisite for subsequent accurate modeling and state estimation. The primary objective of data cleaning is noise suppression and signal smoothing. Since current and voltage sensors are highly susceptible to high-frequency interference caused by switching operations of power electronic devices, directly computing differential features from raw data leads to significant numerical oscillations. In engineering practice, moving average filters or Gaussian filters are commonly employed to eliminate high frequency noise. However, when processing incremental capacity curves where peak features must be preserved, the Savitzky Golay filter, owing to its local polynomial least-squares fitting approach, provides superior performance by effectively maintaining the shape and width of signal peaks while achieving smoothness.

In addition to noise, the detection and removal of outliers are equally critical. During longterm aging tests, data transmission packet loss or poor electrical contact often causes voltage readings to instantaneously drop to zero or spike abruptly, nonphysical transitions that must be identified and eliminated. Common methods include statistical filtering based on the sigma rule or interquartile range analysis using box plots. For data gaps remaining after outlier removal, or signal misalignment caused by differing sampling frequencies, linear or spline interpolation techniques are typically employed for data imputation and resampling, ensuring strict temporal alignment among voltage, current, and temperature data.

Data normalization and standardization are essential steps to meet the input requirements of machine learning algorithms. Due to the significant differences in physical dimensions and magnitudes among voltage (typically ranging from 2.5 to 4.2 V), current (potentially reaching hundreds of amperes), and temperature (usually between −20 and 60°C), directly feeding raw data into a neural network may result in slow convergence of gradient descent or even entrapment in local optima. By applying Min-Max normalization to map all data into the [0, 1] interval, or by using Z Score standardization to transform the data into a distribution with zero mean and unit variance, the influence of different units can be eliminated, thereby significantly enhancing the training efficiency and convergence stability of the model.

8.3.2 Multidimensional data visualization approaches

A battery is a highly complex nonlinear system, and a one-dimensional time domain curve (such as a voltage and time profile) often fails to fully reveal its internal degradation mechanisms. Multidimensional data visualization aims to transform hidden electrochemical states into intuitive visual representations by constructing a high dimensional feature space. Among various approaches, feature curve transformation is the most classical method, particularly incremental capacity analysis and differential voltage analysis. By converting the flat charge/discharge voltage plateaus into differential curves with distinct peaks and valleys, researchers can visually identify electrochemical reaction peaks corresponding to different phase transition processes. The shift in peak positions and their degree of attenuation can directly reflect microscopic aging patterns such as active lithium loss or active material loss, thereby achieving the visual representation of "macroscopic data" for "microscopic mechanisms."

Three dimensional surface plots of multiphysics coupling serve as powerful tools for analyzing the performance boundaries of batteries under dynamic operating conditions. For instance, to evaluate a battery's power capability, a three dimensional surface can be constructed with state of charge as the X axis, temperature as the Y axis, and either maximum pulse power or DC internal resistance as the Z axis. This visualization clearly reveals the battery's "comfort zone" and "danger zone," providing intuitive guidance for the Battery management system (BMS) in formulating safety boundary control strategies. Heatmaps are commonly employed for correlation analysis, using color intensity to represent Pearson correlation coefficients among dozens of feature variables such as voltage, current, and temperature, thereby assisting data scientists in rapidly identifying key features most sensitive to State of health (SOH) estimation.

In long term life assessment, dimensionality reduction visualization of aging trajectories has gained increasing attention. By employing Principal component analysis (PCA) or t SNE algorithms, high dimensional sampling data from hundreds or even thousands of cycles can be projected onto a two dimensional plane. In such visual representations, batteries of the same type or those exhibiting similar degradation patterns cluster together, while outlier batteries that fail prematurely remain isolated from these clusters. This approach enables engineers to comprehensively observe the evolution of consistency across an entire battery pack within a single chart and rapidly identify potential "bottleneck" cells.

8.3.3 Visualization tools and application scenarios

In the modeling, state estimation, and intelligent decision-making processes of lithium-ion battery management systems, the scale and complexity of experimental data continue to increase. Voltage, current, temperature, strain, internal resistance, and multisource derived features accumulate across different time scales and operating conditions, making it increasingly difficult to fully characterize the evolution patterns of battery states using numerical analysis or statistical metrics alone. Under this context, data visualization tools have gradually evolved from auxiliary presentation methods into indispensable analytical and decision-support modules within battery management systems. Their role spans the entire workflow, from data understanding and model evaluation to system operation, as illustrated in Fig. 8.15.

From the perspective of data organization, battery experimental and operational data typically exhibit pronounced multidimensionality and temporal characteristics. Signals acquired from different sensors differ significantly in sampling frequency, dynamic range, and noise levels, and are simultaneously influenced by varying operating conditions, environmental factors, and aging states. Visualization tools enable effective representation of complex information within limited visual space by performing dimensionality reduction and structured data mapping. For instance, timeseries visualization can reveal battery response characteristics under dynamic

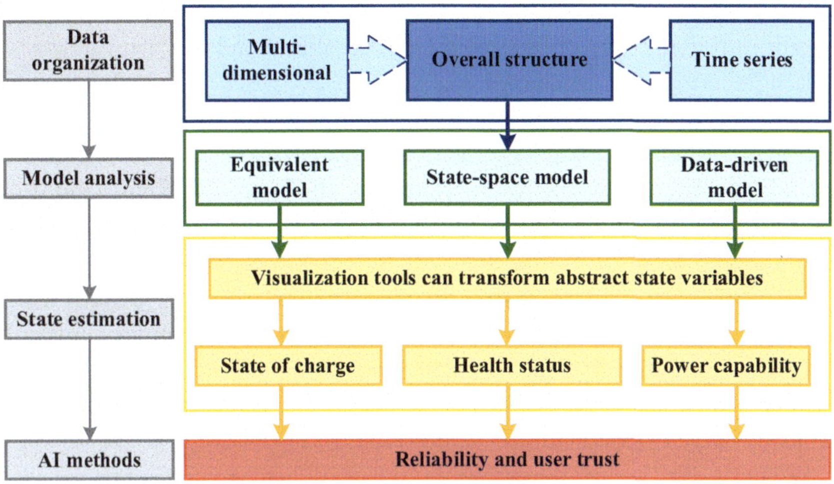

Figure 8.15 Data visualization tools.

loading, while state-space projections facilitate the observation of distributional differences across distinct operational phases or between different samples. This intuitive representation approach provides researchers and engineers with rapid insight into the overall structure of the data.

During the model analysis and validation phase, visualization tools also play a critical role. Regardless of whether the models are mechanism based equivalent models, statespace models, or data driven machine learning and deep learning models, their performance evaluation relies fundamentally on the comparison between predicted results and actual measured data. Through overlaid curves, error distribution plots, and multidimensional metric visualizations, the adaptability and limitations of models under various operating conditions and different aging stages can be clearly illustrated. This visualization based assessment approach not only facilitates the identification of model mismatches or parameter drift issues but also provides intuitive guidance for model structure optimization and feature selection.

Furthermore, in battery state estimation and health management applications, visualization tools can transform abstract state variables into comprehensible information formats. Key indicators such as state of charge, state of health, and power capability are not directly measurable; their credibility often relies on algorithmic inference. By visually representing the trends of these states over time or under varying operating conditions, one can assist in evaluating the reasonableness and stability of estimation results, while enhancing the interpretability of system outputs. This is of significant importance for improving system reliability and user trust in practical engineering applications.

With the widespread application of artificial intelligence methods in battery management systems, the role of visualization tools has further expanded to encompass the interpretation of algorithmic behavior and decision support. High dimensional feature spaces, latent states, and internal weight distributions of models are partially visualized, enabling researchers to analyze key features that models focus on and their response patterns under varying operating conditions. This "visualization driven understanding" approach helps mitigate the opacity associated with blackbox models, providing essential support for the application of artificial intelligence algorithms in safety critical systems, as illustrated in Fig. 8.16.

Data visualization tools in lithium ion battery management systems have evolved from mere result presentation instruments into essential links connecting experimental data, model algorithms, and engineering applications. Through thoughtful design of visualization formats tailored to specific application scenarios, not only can data analysis efficiency be enhanced, but also the interpretability of system operations and the reliability of decision making can be improved, thereby laying a foundation for the development of more complex application scenarios.

At the practical application level, data visualization tools in lithium-ion battery management systems assume context dependent functional roles, with their design priorities shifting according to varying application

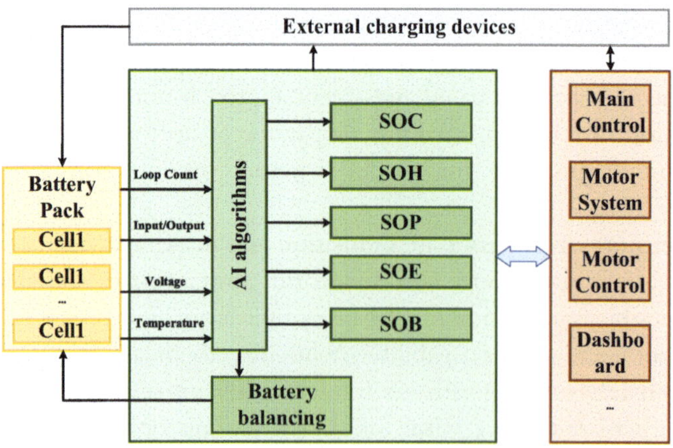

Figure 8.16 Application of artificial intelligence algorithms in BMS.

objectives. During the experimental research phase, visualization emphasizes understanding the underlying mechanisms of battery behavior and the evolution patterns of battery states; in contrast, during engineering testing and real world operational environments, visualization gradually shifts toward monitoring system operational status, identifying anomalies, and supporting decision making.

In laboratory research and performance evaluation scenarios, visualization tools are primarily employed to assist in analyzing the response characteristics of batteries under controlled operating conditions. By enabling multidimensional, interactive visualization of fundamental signals such as voltage, current, and temperature, these tools provide intuitive insights into how varying operational conditions influence the dynamic behavior of batteries. For instance, in pulse load or dynamic condition experiments, combining time-series plots with phasespace diagrams facilitates the identification of transient response features and hysteresis effects. Furthermore, presenting multiple sets of experimental data in parallel or overlaid formats enables effective comparison of differences among various battery samples, testing conditions, or aging stages, thereby supporting the synthesis of experimental conclusions.

During the model development and algorithm validation phases, visualization tools are frequently employed to bridge the gap between experimental data and model outputs. By presenting measured data, model predictions, and error distributions within a unified coordinate system, the applicable range of the model across different operating regimes can be clearly revealed. This visualization based analytical approach is particularly crucial during model tuning and structural optimization, as it facilitates rapid identification of performance bottlenecks and reduces reliance on single evaluation metrics. In the context of data driven models and artificial intelligence algorithms, visualization of the training process further enables monitoring of convergence behavior, helps prevent overfitting, and enhances the efficiency of algorithm development.

In engineering testing and system integration scenarios, visualization tools have gradually shifted from an analytical focus to a monitoring orientation. In such contexts, systems typically require prolonged stable operation, making realtime performance and reliability paramount. Through instrumented interfaces and status dashboards, visualization systems can provide realtime insights into battery operational status, trends in critical parameters, and system alarm information. This approach not only enables engineering personnel to rapidly grasp the overall system status but also delivers intuitive feedback at the earliest signs of anomalies, thereby reducing fault response times.

In practical battery management systems, visualization primarily serves decision support and information dissemination functions. For complex battery systems, management algorithms typically operate automatically in the background, while the visualization interface acts as a crucial human machine interaction window, presenting system status and decision outcomes in a concise and comprehensible manner. Through trend charts, status bars, and graded alarm indicators, users can quickly assess whether the system is operating within safe and efficient parameters. This application oriented visualization design holds significant importance in enhancing system usability and safety.

As application scenarios continue to expand, data visualization tools are progressively integrated into multi-layered system architectures, exhibiting distinct hierarchical characteristics in form and functionality, from laboratory analysis software and embedded monitoring interfaces to cloud-based data management platforms. Targeted visualization design tailored to specific application requirements represents a crucial component in achieving efficient operation and intelligent decision making within battery management systems.

Meanwhile, the continuously increasing scale and system complexity of lithium ion battery applications have made single-scale or single level visualization approaches inadequate for meeting the comprehensive requirements of battery management systems in monitoring, analysis, and decision-making. In response, a multiscale, multilayered visualization framework has emerged. Its core principle lies in hierarchically organizing and differentially presenting information according to variations in data sources, time scales, and user needs, thereby enabling efficient representation and comprehension of complex system states, as illustrated in Fig. 8.17.

In terms of spatial scale, visualization systems typically need to cover multiple levels, ranging from individual cells to modules and ultimately to system level integration. Visualization at the cell level focuses on subtle variations in voltage, temperature, and state variables, enabling analysis of individual disparities and localized anomalies. Module level visualization emphasizes consistency and distribution characteristics among cells, facilitating the identification of imbalances. System level visualization further abstracts these aspects into overall performance metrics and safety status, supporting high level decision making. Through hierarchical representation, management systems can flexibly switch perspectives according to

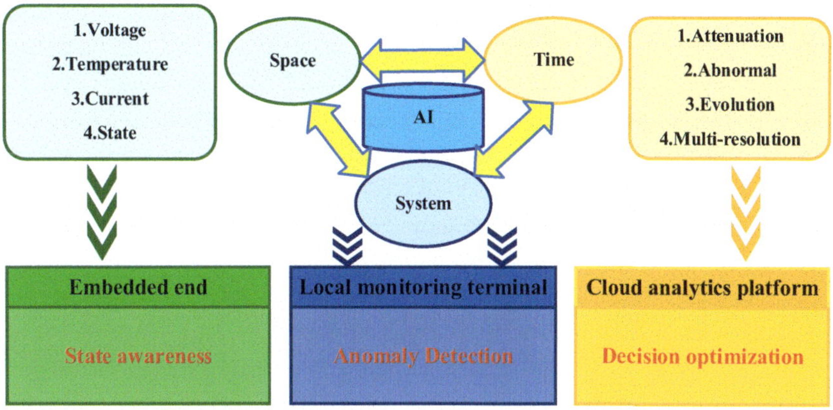

Figure 8.17 Application of multiscale AI in BMS.

varying precision requirements, thereby avoiding information redundancy or the concealment of critical details.

In terms of temporal scale, multiscale visualization requires simultaneous consideration of transient behaviors and longterm evolutionary processes. On short time scales, visualization focuses on capturing rapid dynamic responses and sudden anomalies, commonly applied in realtime monitoring and security protection; whereas on longer time scales, trend oriented presentation proves more effective for analyzing performance degradation and state evolution patterns. By introducing a scalable timeline and multiresolution representation within a unified visualization framework, information across different temporal scales can be systematically managed, thereby enhancing the system's capability to comprehend complex operational processes.

At the system architecture level, visualization tools are typically distributed across different layers, including embedded systems, local monitoring stations, and cloud-based analytics platforms. Visualization at the embedded level emphasizes simplicity and realtime performance, primarily displaying critical operational parameters and alert information. The local monitoring layer offers more sophisticated graphical interfaces for engineering debugging and system maintenance. The cloud platform focuses on integrating historical data, enabling cross system comparisons, and presenting advanced analytical results. This layered deployment approach allows users at different levels of interest to access relevant information while minimizing unnecessary data transmission and computational overhead.

When designing a multiscale, multilevel visualization system, attention must also be paid to issues of information consistency and interactivity. Visualization results across different levels should maintain logical coherence, ensuring that key metrics remain traceable across varying scales of presentation. Furthermore, through interactive design, users can rapidly switch between system level and detailed views, enabling a seamless transition from macroscopic monitoring to microscopic analysis. This interaction driven visualization approach enhances both the usability of the system and analytical efficiency.

Overall, the multiscale and multilevel visualization framework provides a structured approach to information organization and representation for complex lithium-ion battery management systems. Through rational segmentation and collaborative design across spatial, temporal, and system levels, visualization tools effectively support state awareness, anomaly detection, and decision optimization, laying a crucial foundation for the realization of intelligent battery management systems.

Although data visualization has demonstrated significant value in lithium ion battery management systems, its application still faces numerous challenges as system scale expands and algorithmic complexity increases. First, the rapid growth in data volume imposes higher demands on the realtime performance and scalability of visualization systems. Largescale data generated by high sampling rates, multiple sensors, and longterm operation render traditional static visualization methods inadequate for realtime monitoring and interactive analysis. Achieving efficient rendering and responsiveness while preserving information integrity has thus become a critical issue in the design of visualization systems.

Second, the complexity of visualization continues to increase due to data heterogeneity. Lithium-ion battery management systems often involve physical measurement data, model estimation results, and outputs from artificial intelligence algorithms, which exhibit significant differences in dimensionality, temporal scale, and uncertainty characteristics. A major challenge in current visualization applications lies in the rational integration of multisource heterogenous information within a unified visualization framework while avoiding information conflicts or misinterpretations. The absence of a standardized data representation protocol may lead to uninterpretable visualizations or even erroneous conclusions.

Meanwhile, with the deepening application of artificial intelligence methods in battery management systems, visualization faces the challenge of insufficient algorithmic interpretability. Deep learning models often possess highly nonlinear internal structures, making their decision making mechanisms difficult to express directly through conventional graphical representations. Existing visualization techniques still have limitations in revealing internal feature correlations, weight distributions, and decision pathways within models, which to some extent hinders the broader adoption of artificial intelligence algorithms in safety critical systems. Therefore, transforming internal model information into intuitive and reliable visual representations remains a key direction for future research.

From an engineering application perspective, a conflict also exists between generality and customization in visualization systems. On one hand, battery management systems exhibit significant differences in visualization requirements across various application scenarios; on the other hand, excessive customization may increase system development and maintenance costs. The challenge of achieving flexible configuration within a general framework, so that the visualization system remains both scalable and capable of meeting specific application requirements, represents an urgent issue in practical deployment.

Looking ahead, data visualization in lithium-ion battery management systems is expected to evolve toward greater intelligence and interactivity. By integrating automated feature extraction and anomaly detection algorithms, visualization systems will transition from passive display to proactive alerting, automatically highlighting critical information and guiding users' attention toward potential risks. Meanwhile, with the advancement of cloud computing and edge computing technologies, multilevel collaborative visualization will become increasingly mature, enabling efficient data flow and unified representation across different system layers. Explainable visualization methods tailored for artificial intelligence are anticipated to emerge as a key research direction. By visually representing model inference processes, feature contribution levels, and uncertainty information, such approaches will enhance algorithmic transparency and system credibility. As theoretical foundations and supporting tools continue to improve, visualization will transcend mere data comprehension and become an essential component underpinning intelligent decision-making and safe system operation.

Overall, the role of data visualization in lithium ion battery management systems is continuously deepening, with its development direction becoming increasingly integrated with modeling, estimation, and artificial intelligence technologies. By addressing existing challenges and introducing novel visualization concepts, further enhancement of the intelligence level and engineering application value of battery management systems is anticipated.

References

[1] J. Xie, Z. Jia, T.H. Sheng, et al., The in-situ electrochemical impedance spectroscopy unveils the microscopic evolution of vanadium-based cathode materials in aqueous zinc-ion batteries, Chem. Eng. J. 518 (2025) 164660.

[2] Y. Liu, L. Yang, R. Liao, et al., Degradation mechanism of sodium-ion batteries and state of health estimation via electrochemical impedance spectroscopy under temperature disturbances, Energy 332 (2025) 137064.

CHAPTER NINE

Future directions and societal impact

9.1 Technical future directions in battery management

9.1.1 AI-driven advanced modeling

The future Battery management system (BMS) will evolve into an integrated platform combining multiscale sensing, real-time analysis, and intelligent decision-making. This evolutionary trajectory will primarily be driven by the following three technological pillars.

9.1.1.1 AI-driven advanced modeling advancing toward the digital twin paradigm

The battery modeling field has long been characterized by a dichotomy between mechanistic models and data-driven models. The former offers high computational efficiency but struggles to capture complex degradation mechanisms, whereas the latter achieves high accuracy yet is frequently criticized for lacking physical interpretability, being regarded as a "black box." The key future breakthrough lies in overcoming this divide, moving toward a digital twin paradigm [1].

The physics-informed battery digital twin framework illustrated in Fig. 9.1 constructs a digital model by acquiring data such as battery voltage, current, and temperature, enabling visualization of internal states including lithium-ion distribution and concentration, thereby achieving a digital mapping of the battery's physical state.

Digital twins are not merely static virtual replicas of physical entities but dynamic systems capable of evolving in synchronization with their physical counterparts. The core enabler lies in the integration of Physics-informed machine learning (PIML). By embedding partial differential equations that describe processes such as lithium-ion transport and electrode reaction

Modelling, Estimation and AI Applications for Lithium-Ion Battery Management Systems
ISBN: 978-0-443-45347-2
https://doi.org/10.1016/B978-0-443-45347-2.00009-1

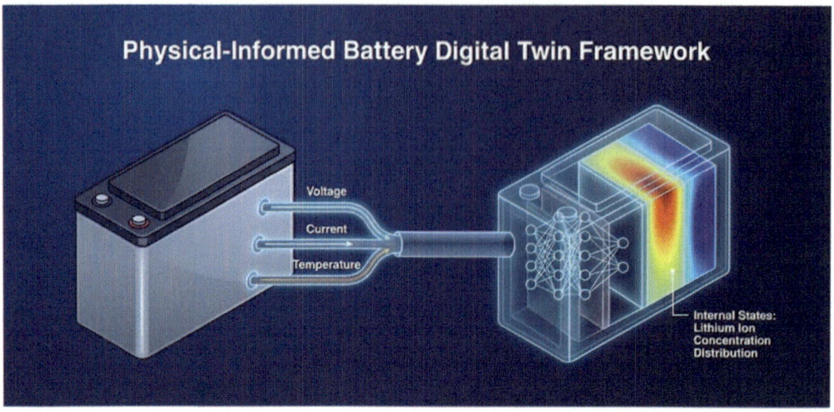

Figure 9.1 Physical information-based battery digital twin framework.

kinetics as constraints within the loss function of neural networks, a "gray-box" model is constructed that combines mechanistic robustness with data-driven flexibility. This fusion not only enhances the predictive reliability of models under extreme operating conditions but also significantly reduces dependence on large volumes of annotated data, thereby opening new avenues for understanding the internal microstates of batteries.

9.1.2 Adaptive parameter reconstruction

Lithium-ion batteries, serving as the core carriers for electrochemical energy storage and conversion, inherently constitute a typical time-varying nonlinear system. This time-varying nonlinear behavior arises from the complex electrochemical reactions and mass transport processes occurring within the battery, manifesting as continuous dynamic drift in key performance parameters with increasing aging, temperature fluctuations, and varying operational conditions. From an aging perspective, prolonged charge/discharge cycling leads to irreversible degradation processes such as progressive loss of active electrode materials, gradual decomposition of the electrolyte, and continuous growth of the Solid electrolyte interphase (SEI) layer—collectively resulting in capacity fade and increased internal resistance. From a thermal standpoint, low temperatures significantly reduce lithiumion mobility and electrolyte conductivity, while high temperatures accelerate parasitic reaction kinetics; both extremes alter the battery's charge/discharge characteristic curves. From the operational condition perspective, varying charge/discharge rates and fluctuating load intensities directly influence internal polarization levels within the battery, thereby inducing dynamic variations in voltage response and energy output efficiency.

Owing to the aforementioned characteristics, conventional static models calibrated at the factory exhibit significant limitations and can no longer meet the demands for precise battery management throughout the entire lifecycle. The fundamental drawback of static models lies in their fixed parameters, which are only suitable for specific temperature conditions and operating scenarios at the initial stage after manufacturing, and cannot dynamically adapt to the time-varying characteristics of the battery.

Therefore, next-generation BMS algorithms must transcend the limitations of static models and possess robust Adaptive parameter reconstruction (APR) capabilities. The core objective is to achieve real-time alignment between model parameters and the actual state of the battery. By dynamically updating model parameters, these algorithms can compensate for characteristic drift caused by aging, temperature variations, and changing operating conditions, thereby ensuring high estimation accuracy and control reliability throughout the battery's entire lifecycle. This provides essential algorithmic support for safe battery operation, extended lifespan, and efficient utilization.

From a technical perspective, future mainstream solutions will integrate efficient online identification algorithms. Among these, the Forgetting-factor recursive least squares (FF-RLS) algorithm and the more advanced Dual extended Kalman filter (DEKF) represent the two most promising core algorithms. The FF-RLS algorithm incorporates a forgetting factor that dynamically adjusts the weighting between historical and current data, thereby reducing the influence of outdated data on parameter identification. This enables rapid detection of subtle variations in voltage and current responses under dynamic vehicle operating conditions, allowing real-time updating of key internal model parameters, such as the increase in ohmic internal resistance reflecting battery aging, and changes in polarization capacitance and resistance characterizing polarization effects. The FF-RLS algorithm offers low computational complexity and strong real-time performance, making it well-suited for embedded systems with limited computational resources on-board vehicles. In contrast, the DEKF algorithm integrates both battery model parameters and state variables within an extended Kalman filtering framework, enabling concurrent parameter identification and state estimation. Compared to FF-RLS, DEKF achieves higher identification accuracy under complex operating conditions, particularly meeting stringent accuracy requirements in high-end electric vehicles and energy storage power stations.

At the fundamental research level, deeper exploration will focus on achieving precise decoupling of aging mechanisms, representing an advanced direction for adaptive parameter reconfiguration technology. Battery performance degradation typically results from the combined effects of multiple factors, including temporary and reversible performance decline induced by environmental temperature fluctuations, as well as permanent and irreversible aging caused by SEI layer growth, active material loss, lithium dendrite deposition, and similar phenomena. These two types of degradation exhibit fundamentally distinct mechanisms and influence patterns, necessitating markedly different mitigation strategies. Future research will integrate electrochemical mechanism models with data-driven approaches to develop adaptive multi-physics coupled models capable of intelligent differentiation and quantitative characterization of distinct degradation mechanisms. This mechanistic-level decoupling capability will provide precise foundations for formulating differentiated battery maintenance and management strategies, for instance, addressing transient degradation through adjustments in charging/discharging protocols, while initiating equalization maintenance or replacement procedures promptly in response to irreversible aging, thereby maximizing both the utilization value and service life of batteries.

9.1.3 Real-time learning and edge intelligence collaboration

With the continuous advancement of lithium-ion battery management requirements, deep learning models are increasingly applied in BMS. Although these models can enhance the accuracy of state estimation and risk prediction, they typically feature complex network architectures and large parameter scales. Meanwhile, in-vehicle edge computing chips face significant constraints in terms of physical space, power budget, and cost control, resulting in pronounced computational bottlenecks. The conflict between these two factors has become increasingly pronounced. A single computational architecture can no longer simultaneously satisfy model accuracy and real-time response demands: deploying complex deep learning models directly on edge devices leads to excessive computation latency due to insufficient computing power, failing to meet millisecond-level safety monitoring requirements; conversely, relying solely on lightweight edge-based models sacrifices estimation accuracy and the foresight of risk warnings, rendering them inadequate for fine-grained management under complex operating conditions. Under this context, constructing a hierarchical intelligent architecture based on "cloud-edge collaboration" has emerged as an inevitable solution to resolve this contradiction and is expected to gradually become an industry standard in the future.

The core logic of the "edge-cloud collaboration" architecture lies in differentiated allocation of computational resources based on task characteristics, achieving the dual objectives of "ensuring real-time responsiveness and safety" as well as "enhancing accuracy through massive computation" via functional complementarity and efficient coordination between edge and cloud. Here, the "edge side" centers on the vehicle-mounted BMS, whose primary role is defined as "real-time response and safety fallback," requiring priority to be given to low latency and high reliability. Consequently, the edge side primarily handles millisecond-level critical tasks: first, safety monitoring and rapid fault isolation, this includes real-time surveillance of core parameters such as voltage, current, and temperature to detect whether they exceed safety thresholds; upon anomaly detection, protective mechanisms are immediately triggered to rapidly disconnect charging/discharging circuits, preventing escalation of safety incidents. Second is real-time state estimation based on lightweight models, complex deep learning models are compressed using techniques such as model pruning, quantization, and knowledge distillation to produce lightweight models suitable for edge-side computational capacity, enabling rapid estimation of key states such as State of charge (SOC) and State of health (SOH), thereby ensuring timely adjustments to energy management strategies during driving. These tasks typically require processing delays maintained within 10–100 ms, well within the capability of local edge computing, and simultaneously eliminate latency and security risks associated with data transmission.

In contrast to the "real-time response" role fulfilled by the edge side, the cloud assumes the function of the system's "central brain" by leveraging powerful computational resources provided by its distributed computing clusters. It focuses on handling high-computational-demand, non–real-time complex tasks. Specifically, the cloud's core responsibilities encompass three main aspects: first, deep mining of massive historical operational data, aggregating full-life-cycle operational data from lithium-ion batteries across diverse vehicles and application scenarios, and employing big data analytics techniques to uncover battery aging patterns and correlation features between operating conditions and performance degradation, thereby providing rich data support for model optimization; second, deep training and iterative refinement of complex models, utilizing vast datasets to train high-accuracy deep learning models such as transformer-based models for long-term SOH prediction and multimodal fusion models for thermal runaway risk early warning. These model training processes typically require substantial computational power that edge devices cannot meet; however, the cloud can significantly enhance training efficiency through parallel computing; third, long-term health feature extraction and early detection of potential risks, applying well-trained high-accuracy models to forecast future trends in battery health status over extended periods, precisely identifying latent risks such as thermal runaway or sudden capacity drop, and generating targeted early-warning strategies.

Collaborative interaction between the edge and the cloud is achieved through Over-the-air (OTA) technology, enabling closed-loop iteration and ensuring the continuous evolution of the BMS. After model training and optimization in the cloud, updated model weights, revised control strategies, or adjusted risk alert thresholds are compressed and packaged before being transmitted to edge-side BMS units via vehicle networking. The edge devices receive these updates and perform local upgrades without disrupting normal operation, thereby migrating core capabilities of high-precision models into lightweight models deployed at the edge, enhancing estimation accuracy and early warning performance. Simultaneously, the edge periodically uploads locally collected operational data to the cloud, providing up-to-date data support for ongoing model training and strategy optimization in the cloud. This establishes a virtuous cycle: "data upload → cloud training → model delivery → edge application." This collaborative paradigm fully leverages the cloud's superior computational power and advanced data processing capabilities while ensuring real-time responsiveness and operational reliability at the edge. It effectively resolves the conflict between complex models and limited onboard computing resources, providing a foundational architectural framework for the practical deployment of next-generation intelligent BMS systems.

9.2 Societal impact and industrial transformation

9.2.1 Economic impacts on energy ecosystems

In the new energy vehicle industry ecosystem, batteries serve as the core power components, accounting for 30%–40% of the total cost, making them a critical determinant of end-user pricing and industrial profitability. However, for a long time, the uncertainty in battery valuation has represented a major bottleneck hindering high-quality industry development. Battery value is closely tied to its health status throughout its entire lifecycle; yet traditional evaluation methods fail to accurately quantify remaining service life, degradation trends, or potential failure risks. As a result, battery valuation remains in a state of "ambiguous pricing," causing significant negative impacts across multiple stages, including used vehicle transactions, battery second-life applications, and leasing services, thereby severely impeding the smooth operation of the industry's economic value chain. The breakthrough and application of intelligent modeling technologies now offer precise and transparent assessment of battery health across their entire lifecycle, fundamentally resolving this challenge and driving the reconstruction and upgrading of the energy ecosystem's economic value chain.

Leveraging intelligent modeling technologies, service providers can achieve fine-grained monitoring and full-cycle management of batteries: during the battery usage phase, real-time state monitoring enables early detection of potential faults, allowing for timely maintenance and preservation efforts that slow down battery degradation. When degradation reaches a certain threshold, the system can accurately determine whether the battery remains suitable for continued use in new energy vehicles; if not, it is promptly transferred to secondary utilization. Large-scale energy storage power stations represent the core application scenario for second-life batteries. This alignment not only reduces the initial investment cost of energy storage power stations but also ensures operational stability and safety through precise health status monitoring of second-life batteries via intelligent modeling technologies, thereby enhancing the return on investment for energy storage projects. Once batteries no longer meet secondary utilization requirements, they are systematically recycled and dismantled to enable material regeneration. This full-cycle operational model transforms manufacturers from mere "battery sellers" into "full-cycle energy service providers." On one hand, this significantly reduces consumers' initial purchase costs and enhances the market competitiveness of new energy vehicles; on the other hand, by extending battery asset lifespans, it maximizes the lifecycle economic benefits of batteries and establishes a multitiered, full-cycle energy ecosystem value chain, providing sustained momentum for high-quality industrial development.

9.2.2 Environmental and sustainability implications

Against the global backdrop of active efforts to combat climate change and advance the implementation of dual-carbon goals, the green attributes of the new energy industry extend beyond clean end-use energy consumption to encompass environmental sustainability throughout the entire lifecycle of core components. At the heart of this green value lies the improvement in lifecycle utilization efficiency of lithium-ion batteries. During conventional usage, lithium-ion batteries often suffer premature degradation and early retirement due to suboptimal charging strategies, delayed thermal management, and misuse under harsh operating conditions. This not only reduces resource utilization efficiency but also intensifies pressure on upstream mineral extraction and increases carbon emissions from manufacturing processes due to frequent battery replacements. Intelligent BMS, leveraging precise state perception and adaptive control capabilities, have emerged as a critical enabler for enhancing lifecycle utilization efficiency and reinforcing the green characteristics of batteries. Through multidimensional technological approaches, intelligent BMS significantly prolongs the average service life of batteries, thereby demonstrating substantial ecological and environmental benefits.

The significant extension of battery average service life brings multiple ecological and environmental benefits: under the premise of meeting equivalent transportation or energy storage demands, battery replacement frequency can be substantially reduced, thereby decreasing the demand for mining of scarce upstream mineral resources such as lithium, cobalt, and nickel. Meanwhile, battery manufacturing involves multiple energy-intensive and high-carbon-emission processes, including raw material processing, cell assembly, formation testing, and others. Reducing the number of batteries produced directly lowers the carbon footprint across the entire manufacturing lifecycle, providing critical support for low-carbon development in the new energy industry.

Moreover, the AI-powered big data early-warning technology integrated into intelligent BMS can mitigate environmental pollution risks from a safety protection perspective, further reinforcing the green characteristics of batteries throughout their entire life cycle. Among lithium-ion battery safety incidents, thermal runaway leading to fire or explosion represents one of the most severe scenarios. Such events not only cause significant loss of life and property but also release toxic gases such as hydrogen fluoride and carbon monoxide during combustion. Additionally, leakage of electrolytes and electrode materials can contaminate soil and water bodies, creating long-term environmental hazards. AI-driven big data early-warning technology enables deep analysis of battery operational data across the entire life cycle, establishing predictive models for thermal runaway risk. It can detect subtle precursors to thermal runaway, such as abnormal increases in internal resistance, sudden local temperature gradients, or voltage platform fluctuations—hours or even days in advance, issuing early warnings that guide maintenance personnel to promptly implement intervention measures such as cooling, power disconnection, or isolation, thereby preventing safety incidents at an early stage.

In summary, intelligent BMS has profoundly reshaped the green characteristics of lithium-ion batteries through multiple pathways, including extending battery lifespan, optimizing resource utilization, and mitigating environmental risks. As a result, lithium-ion batteries have evolved from end-point clean energy carriers into core components of low-carbon and environmentally sustainable systems across their entire life cycle. In the global context of addressing climate change, this technology-driven green transformation not only aligns with the dual carbon targets but also promotes the synergistic integration of economic and ecological values within the new energy industry, laying a solid foundation for building a sustainable energy ecosystem.

9.2.3 Promoting the dual-carbon objectives and the transition to green energy

Against the backdrop of the rapid global development of the new energy vehicle industry, the number of Electric vehicles (EVs) has experienced explosive growth, a trend profoundly reshaping the landscape of traditional energy systems. Electric vehicles are no longer merely grid loads consuming electrical energy; instead, leveraging their vast battery storage capacity, they have evolved into large-scale, widely distributed mobile energy storage resources. Intelligent BMS, serving as the core control hub for EV battery management, have become a critical technological enabler for unlocking this mobile energy storage potential and achieving efficient Vehicle-to-grid (V2G) interactions through precise state perception, intelligent decision-making, and coordinated control capabilities.

The core logic underlying intelligent BMS for efficient V2G interaction lies in constructing a coordinated decision-making framework integrating "travel demand, battery state, and grid conditions" through AI algorithms, enabling precise control over charging and discharging behaviors.

Specifically, its technical approach comprises three key components: first, accurate prediction of user travel demand. By integrating historical travel data, real-time traffic information, and meteorological data, the intelligent BMS employs deep learning models to forecast the probability of future travel, required mileage, and corresponding battery energy demands over a given time horizon. This enables determination of the "minimum available capacity" required to ensure that V2G operations do not compromise users' normal mobility needs. Second is dynamic assessment of battery availability. Based on the high-accuracy SOH estimation model previously discussed, the intelligent BMS continuously monitors parameters such as remaining charge level, battery health status, charge/discharge efficiency, and temperature. By combining these real-time measurements with travel demand forecasts, it dynamically calculates the maximum amount of energy that can be safely allocated for V2G participation while avoiding excessive discharge that could degrade battery lifespan. Third is real-time response to grid conditions. The intelligent BMS exchanges information with grid dispatch centers via vehicle networking technologies to obtain up-to-date data on load peak/off-peak periods, electricity price fluctuations, and renewable energy generation output. This enables the system to formulate optimal charging and discharging strategies in alignment with grid dynamics.

Distributed mobile energy storage resources based on intelligent BMS and V2G technologies can be integrated into a "virtual power plant" model, providing core support for renewable energy integration and energy system optimization. Renewable energy sources such as wind and solar power inherently suffer from intermittency and volatility, with their output significantly influenced by weather, seasonal variations, and other environmental factors. Large-scale grid integration of these sources poses challenges to the stable operation of power systems. However, a virtual power plant composed of vast numbers of electric vehicles can serve as a "buffer" and "regulator" for renewable energy through unified coordination and scheduling enabled by intelligent BMS: when renewable generation exceeds demand, the intelligent BMS directs electric vehicles within the region to charge collectively, storing surplus electricity; conversely, during periods of insufficient renewable output, it triggers collective discharging from electric vehicles to compensate for grid supply deficits. Data indicate that the V2G-based virtual power plant model can increase the regional integration rate of wind and solar power by 15%–25%, significantly reducing curtailment of wind and solar energy.

Moreover, the V2G virtual power plant model can significantly enhance the flexibility and stability of the entire energy system. Conventional energy systems primarily rely on dispatchable sources such as thermal and hydro power, which offer strong flexibility but high carbon emissions. As the share of renewable energy increases, the system's flexibility gradually declines, leading to an increasingly urgent demand for dispatchable energy storage resources. Electric vehicle (EV) virtual power plants exhibit widespread distribution and rapid response characteristics; intelligent BMS enable millisecond-level switching between charging and discharging, allowing for swift mitigation of grid frequency fluctuations and compensation for voltage deviations, thereby enhancing the grid's resilience to disturbances. Furthermore, virtual power plants can participate in ancillary services markets by providing services such as peak shaving, frequency regulation, and backup power supply, generating additional revenue and further improving the economic viability of V2G models. This fosters a virtuous cycle of "grid optimization—user benefits—industrial upgrading."

In summary, intelligent BMS serves as the core hub of V2G technology, not only unlocking the distributed mobile energy storage potential of electric vehicles but also driving the transformation of energy systems from "centralized power supply" to "distributed collaboration." Against the backdrop of global energy transition, this technology-driven integration of vehicles and power grids enhances the grid's capacity to absorb renewable energy, improves the flexibility and stability of energy systems, and generates additional value for users and industries alike. As such, it has become a crucial enabler in building a clean, low-carbon, safe, and efficient energy system.

9.3 Challenges and mitigation strategies

9.3.1 Limited model generalization and challenging transferability

The large-scale application of data-driven approaches and AI models in lithium-ion BMS is currently encountering a core bottleneck characterized by insufficient generalization capability and limited transferability. This issue stems from the complexity of battery systems, the heterogeneity of data distributions, and practical constraints in technological implementation. A detailed analysis can be conducted from the following dimensions.

9.3.1.1 Origins and manifestations of insufficient generalization capability

The lack of model generalization capability fundamentally stems from "distribution shift" between training data and real-world application scenarios, with its core causes and manifestations exhibiting multidimensional characteristics. On one hand, there is an inherent constraint imposed by data distribution bias: the assumption of independent and identically distributed training and testing data, central to data-driven modeling is difficult to satisfy in lithium-ion battery applications. Laboratory training datasets are typically constructed under standardized conditions such as constant temperature and fixed charge/discharge rates, whereas real-world conditions involve significant deviations, including diverse driving behaviors in electric vehicles, stochastic charge/discharge loads in energy storage power stations, and extreme temperature fluctuations in harsh climatic regions, all of which induce substantial distributional discrepancies between test and training data. On the other hand, battery-to-battery variability and batch-to-batch variation introduce further interference. Minor initial variations arising from manufacturing processes, such as material synthesis precision and electrode coating uniformity can manifest early as divergent charge acceptance voltages, thereby affecting long-term degradation trajectories. Furthermore, there exists a fundamental challenge in cross-chemical-system adaptability: different cathode material systems exhibit intrinsic electrochemical differences. For instance, Nickel–cobalt–manganese (NCM) lithium batteries exhibit a nominal voltage plateau around 3.6 V and are highly sensitive to temperature variations; in contrast, Lithium iron phosphate (LFP) batteries feature a flat voltage plateau with superior safety but inferior low-temperature performance. These material-level disparities hinder model transferability across different chemical systems.

9.3.1.2 The core obstacle of limited mobility

The challenge of model transferability manifests in the difficulties of adaptation across different scenarios, devices, and lifecycle stages, rooted in inconsistencies in feature space and the absence of physical mechanisms: cross-scenario transfer encounters feature misalignment, where the characteristic representations of core battery monitoring parameters significantly differ across application contexts, and even within the same scenario, varying operational modes may lead to feature degradation; this misalignment in feature space renders models optimized for a single scenario ineffective when directly transferred to other contexts. Concurrently, there exists a "black-box" dilemma due to missing physical mechanisms, pure data-driven models, lacking constraints from electrochemical principles, struggle to capture commonalities across domains and their opaque nature prevents interpretation of cross-domain error sources, further limiting the precision of transfer optimization. Moreover, limitations imposed by small sample sizes and privacy constraints form dual barriers to practical deployment of model transferability: in electric vehicle applications, individual user battery data are inherently limited and subject to privacy protection regulations, making it difficult to obtain sufficient labeled data for model fine-tuning; in energy storage power stations, battery data from different manufacturers cannot be shared due to commercial barriers, while full-parameter fine-tuning demands substantial computational resources, creating a conflict with existing computational constraints.

9.3.1.3 Mitigation strategies and technical pathways

To address the aforementioned bottlenecks, multidimensional solutions must be developed across data, model, engineering, and industry collaboration dimensions, enabling dual improvements in generalization and transferability. By leveraging knowledge reuse, the rigid constraint of "training-test distribution consistency" can be broken, thereby reducing dependency on target-domain data.

Hierarchical transfer strategy: Select an appropriate adaptation method based on the degree of discrepancy between the source and target domains. When the discrepancy is small, fine-tuning transfer is adopted by fixing the parameters of the lower-level feature extraction layers in the model and updating only the top-level prediction layers. When the discrepancy is significant, adversarial training transfer is employed, leveraging a game-like interaction between a domain discriminator and a feature extractor to align cross-domain distributions.

Federated transfer fusion architecture: To address issues of data privacy and scarcity, a two-stage framework of "federated training + personalized fine-tuning" is adopted. In the first stage, federated learning aggregates multi-source distributed data to train a global general model, avoiding the sharing of raw data. In the second stage, model parameters are fine-tuned using only a small amount of data from the target domain to construct personalized models. The schematic diagram of the federated transfer joint architecture is shown in the following.

Fig. 9.2 illustrates the federated transfer fusion architecture for cross-scenario and cross-manufacturer migration of lithium-ion battery BMS

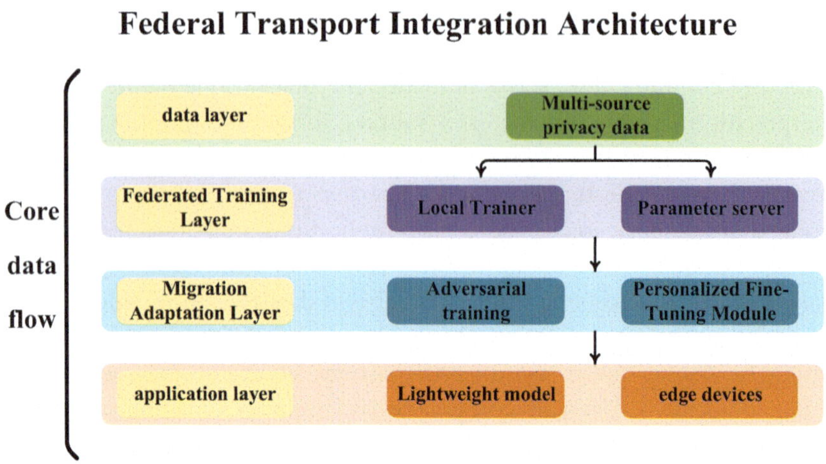

Figure 9.2 Integrated architecture diagram of federated migration.

models. This architecture, designed through a four layer collaborative framework, "data layer, federated training layer, transfer adaptation layer, application layer", addresses the dual challenges of "data privacy protection" and "small-sample cross-domain adaptation." From a logical standpoint, the data layer aggregates multisource smallsample data, with all raw data remaining locally stored; only model parameters are transmitted to the federated training layer, thereby eliminating privacy risks associated with cross-entity data sharing. The federated training layer aggregates local model parameters from multiple sources via a parameter server, generating a global general model that spans diverse scenarios and lays the foundation for cross-domain transfer. The transfer adaptation layer aligns source-domain and target-domain data distributions through an adversarial training module, followed by fine-tuning via a personalized adjustment module to optimize top-layer parameters. The application layer employs knowledge distillation and pruning techniques to lightweight the model, enabling adaptation to device-level computational constraints, while leveraging an OTA update module to establish a continuous iteration loop of "edge feedback–cloud optimization–parameter dissemination." This integrated architecture not only overcomes the traditional transfer learning dependency on full-scale datasets but also achieves a balanced trade-off among model accuracy, privacy preservation, and engineering deployability through its layered design. It thus provides a practical technical solution for collaborative management of multi-manufacturer battery packs and reusable BMS models across diverse operational scenarios.

9.3.2 Underdeveloped continual learning mechanisms

Current mainstream BMS algorithms universally suffer from a core limitation of "static deployment and parameter fixation," rendering them inadequate for adapting to the dynamically evolving operating environments and aging characteristics of lithium-ion batteries throughout their entire lifecycle. Most existing algorithms are trained and their parameters calibrated prior to shipment, based on standardized laboratory data; once deployed, the model parameters remain fixed, allowing adaptation only to the limited scenarios covered by the training data. However, real-world battery operation involves significant dynamic variability, experiencing wide-range temperature fluctuations due to seasonal changes, regional climatic differences, diverse user driving patterns, and variable charge/discharge conditions. Moreover, over a lifespan exceeding 10 years, batteries undergo continuous degradation processes such as active material loss, SEI layer growth, and increasing internal resistance, leading to ongoing evolution of electrochemical properties and dynamic drift in data distributions across life stages. Static BMS algorithms, lacking mechanisms for dynamic parameter updating and knowledge accumulation, fail to continuously adapt to these evolving conditions, resulting in progressively deteriorating estimation accuracy throughout the battery's lifecycle.

More critically, when static BMS algorithms attempt to adapt to new operating modes through online learning, they are highly prone to triggering the classic "catastrophic forgetting" phenomenon prevalent in the field of artificial intelligence. Catastrophic forgetting is characterized by the fact that, during the learning process of new tasks or data patterns, the model significantly alters its network parameters to accommodate new information, thereby overwriting and forgetting previously acquired knowledge related to old tasks and patterns. This results in a substantial decline in the model's ability to adapt to prior scenarios. The inherent "learning new while forgetting old" trait prevents static BMS algorithms from achieving stable and accurate management across all scenarios and throughout the entire lifecycle, representing a key bottleneck that hinders advancements in BMS performance.

To overcome the limitations of static algorithms and the challenge of catastrophic forgetting, advancing lifelong learning (lifelong learning/continual learning) algorithms has become a core technological direction for enabling comprehensive lifecycle precision management of BMS. The central objective of lifelong learning is to equip BMS with the capabilities of "continuous learning, memory retention, and knowledge evolution," allowing them to continually acquire new data and patterns in dynamically changing environments while firmly preserving previously acquired knowledge. This results in an expanding knowledge base and enables sustained model performance optimization rather than phase-wise degradation. This technical requirement aligns closely with the demands of lithium-ion battery full lifecycle management. The core strategies can be categorized into two key technical pathways, both of which can be seamlessly integrated with edge-cloud collaborative architectures for practical implementation.

The first core technical approach is the elastic Weight consolidation (EWC) algorithm, which protects prior knowledge through quantification of parameter importance and constraint mechanisms. The fundamental principle of the EWC algorithm is as follows: after the model has learned an old task, critical network parameters essential to the performance of the previous task are identified via quantitative analysis, and "elastic constraints" are imposed on these parameters, specifically, threshold limits and penalty terms are set on weight updates. When the model subsequently learns a new task, parameter updates prioritize those with minimal impact on prior tasks, while strict restrictions are enforced on modifications to core parameters to prevent catastrophic forgetting caused by significant parameter adjustments. This parameter constraint mechanism requires no additional storage of old data and exhibits low computational complexity, making it well-suited for real-time deployment in BMS under constrained onboard computational resources.

The second core technical approach is the replay mechanism, which achieves knowledge retention and incremental learning through a "learning new data + reviewing old data" paradigm. This mechanism mimics human learning and memory patterns: when the model learns data from new scenarios or new life stages, it draws a small number of representative samples from old scenarios and previous life stages from a historical data pool stored in the cloud. These old samples are then jointly input into the model alongside new data for training, enabling the model to "review" prior knowledge while acquiring new knowledge, thereby mitigating catastrophic forgetting. Depending on differences in data storage and utilization methods, the replay mechanism can be categorized into two types: real-sample replay and generative-sample replay. Real-sample replay directly stores historical representative data, ensuring high data fidelity but requiring substantial storage resources. In contrast, generative-sample replay employs models such as

Generative adversarial networks (GANs) to learn the distribution characteristics of old data and generate virtual samples of past scenarios in realtime, eliminating the need for large-scale storage and thus meeting lightweight requirements on vehicle-mounted edge devices. Under an edge-cloud collaborative architecture, the cloud can establish a large-scale historical data pool and periodically deliver representative old samples to edge devices. The edge devices then combine these with real-time new data to perform replay-based training, forming a lifelong learning loop characterized by "cloud-based data support + edge-device real-time learning."

Constructing a lifelong learning AI-BMS with memory retention and continuous evolution capabilities is the essential pathway toward achieving precise, 10 years plus lifecycle management of lithium-ion batteries, offering significant industrial value. From a safety perspective, the system can continuously adapt to varying operating conditions and aging stages of batteries, preventing the expansion of estimation errors caused by model forgetting and thereby reducing safety risks such as overcharging, overdischarging, and thermal runaway. From an economic standpoint, it enables accurate tracking of battery health state changes throughout the entire lifecycle, providing more precise valuation foundations for BaaS services, secondary utilization, and recycling processes, thus further enhancing the overall lifecycle economic benefits of battery assets. Technologically, lifelong learning transforms BMS from a "static deployment" model into a "dynamically evolving" system capable of continuously integrating new operational data and technical insights to sustainably optimize model performance, providing core technological support for the iterative advancement of next-generation intelligent BMS. Moreover, lifelong learning synergizes with previously discussed techniques, including transfer learning, domain adaptation, and adaptive parameter reconfiguration, to jointly establish a comprehensive intelligent algorithm framework characterized by "general foundational capabilities + cross-domain adaptability + dynamic update capacity + lifelong memory functionality," thereby ensuring comprehensive support for the engineering implementation of intelligent lithium-ion battery management.

Reference

[1] T. Hussein, B. Ali, J.M. Jaber, et al., Advancing organic photovoltaic cells for a sustainable future: the role of artificial intelligence (AI) and deep learning (DL) in enhancing performance and innovation, Sol. Energy 291 (2025) 113378.

Index

'*Note:* Page numbers followed by "f" indicate figures and "t" indicate tables.'

I

J

K

L

M

N

O

P

R

S

Printed and bound by CPI Group (UK) Ltd, Croydon, CR0 4YY
17/09/2026
02218533-0003